Kehui Sun

Chaotic Secure Communication

Also of interest

Coding and Cryptography
Nataša Živić, 2013
ISBN 978-3-486-75212-0, e-ISBN 978-3-486-78126-7,
Set-ISBN 978-3-486-80581-9

Modern Communications Technology
Nataša Živić, 2016
ISBN 978-3-11-041337-3, e-ISBN 978-3-11-041338-0,
e-ISBN (EPUB) 978-3-11-042390-7

A Course in Mathematical Cryptography
G. Baumslag, B.Fine, M. Kreuzer, G. Rosenberger
ISBN 978-3-11-037276-2, e-ISBN 978-3-11-037277-9,
e-ISBN (EPUB) 978-3-11-038616-5

Computational Methods in Applied Mathematics
Carsten Carstensen (Editor-in-Chief), 4 issues per year
ISSN 1609-4840, e-ISSN 1609-9389

Kehui Sun

Chaotic Secure Communication

Principles and Technologies

DE GRUYTER

清華大學出版社
TSINGHUA UNIVERSITY PRESS

This work is co-published by Tsinghua University Press and Walter de Gruyter GmbH.

Author
Prof. Kehui Sun
School of Physics and Electronics
Central South University
Hunan, China

ISBN 978-3-11-042688-5
e-ISBN (PDF) 978-3-11-043406-4
e-ISBN (EPUB) 978-3-11-043326-5
Set-ISBN 978-3-11-043407-1

Library of Congress Cataloging-in-Publication Data
A CIP catalog record for this book has been applied for at the Library of Congress.

Bibliographic information published by the Deutsche Nationalbibliothek
The Deutsche Nationalbibliothek lists this publication in the Deutsche Nationalbibliografie; detailed bibliographic data are available on the Internet at http://dnb.dnb.de.

Cover image: scanrail/iStock/thinkstock
Typesetting: Integra Software Services Pvt. Ltd.
Printing and binding: CPI books GmbH, Leck
♾ Printed on acid-free paper
Printed in Germany

www.degruyter.com

Preface

Since meteorologist Lorenz discovered chaotic phenomenon in 1963, it has been widely and deeply studied. Chaos theory is known as the third scientific theoretical revolution following the theory of relativity and quantum theory in the 20th century. Especially, Pecora and Carroll reported chaos synchronization in 1990, which aroused an upsurge of study on chaotic secure communications. After more than 20 years of research and development, chaotic secure communication technology is becoming a new growth point in the field of information technology. On one hand, the application of chaos will continue to promote people's profound understanding of the nature of chaos. On the other hand, the new problems in the application of chaos will further lead to the further research of chaos. Although chaotic secure communication technology still has a long way to go, the application of chaos in secure communication field shows us a promising prospect.

This book includes nine chapters. In the first chapter, the history of chaos research is reviewed, and the important characters and stories in the development of chaos are introduced, and the advances in research on chaotic synchronization control technology and chaotic secure communication method are reviewed. In the second chapter, the main analysis methods of the chaotic characteristic of the nonlinear system are presented. The typical chaotic dynamic system models are introduced briefly in the third chapter. In the fourth chapter, the principle and technology of the synchronization control between chaotic systems are investigated. In the fifth chapter, the technology and scheme of secure communication based on chaotic synchronization are discussed. In the sixth chapter, the data encryption technology based on chaotic sequences is discussed. In the seventh chapter, the audio and video real-time encryption communication technology based on chaos is studied. In the eighth chapter, the latest research progress, dynamic characteristic analysis, and simulation methods of fractional-order chaotic system are studied. In the ninth chapter, the design and implementation technology of chaotic circuit is studied.

In order to facilitate readers to grasp the important knowledge points in the book, the author lists a few questions at the end of each chapter. In order to facilitate the search of relevant literature and to publish the research results on time, journals about chaos research are listed as Appendix A of this book. In addition, in order to facilitate readers to quickly grasp the chaos of research methods, the author collected the relevant analysis programs as shown in Appendix B.

Based on the research results in the field and a large number of related literatures, this book is written through repeated modification.

In the field of research of chaos, the author has got the guidance and inspiration of many professors including Professor Sprott J C from the University of Wisconsin Madison in the United States, Professor Qiu S S from South China University of Technology, Professor Yu S M from Guangdong University of Technology, Professor Wang

G Y from Hangzhou University of Electronic Science and Technology, Professor Zhang T S, Professor Sheng L Y, and Professor Zhu C X from Central South University. The author expresses sincere gratitude to all the professors. In addition, in the preparation of this book, previous graduate students gave a lot of support. They are Mou J, Zhou J L, Tan G Q, Ren J, Shang F, Wang X, Bao S Q, Yang J L, Cheng W, Liu X, He S B, Wang H H, Wang Y, Zuo T, Ai X X, Liu W H, Ruan J Y, Zhang L M, Xu Y X, Yu M Y, Yin Y H. Thanks for their contributions!

I would sincerely like to appreciate and acknowledge the comments and suggestions from the readers for this book.

Kehui Sun
January 2015, Changsha

Contents

1 Introduction

After decades of development, people are no longer strange to chaos, but how to correctly understand and study chaos and its application is still the main task of academic circles. As we all know, chaos is nonlinear, while what we know well is linear. What is the relationship between them? The discussion begins here.

1.1 Linear, Nonlinear, and Chaos

Linear and nonlinear are used to explain the relationship between function $y = f(x)$ and an independent variable x. To be specific, linear refers to the proportional relationship between two independent variables. It represents regular and smooth movement in space and time. However, nonlinear refers to the disproportionate relationship between two dependent variables. It represents irregular and mutation movement in space and time. For example, how many times the visual acuity of two eyes is that of an eye? It is very easy to think that it is two times, but actually, it is 6–10 times. This is the nonlinear relationship. Calculation result of one plus one is not two.

Although nonlinear relationship is ever changing and complicated, there is something in common that is different from linear relationship. Linear relationship is unrelated independence, while nonlinear relationship is interaction. It is the interaction that makes the entirety, which is not simply equal to the sum of the parts. On the contrary, unlike linear superposition, gain or loss may occur. For example, the generation of laser is nonlinear. Laser scatters in all directions like lamp when the voltage is small. When the applied voltage reaches a certain value, a new phenomenon applied will appear suddenly. The excited atoms scatter a kind of homogeneous light whose emit phase and direction are consistent as if they received an order to look right. This kind of homogeneous light is laser. For another example, the frequency components of linear signal remain the same, but nonlinear relationship leads to changes of frequency configuration. As long as there is nonlinearity, even if it is a small nonlinearity, it will produce sum frequency, difference frequency, frequency doubling, and other components. Behavioral mutations are caused by nonlinear relationships, but small deviation for the linear relationship doesn't lead to behavioral mutations in general. And according to the original linear conditions, the small deviation can be described and comprehended by modified linear theory. System behavior may change suddenly when nonlinear relationship reaches a certain degree. Mutations often occur at a series of parameter thresholds of the nonlinear system, and each mutation is accompanied by a new frequency component. The system will eventually enter the chaotic state.

According to the characteristics of the nonlinear relationship, it can be concluded that if there exists chaos, the system must be a nonlinear system. There is a mutation behavior when a nonlinear system turns into the state of chaos. How can we know whether a system has turned into the state of chaos? How to judge a system that presents phenomenon for a long period of time? How to distinguish whether a system is subject to external random disturbances? All those questions are important issues in chaos research.

Chaos theory belongs to nonlinear science. Nonlinear scientific research always seems to make people's understanding of "normal" things and "normal" phenomenon to explore the "abnormal" things and "abnormal" phenomenon. For example, solitary wave is not a regular propagation with periodic oscillation. Unconventional edit methods will be taken when a large number of unconventional phenomena occur when using multimedia technology for information storage, compression, transmission, conversion, and control process encountered. Chaos broke the convention that the future movement of a deterministic equation is determined strictly by the initial conditions and led to a so-called strange attractor phenomenon.

In nonlinear science, chaos is not exactly the same as what it means. Chaos phenomenon refers to a determined but unpredictable motion in nonlinear science. The external manifestations of chaos are very similar to purely random motion. That is, they are all unpredictable. Comparing with random motion, chaos motion is determinate. And its unpredictability is derived from the instability of the system of internal movement. Chaos system is sensitive to initial conditions and little change, no matter how small they are. Therefore, after a period of time evolution, the system will deviate from its original direction of evolution. Chaos is a common phenomenon in nature, such as the change of weather. An important feature of chaos is butterfly effect. So chaos means a nonperiodic motion generated by a deterministic system. Chaos, fractals, and soliton are the three most important nonlinear science concepts. Chaos theory is a part of nonlinear science, and only nonlinear system can produce chaotic motion.

In 1977, the first international chaos conference was held in Italy. It marks the official birth of the science of chaos. The physicist Ford who was one of the assembly moderators considered that chaos theory was the third revolution of physics following relativity theory and quantum mechanics in the twentieth century. He said, relativity theory eliminated illusions about absolute space and time, and quantum mechanics broke Newton's dream about controllable measurement process; and chaos burst the bubble of determinism of predictability by Laplace [1].

1.2 Development of Chaos

In the latter half of the twentieth century, nonlinear science has been developing rapidly, which accounted for a great proportion of the study on chaos. Chaos, as a remarkable frontier project and academic hot topic, reveals the unity of simplicity and complexity, order and disorder, and certainty and randomness in nature and human society, and broadens people's view greatly, and deepens people's understanding of the objective world. In the field of natural science and social science, chaos is striking and changing the field of science and technology, and it presents a huge challenge to people.

In the early 1960s, scientists began to explore some elusive phenomenon in nature. In 1963, Lorenz proposed deterministic nonperiodic flow model [2], and later he proposed the butterfly effect. In 1975, Li and Yorke published “Period three implies chaos” [3] in *American Mathematical Monthly*, which revealed the evolution process from order to chaos. This famous article also defined “chaos” as a new scientific term that officially appeared in the literature. In 1976, American biologist May published “Simple mathematical models with very complicated dynamics” [4] in *Nature*. This article pointed out that a very simple one-dimensional iterative map can produce complex period-doubling and chaotic motion. He revealed to the people that some simple deterministic mathematical model in ecology can generate chaotic behavior. In 1978 and 1979, Feigenbaum independently discovered scaling property and universal constant [5, 6] in the phenomenon of period-doubling bifurcation, which lays a solid theoretical foundation for chaos in modern science. Lorenz defined that chaos is a science that uses fractal geometry to analyze and study the nonlinear dynamics problem from butterfly effect, aperiodicity, and so on.

In the 1980s, chaos science has been further developed. Many researchers have focused on how an order system turns into the chaotic state and on the characteristics of chaos. In 1981, Takens proposed experimental method of determining strange attractors [7]. In 1983, Glass published an article to set off a wave of computing time series dimension [8]. In 1986, Grassber proposed dynamical systems theory and methods of reconstruction [9]. By calculating chaotic characteristics, such as fractal dimension, Lyapunov exponent from the time series, chaos theory entered into the stage of practical application.

In the 1990s, chaos science got mutual infiltration, promotion, and wide application with other subjects, such as synchronization, secure communication, chaos cryptography, chaotic neural networks, and chaotic economics research areas. Exploring the unique nature, role, and function of chaotic systems to benefit humanity becomes a very great task.

Research on deterministic chaotic system has undergone three stages. The first stage is to study how an order system turns into chaos, which focuses on the conditions, mechanisms, and generated pathways. The second stage is to study what is the order of chaotic system, which focuses on universality, statistical characteristics, and the fractal structure in chaos. The third stage is to study how a chaotic system turns into an order system, which focuses on controlling chaos to achieve order. The current study of chaos focuses on how to control and utilize chaos.

Due to being sensitive on initial conditions, even two identical chaotic systems evolve from almost the same initial conditions, and their orbits also quickly become uncorrelated over time, which makes the chaotic signal become unpredictable and with long-term anti-interception capability. At the same time, chaotic system is deterministic, which is completely determined by the equations, nonlinear system parameters, and initial conditions, and therefore chaotic signal is easy to be generated and replicated. Due to these features, chaos has excellent prospects in secure communications and becomes a hot research topic in chaotic application.

Internationally, the study of secure communications originated in the early 1990s, and it becomes one of the central issues in the field of information science. In 1990, Pecora and Carroll proposed a chaos synchronization method and first observed the phenomenon of chaotic synchronization [10, 11] on electronic circuits. In the same year, Ott, Grebogi, and Yorke realized the control to the chaotic attractor unstable periodic orbits [12]. These works have greatly aroused people's interest in chaos mechanism and applications, and quickly pushed the theoretical and experimental study on chaos synchronization and control of chaos, and opened the prelude to apply chaos in engineering field. In fact, chaotic encrypted communication will play an important role not only in the field of military, national security, and communications countermeasure, but also in the business world and people's lives, which is the new trend of future information security technology.

1.3 Famous Scientists and Important Events

1.3.1 Lorenz and Butterfly Effect

Edward Norton Lorenz (1917–2008) was born in Hartford, Connecticut, USA. He loved science from a young age and graduated from Harvard University in 1940. During World War II, Lorenz served as a forecaster in the United States Army Air Corps. After the war, he received his master's and doctorate in meteorology. Then he taught at the Massachusetts Institute of Technology. In 1972, he put forward the famous "butterfly effect" theory. In 1975, he became an academician in the US National Academy of Sciences. He won the "Crafoord Prize," which is called Ecology "Nobel Prize" in 1983 and won the "Kyoto Prize" in 1991. The judges thought his chaos theory "brought the most dramatic change to the human nature view after Newton's theory."

The butterfly effect is one of his important scientific contributions, and its discovery is interesting and enlightening. In 1961 winter, Lorenz used computer for numerical prediction calculation. In order to save time, he selected a row data (equivalent to one day's weather conditions) from the original calculation results as an initial value to calculate again, and then left his office to drink a cup of coffee. After an hour, when he returned to the laboratory, the computer had already calculated the predicted results of the next two months. He was shocked by the new results, which were quite different from the original results. Minor deviation doubled every four days until the similarity between the old and the new data was completely lost. The difference of initial value was less than 1/1,000, but it causes the second calculation result to be completely different from that of the first one. This is the "butterfly effect," which shows the initial value sensitivity.

In 1979, Lorenz delivered a speech entitled "Butterfly Effect" on the 139th meeting of the American Academy of Science Development and proposed a seemingly preposterous assertion: a butterfly flapping its wings in Brazil, there may be a tornado in Texas, United States, a month later. Thus, he pointed out that it is very difficult to forecast weather accurately. Today, people still talk about this assertion. More importantly, it stimulated people's interest in chaos. With the rapid development of computer technology, chaos has become a frontier field with far-reaching influence and rapid development. "Butterfly Effect" is fascinating, exciting, and thought provoking, so it not only is a bold imagination and charming color aesthetics, but also lies in the profound scientific connotation and inherent philosophical charm. Lorenz's discovery laid the foundation for chaos theory. The chaos theory not only affects meteorology, but also profoundly affects every branch of science.

1.3.2 Li Tien-Yien and the Concept of Chaos

Li Tien-Yien (1945–) was born in Shaxian, Fujian Province, China. He and his family moved to Taiwan when he was three years old. He received education there until he graduated from university. He is the 68th graduate of mathematics from Tsinghua University, Hsinchu, Taiwan province, China. In 1969, he went to the Department of Mathematics, University of Maryland, United States, for doctoral degree, under Professor York. He was a lecturer in the University of Utah in the United States from 1974 to 1976. Since 1976, he taught at Michigan State University in the United States. Li Tien-Yien has made several important pioneering works and extraordinary achievements in the field of applied mathematics and computational mathematics. The paper "Period Three Implies Chaos" was written by him and York, which first proposed the concept of chaos in science and opened a new era in the scientific community for research in chaotic dynamical systems.

In the past few decades, Professor Li Tien-Yien is suffering from a long-term illness. So far, he has undergone major surgeries under general anesthesia more than ten times and countless surgeries under local anesthesia. Innumerable wounds are on his body. However, he struggled in the face of adversity and overcame the disease with an indomitable spirit of optimism again and again. He sought breakthrough in adversity and fought the disease optimistically. He overcame all the difficulties with a surge of strong determination and ultimately made a first-class research work under difficult environments.

Li Tien-Yien has had rigorous scholarship for decades [13]. He believes that the only way to success is insistence. He said to his students that he was not smart and that it is not really important to be smart. The most important issue is to get to the bottom. He often emphasized to think on the problem just a minute more than others. He believed the precious one minute paves the road for success. He often exhorts his students that with tremendous effort, one must persist till the end of everything and should never give up. He also said that being engaged in researches they must have a thorough understanding, especially on mathematical logic.

Li Tien-Yien witnessed interesting research experience in the birth of chaos. The famous paper, "Period Three Implies Chaos," also underwent topic selection, research, submission, rejection, and modification, and finally was published in *American Mathematical Monthly*. It focused on the meaning of section iterative, definitions, and main results of some basic concepts. And some courses are provided as an appendix. A mathematician, who most values clarity and precision, had even talked about mess, which has attracted many readers. Physicists and biologists have also quickly accepted the mathematical articles. Since then, "chaos" which was put forward first by Li and York, has become a technical term in chaos theory and has even become synonymous with chaos theory [14].

1.3.3 Feigenbaum and Feigenbaum Constant

Feigenbaum M. J. (1944–), a professor of physics at Cornell University, entered the Graduate School of MIT after graduating from New York College in 1964. He received his PhD in basic particle physics in 1970 and went to Cornell University to engage in postdoctoral research in the same year, but there was no progress for two years. Then he went to Virginia Polytechnic Institute for two years without success. Carruthers P., a former professor in Cornell University, recognized his excellence and selected him as his assistant in Los Alamos (affiliated to California campuses) for free research. Then, after two years, Feigenbaum found a wonderful constant 4.6692... [5]. He published the important result in the *Journal of Statistical Physics* in 1979 [6] and became famous overnight.

Feigenbaum studied about spacing ratio changes in period-doubling bifurcations, which are shown in Table 1.1. He found that bifurcation point of two adjacent spacing Δk forms a geometric sequence with the increase in the number k of bifurcation

$$\delta = \lim_{n\to\infty} \frac{k_n - k_{n-1}}{k_{n+1} - k_n} = 4.6692016091\ldots. \tag{1.1}$$

Bifurcation width w_i also forms a geometric sequence

$$\alpha = \lim_{n\to\infty} \frac{w_n}{w_{n+1}} = 2.5029078750\ldots. \tag{1.2}$$

Calculation of Feigenbaum constant is shown in Figure 1.1.

The discovery of Feigenbaum constant was one of the twentieth century's greatest discoveries. It found a deeper regularity in the periodic-doubling bifurcations and

Table 1.1: Spacing ratio changes of period-doubling bifurcation ($f(x) = \lambda x(1-x)$).

k	Bifurcation	Bifurcation value λ_k	Spacing ratio $(\lambda_k - \lambda_{k-1})/(\lambda_{k+1} - \lambda_k)$
1	1 to 2 = 2^1	3	
2	2 to 4 = 2^2	3.449489743	4.751466
3	4 to 8 = 2^3	3.544090359	4.656251
4	8 to 16 = 2^4	3.564407266	4.668242
5	16 to 32 = 2^5	3.468759420	4.66874
6	32 to 64 = 2^6	3.569691610	4.6691
⋮	⋮	⋮	⋮
∞	Periodic solution → chaos	3.569945972	4.669201609

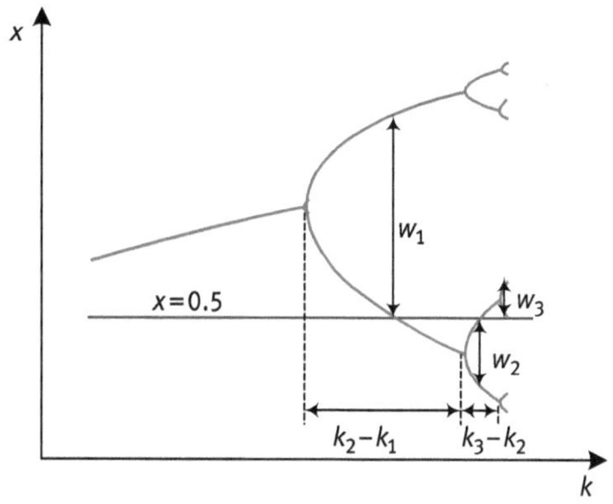

Figure 1.1: Calculation of Feigenbaum constant.

revealed the secret of nonlinear systems from order to chaos. The discovery of Feigenbaum constant greatly changed humans' understanding of the universe and promoted chaos research from qualitative analysis to quantitative calculation. It became an important milestone in the study of chaos. As we all know, the birth of a new theory is often accompanied with the emergence of new constant in the development of physical theory, such as gravitational constant G in Newton mechanics, Planck's constant h in quantum mechanics, and the speed of light c in relativity. So we can easily understand the importance of Feigenbaum constant in chaos theory research and development.

1.3.4 Leon Ong Chua and Chua's Circuit

Leon Ong Chua (1936–) is a professor at Department of Electrical Engineering and Computer Science, University of California, Berkeley. He received his bachelor's degree in electrical engineering in the Philippines Mapúa Institute of Technology in 1959.

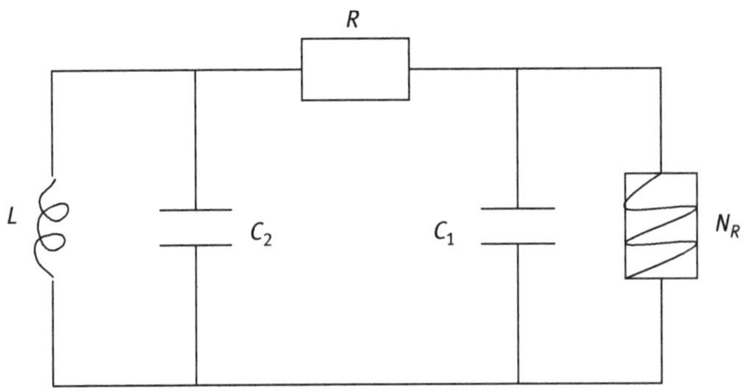

Figure 1.2: Chua's circuit.

He received his master's degree from MIT in 1961 and his doctoral degree from the University of Illinois, Urbana – Champaign in 1964. He taught at the Purdue University in 1964–1970 and became a professor and an IEEE fellow at the University of California, Berkeley, in 1971. He proposed the memristor, Chua's circuit, and cell-type neural network theory. He is considered the father of the theory of nonlinear circuit analysis. He predicted the existence of the memristor. Thirty-seven years later, the research team of Williams R. S. in Hewlett-Packard developed a solid memristor.

Chua's circuit provides a hardware foundation for chaos theory and applications research. When Professor Chua was a visiting scholar at Waseda University in 1983, he designed a simple nonlinear electronic circuit, which is shown in Figure 1.2 [15]. The circuit consists of an energy-storage component (capacitors C_1 and C_2), an inductor L_1, an active resistance R, and a nonlinear resistor N_R. Nonlinear resistor consists of three linear resistors and an operational amplifier. The circuit can exhibit standard chaotic behavior. Its simple manufacturability makes it an example of chaotic system in real world and becomes a typical model of chaotic systems.

1.3.5 Guanrong Chen and Chen's Attractor

Guanrong Chen (1948–) is a professor in the Department of Electronic Engineering in the City University of Hong Kong. During the Cultural Revolution (for about 10 years), he studied on his own and was directly admitted to the Department of Mathematics of Sun Yat-Sen University at the end of the Cultural Revolution. He graduated in 1981 and received his master's degree in computational mathematics. Later, he studied applied mathematics for his doctoral degree in Texas A & M University in 1987. He worked as a visiting assistant professor at the Rice University in the United States from 1987 to 1990, and then as an assistant professor, associate professor, and professor at the Houston University from 1990 to 2000. In 1996, Professor Chen became an IEEE Fellow

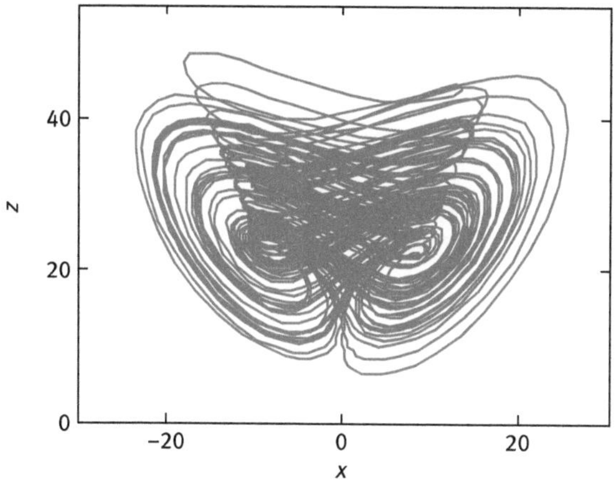

Figure 1.3: Chen attractor (x–z plane).

because of his contributions in chaos control and bifurcation theory and application. Since 2000, he has been serving as a professor of electronic engineering in the City University of Hong Kong and the director of the chaotic and complex network of academic research center. His research focused on related applications areas of control theory of nonlinear systems and dynamics analysis and complex network. In 1999, he proposed a new chaotic attractor [16], known as Chen's attractor, which is shown in Figure 1.3. It has become an important model in chaos theory and applied research.

1.4 Definition and Characteristics of Chaos

Due to the singularity and complexity of chaotic systems, there is no unified definition for chaos. Different definitions reflect the nature of chaotic motion from different angles. The larger impact is the definition of Li–Yorke chaos [3], which is defined from the interval mapping, and it is described as follows.

Li–Yorke theorem: Let $f(x)$ be a continuous map on $[a, b]$. If there are three cycle points, then for any positive integer n, there are n periodic points for $f(x)$.
Li–Yorke chaos definition: For continuous map $f(x)$ on closed interval I, if the following conditions are met, it is confirmed that there is chaos phenomenon: (1) period of periodic points of $f(x)$ is unbounded and (2) the uncountable set S, on closed interval I, meets the following conditions:

1. $\forall x, y \in S,\ \lim_{n\to\infty} \sup\left|f^n(x) - f^n(y)\right| > 0$ when $x \neq y$,
2. $\forall x, y \in S,\ \lim_{n\to\infty} \inf\left|f^n(x) - f^n(y)\right| = 0$,
3. $\forall x \in S$ and any periodic point y of f, $\lim_{n\to\infty} \sup\left|f^n(x) - f^n(y)\right| > 0$.

Chaos has aperiodic bounded dynamic behaviors in deterministic nonlinear dynamical systems, which is sensitive to initial conditions. Chaotic motion is random-like phenomenon that occurs in deterministic systems. Chaos has the following main features:

1. Boundedness. Chaos is bounded, and its track of movement has always been confined to a certain area that is called the chaotic attracting basin. No matter how instable the chaotic system is, its trajectory will not be out of the chaotic attractor basin. So chaotic systems are stable in general.
2. Ergodicity. Chaotic motion is ergodic in their attractor area, which means chaotic orbits experience every status point within a limited time in the chaotic region.
3. Randomness. Chaos is uncertain behavior generated by a certain system, and it has the intrinsic randomness, regardless of external factors. Although the equations of a system are deterministic, its dynamic behavior is difficult to determine. The probability density function of any area is nonzero in its attractor. In fact, unpredictability and sensitivity to initial conditions of chaos lead to internal randomness, but it is also proved that chaos is locally instable.
4. Fractal dimension. Fractal dimension character refers to the behavior characteristics of chaos trajectories in phase space. The dimension is a quantitative description for the geometry complexity of an attractor. It shows chaotic motion features with multi-leaf and a multilayer structure, and each leaf and layer is fine, performing infinite hierarchy self-similar structure.
5. Scaling property. Scaling property refers that the chaotic motion is ordered in the disordered state. It can be considered that as long as the precision of value or laboratory equipment is high enough, you can always see the orderly movement in small-scale chaotic region.
6. Universality. It refers to the different systems that show some common features in the route to chaos, and it does not change with systems equations or parameters. Specific performance is universal constant in chaotic systems, such as the famous Feigenbaum constant. Universality is a reflection of the inherent regularity in chaos.
7. Positive Lyapunov exponent. Lyapunov exponent is the quantitative characterization of approaching or separating trajectories generated by nonlinear system from each other. Positive Lyapunov exponents indicate that tracks are unstable in each locality and adjacent tracks separated exponentially. Meanwhile, the positive Lyapunov exponent also indicates the loss of the adjacent point information – the greater the value, the more serious the loss of information, the higher the degree of chaos.

1.5 Overview of Chaos Synchronization Method

Chaos is so sensitive to initial conditions that people once thought that chaos synchronization is almost impossible. Until 1990, Pecora and Carroll proposed chaos self-synchronization method [10]. Synchronization between two chaotic systems was achieved by using the drive-response method for the first time. This breakthrough research progress broke the traditional idea that movement pattern of chaos is dangerous and uncontrollable. The research results show that chaos not only can achieve control and synchronization, but also can serve as the dynamic basis of information transmission and processing. It is possible to apply chaos to secure communication.

Generally speaking, synchronization belongs to the category of chaos control. Synchronization and control have been combined into one issue by Chua et al. [17]. That is to say, the problem of chaotic synchronization can be considered as a kind of control problem of the chaotic orbit of the controlled system moving according to the orbit of the target system. Traditional chaos control is generally to control a system on unstable periodic orbit, while chaos synchronization is to realize complete reconstruction of two chaotic systems.

1.5.1 Chaos Synchronization Methods and Characteristics

Based on chaos synchronization theory and chaos control methods, researchers have put forward many chaos synchronization schemes [18] as shown in Table 1.2.

Drive-Response Synchronization: In 1990, Pacora-Carroll first realized synchronization between two chaotic systems by using drive-response synchronization method. Later, Cuomo and Oppenheim also successfully simulated the synchronization of Lorenz system by using electronic circuit [19]. Soon, Carroll and Pecora promoted drive-response chaos synchronization method to higher-order cascaded chaotic system [20]. The biggest feature of drive-response synchronization is that there is a relationship of drive and response between two nonlinear dynamic systems. Behavior of the response system depends on the drive system, while the behavior of the drive systems has nothing to do with the response system. However, for some actual

Table 1.2: Chaos synchronization methods.

Time chaotic synchronization	Spatiotemporal chaos synchronization	Hyperchaos synchronization
Drive-response synchronization	Feedback technology	Drive-response synchronization
Active-passive synchronization	Driven variable feedback	Coupled synchronization
Mutual coupling synchronization	Other	Variable feedback synchronization
Continuous variable feedback		Hybrid synchronization
Adaptive synchronization		Drive and variable feedback
Pulse synchronization		Active-passive feedback
Projective synchronization		
D-B synchronization		

nonlinear systems, they cannot be decomposed due to physical nature or natural characteristics or other reasons. Then, the drive-response synchronization method will fail.

Active-Passive Synchronization: In 1995, Kocarev and Parlitoz proposed active-passive synchronization method [21]. The advantage of this method is that it can choose the function of drive signal without limitation. Therefore, it is very flexible and has greater universality and practicality, and it includes the drive-response synchronization method as a special case. Active-passive synchronization method is particularly suitable for applications in secure communications. The method can also be used to synchronize control between hyperchaotic systems.

Coupling Synchronization: In 1990, Winful and Rahman theoretically studied the possibility of laser chaos synchronization. In 1994, American scholars Roy and Thornburg and Japan scholars Suga-Wara and Tachikawa independently observed synchronization between two laser chaotic systems in experiments. Study results show that mutually coupled chaotic system can achieve chaotic synchronization under certain conditions. Kapitaniak and Chua achieved synchronization between two Chua's chaotic circuits with mutual coupling method [22]. Dynamical behaviors of mutually coupled nonlinear systems are very complex. So far there is no universal theory. However, mutually coupled nonlinear systems widely existed in nature, and therefore, it is very important to study this synchronization method.

Continuous Variable Feedback Synchronization: In 1993, German scholars Pyragas and Tamasevicius proposed a method to control continuous nonlinear chaotic systems [23]. Later, this idea was used to study the synchronization between two chaotic systems, and it is called continuous variable feedback synchronization method. Variable feedback synchronization method is simple and easy to implement. According to different systems, the single variable, multivariables, or even all variables of the system can be flexibly used to feedback control. Research results show that the minimum number of feedback variables should be equal to the number of positive Lyapunov exponent of the system without perturbation. Meanwhile, multivariable feedback is more effective than single-variable feedback.

Adaptive Synchronization: In 1990, Huberman and Lumer employed adaptive method to control chaos [24]. John and Amritker applied this principle to chaos synchronization, and the phase space trajectory of the chaotic system is synchronized with the desired unstable orbit [25]. Specifically, the adaptive synchronization method is to achieve chaos synchronization by automatically adjusting certain parameters of system with adaptive control technology. There are two prerequisites for the application of this method. (1) At least one or more parameters of the system can be obtained. (2) For the desired orbit, the values of these parameters are known. Adjustment of system parameters depends on two factors: (1) the difference between output variable of the system and corresponding variable of the desired orbit and (2) the difference between the controlled parameter value and the corresponding variable value of desired system.

Pulse Synchronization: In 1997, Yang and Chua put forward pulse synchronization control method [26, 27]. For pulse synchronization, the response system is driven by single pulse transferred from drive signal. The transmitted signal is incomplete chaotic signal, so it is more secure. Pulse synchronization has strong noise immunity and robustness, so it is a promising method of chaos synchronization. According to the basic theory of impulsive differential equations, the stability of synchronization error system depends on its comparative systems. When the comparative system approaches stable state asymptotically, pulses synchronization becomes stable, and the pulse intervals can be obtained.

Projective Synchronization: In 1999, Mainieri and Rehacek observed a new phenomenon – projective synchronization in the research of some linear chaotic systems [28]. By employing this synchronization, for some linear chaotic systems, when an appropriate controller is selected, the output phase of the drive system and response system will be locked, and the amplitude of each corresponding output is also evolved according to a certain proportional relation. This new synchronization method caught researchers' attention, and a lot of projective synchronization schemes are proposed. Recently, a function projective synchronization method is proposed [29]. Compared with the general projective synchronization, function projective synchronization means that the drive system and response system can be synchronized according to a certain function proportion, which has important significance for the realization of chaotic secure communication. In engineering practice, the parameters of a system may be unknown. In order to realize parameter identification of chaotic systems with unknown parameters, people began to apply adaptive synchronization method to the synchronization research of chaotic systems with unknown parameters. Obviously, combining adaptive control and function projective synchronization to research the synchronization of chaotic systems with unknown parameters has more theoretical value and practical significance [30].

Dead-Beat Synchronization: In 1995, Angeli et al. proposed a synchronization method for discrete chaotic systems, called dead-beat synchronization method [31]. The most important feature of this method is that as long as there are several steps of iteration, chaos synchronization can be achieved accurately. Although several synchronization methods described above can be extended to a discrete system, they are different from dead-beat synchronization method. First, dead-beat chaotic synchronization is accurate synchronization, and other synchronization is asymptotic synchronization in a certain sense. Second, the synchronization time for dead-beat synchronization method is the time for iteration N (N is the dimension of the state space). The synchronization time for other synchronization methods is not only closely related with system parameters, but also depends on the given accuracy range.

Performance of different synchronization methods is different. Based on Lorenz system, researchers have conducted simulation experiments about synchronization time and sensitivity to parameter's change with the drive-response synchronization, mutual coupling synchronization, feedback perturbation synchronization, adaptive control synchronization, and pulse synchronization. Results show that for feedback synchronization between Lorenz systems, the robustness to parameter changes is the best, and the establishment of synchronization is the fastest. Setup time for adaptive synchronization and pulse synchronization is longer. Weak robustness to parameter change means relatively complex, but its security is better. Therefore, from the synchronization setup time and robustness to parameter changes, the feedback synchronization method is the optimal synchronization method.

1.5.2 Other Synchronization Methods and Problems

In the twenty-first century, chaos synchronization study shows a new trend. First, new synchronization methods have been put forward. In addition to improving the existing synchronization methods, the main work is to introduce advanced control theory and technology into chaos synchronization research, such as fuzzy control [32], genetic algorithm [33], and state observer method [34]. All this methods have achieved good results. Second, the object of study has turned to discrete chaotic systems from continuous chaotic system, from low-dimensional chaotic system to high-dimensional hyperchaotic system. Third, how to improve the performance of chaotic synchronization began to draw more attention [35], such as to improve the synchronization system performance with system identification and neural network technology.

Synchronization is the key to achieve chaotic security communication. The most recent and the most competitive studies are to achieve chaotic secure communication based on chaos synchronization. As the foundation of chaos theory and technology, chaos synchronization also is an important aspect of a chaotic mechanism. Many chaos synchronization methods perhaps will not soon get practical value, but each new proposed method will give a new inspiration to open up a new avenue of research. Although many chaos synchronization methods have been proposed, they have not reached the mature stage. There are still many theoretical and technical issues to be resolved in practical applications.

1. Establishing chaos synchronization theory. At present, some new chaotic synchronization methods, such as genetic algorithm, neural network, and fuzzy control, still lack universal theory. The synchronization theory needs further establishment and improvement.
2. Improving synchronization performance. Research on the performance of synchronization system is directly related to the engineering application of chaos. Carrying out this research is of great significance. In the case of mismatch between noise and parameter, chaos synchronization may be lost. It is the key technology

of chaotic secure communication to improve the stability of synchronization and to increase the anti-disturbance ability of the synchronization system by using the robust control method and the stability of dynamic system. We will discuss this in Chapter 4 in detail.

3. Developing chaotic network synchronization method. With the development of computer technology and network technology, the Internet has become the main means to transmit information for people. It is convenient, and meanwhile it has brought huge information security risks. How to achieve information-secure transmission becomes a hot spot. It is an effective way to realize the information security by exploring the real-time information encryption transmission based on chaotic network synchronization and network transmission protocol. It has broad application prospects.
4. Combining other chaos communication methods to promote the development of chaotic communications. As described above, besides chaos synchronous communication, there is chaos-coded communication. The two methods can be combined together. If an information signal is encoded first, and it is transmitted based on chaotic synchronization, then it effectively increases the difficulty of deciphering. The fifth, sixth, and seventh chapters of the book will do in-depth research on real-time chaotic secure communication based on chaotic encryption and chaotic synchronization.

1.6 Summary of Chaos Secure Communication

Since 1990, chaotic secure communication and chaotic encryption technology have become a hot topic in the field of international electronic communications [36–39]. So far, the chaos secure communication applications are roughly divided into three categories [18]. The first is to achieve secure communication using chaos directly. The second is to achieve secure communication based on chaotic synchronization signal. The third is digital encrypted communication based on chaos sequence. At present, the second class, chaos synchronous communication, is an international research focus. It has become a new high-tech field. With the development of network technology and computer technology, the third class of chaotic encrypted communication has captured more attention.

So far, four chaos synchronous secure communication technologies have been proposed and developed, including chaotic masking [40], chaos shift keying [41], chaotic modulation [42], and chaotic spread spectrum [43, 44]. The first category belongs to analog communications, and the other three belong to digital communications. At present, the four secure communication schemes are the most competitive technologies. There are three main ways to realize chaos secure communication, such as electrical systems, laser systems, and computer networks. The most mature technology is electrical systems. Here, the overview will focus on the basic principles and study progress of chaotic synchronization secure communication technology based on circuit systems and chaotic encryption technology based on computer network.

1.6.1 Chaotic Analog Communication

The typical technique of chaotic analog communication is chaotic masking. Chaotic masking is to superpose the message signal into the chaotic signal directly, and the signal is masked by the randomness of chaotic signal. Cuomo and Oppenheium constructed chaotic masking secure communication system based on Lorenz system [19, 45], combining two response subsystems into one complete response system, which the structure is identical with drive system. The receiver can replicate all states of the sender and achieve synchronization. At the sender, the message signal overlaps with larger amplitude chaotic signal to form noise-like signal. At the receiver, the chaotic signal is obtained from the response system, and the message signal is recovered by subtracting the chaotic signal generated by the response system from the received signal.

Chaotic masking requires that parameters of chaos circuit of the sender and the receiver are accurately matched. Minor differences of parameters may cause synchronization fail. It is very difficult for an attacker to obtain the key parameters, but the strict matching of parameters also puts forward a very high request to the circuit design.

1.6.2 Chaotic Digital Communication

Chaos digital communications include chaos shift keying, chaotic parameter modulation, and chaotic spread-spectrum communication. Chaos shift keying is a chaos communication method that is suitable for digital communications. It can code a binary signal using different attractors produced by different parameters. For example, code 1 indicates parameter μ_1, and its corresponding chaotic attractor is A_1. Code 0 indicates parameter μ_2, and its corresponding chaotic attractor is A_2. The behaviors of chaos switch between A_1 and A_2. Response time of the system is controlled by changing the parameter. A parameter at the transmitter is modulated by using a binary signal, and then the chaotic signal is sent to the receiver. The modulated signal can be detected at the receiver by using the synchronization error. Thereby the real signal is recovered. Chaos shift keying has better robustness than that of chaotic masking method, and the anti-interference performance is better, but it has lower information transmission rate. The original signal may be recovered by using the short-time zero crossing rate method, but the security will be reduced. So researchers have proposed an improved chaos shift keying digital communication schemes, including chaotic on-off keying (COOK), differential chaos shift keying (DCSK), frequency modulation differential chaos shift keying (FM-DCSK) modulation, etc. [46–48].

Chaos parameter modulation is to modulate certain parameters of chaotic system by using the original signal in chaos oscillator in the sender. The changes of this parameter reflect changes of the original signal. There is a synchronous chaotic system in the receiver. Because the changes of chaotic state also include the changes in the original signal, the original signal can be detected and recovered by a nonlinear filter. Yang and Chua proposed a chaotic parameter modulation scheme in 1996, which is suitable for general signal modulation [42] and comprehensively analyzed several methods for parameter modulation based on Chua circuit. The simulation results show that, when modulating the different parameters of Chua circuit, the signal recovery accuracy is different. If we modulate capacitor and resistor, the recovery accuracy is higher.

Chaotic spread-spectrum communication is to achieve spread-spectrum communication by using chaotic sequence instead of spreading codes. The traditional code division multiple access (CDMA) technology is mainly limited by the periodicity of the *PN* code and the available number of orthogonal *PN* code address. Because chaotic systems are sensitive to initial conditions, a large number of spreading sequences with good correlation properties can be generated through the evolution of chaotic system. Based on the statistical characteristics of the chaotic sequence, it is very effective for the realization of the chaotic spread-spectrum code division multiple access communication. Chaotic sequence can be generated by an initial value and a mapping formula, without having to store values of the sequence. Therefore, it is more suitable for spreading code in spread-spectrum communication, and it has broad application prospects.

In all, chaotic communication has the following advantages [18]:

1. High security. The security of chaotic communication mainly comes from complex dynamic behavior of chaos system, sensitivity to initial conditions, and long-term unpredictability of dynamic behavior.
2. High capacity of dynamic storage.
3. Low power and low accessibility.
4. Low cost.

Based on the advantages above, applications of chaos in secure communication are still in the initial stage of the study; it has captured a lot of attention and interest from physics, information science, and other interdisciplinary fields. It is foreseeable that chaotic secure communication will play an important role in people's lives, especially in the military and national information security.

1.6.3 Encryption Communication based on Chaotic Sequence

In 1989, British mathematician Matthews proposed a new encryption method – chaotic encryption method [49], and thus the study of chaos cryptography began.

In the past decades, along with the deepening theoretical study of chaos, the application scope of chaos theory is also expanding. Chaos cryptography application has become a hot topic, and a number of chaotic encryption algorithms have been proposed [50, 51]. Chaotic systems can provide pseudorandom sequence with good randomness, correlation, and complexity. The key sequences generated by a chaotic system not only exhibit excellent cryptographic performances, but also have a rich source. In addition, the chaotic encryption method greatly simplifies the design process of traditional sequence cipher. Those attractive features make it possible to be a stream cryptosystem. The unique advantages of chaos in cryptography attracted many scholars and academic communities around the world to do research.

Unlike chaos analog communication, the anti-crack ability of chaotic encrypted communication is strong. Chaos analog communication using small signal modulation cannot resist the attacks of neural network and regression map. While for chaos encrypted communications, the attacker cannot get the complete original data, and the chaotic signals generated by original equation or regression map reconstruction cannot be obtained according to neural network or regression method. It is difficult to decipher the encrypted signals and it can be applied to highly confidential communication systems.

With the development of computer technology, digital technology, and network technology, chaos encryption software has been emerging persistently [52]. Chaos-based engineering applications have also appeared [53], and the hardware achievements based on Digital Signal Processing (DSP) and Field Programmable Gate Array (FPGA) technologies have also been reported [54]. Chaotic encryption with digital technology has many advantages, such as low requirements for accuracy of circuit elements, easy hardware implementation, easy computer processing, less loss of information transmission, versatility, and so on. Especially when it is used in real-time signal processing, the encrypted information is difficult to be deciphered. Therefore, chaotic encryption shows strong vitality in secure communications.

Compared with conventional packet encryption system, chaos encryption system has the following characteristics:

1. Complex key signal generated by simple circuit or iterative equation.
2. Key signal is unpredictable. There is no exclusive corresponding plaintext–ciphertext pair, and the same plaintext corresponds to different and irrelevant ciphertext. So it is similar to one-time pad.
3. Design method and implementation technology of key space need to be further studied.
4. Security of systems can only be tested by limited numerical algorithm.

1.7 Chaos Research Methods and Main Research Contents

Because of the complexity of chaos, the research on the chaotic motion should employ many methods, including theoretical analysis, numerical simulation, experimental study, and so on. At present, the methods on the chaos theory mainly include numerical calculation, symbolic dynamics, Melnikov and Shilikov method, phase space reconstruction method, and so on. The numerical methods for judging the chaotic state are the Lyapunov exponents, fractal dimension, power spectrum, Poincaré section, direct observation, and so on. Among them, the Lyapunov exponents and fractal dimension are quantitative methods, and the others are qualitative descriptions. Power spectral density shows the signal changes with the frequency by Fourier transform. The power spectrum of a chaotic signal is changed continuously. Poincaré section is a section in the phase space. The phase trajectory leaves a point at the cross section when it passes the cross section so that the flow in n-dimension can be reduced to a map with $n - 1$ dimension, and many properties of the original system are preserved. For a chaotic system, these points have a fractal structure and cannot fill up the entire section. To observe whether there exists butterfly effect is a typical direct observation method. Comparing the time series of chaotic system, two very similar initial values after a short-time evolution will be very different. It reflects the sensitivity of chaotic system to initial value. For the chaotic systems with known dynamics equations, the symbolic dynamics method is used to study the dynamics equation. In the second chapter of this book, we will focus on the analysis methods of chaotic characteristics.

Numerical simulation is an important method for the study of chaos theory and application. First of all, numerical simulation method provides the application issues associated with the real system for chaos theory analysis, and it broadens the field of theory research. Simulation system provides ideal plasticity experimental model for theoretical research, and it opens up the new field of chaotic application study. Second, the simulation method can provide the basis for the feasibility of constructing the real and practical system. The third and fourth chapters of this book are based on the Matlab/Simulink numerical simulation method. The third, fourth, and fifth chapters mainly use the numerical simulation method based on Matlab/Simulink to study.

Real experimental study is an important method for the practical application of chaos. Since the 1990s, the results of a large number of theoretical studies are based on real system experiments and applications. Real experiments include circuit experiments and computer experiments. With the development of microelectronics technology, it is possible to design various kinds of high precision chaotic components and systems. The development of network technology and computer technology provides a platform for constructing and testing various chaotic encryption softwares. The fifth and sixth chapters of this book will be mainly based on computer system and network.

Three kinds of chaos analysis methods discussed above are mutual dependence and mutual promotion, and constantly push forward the development of chaos research. In this book, the theoretical analysis, numerical simulation, and real experiments are combined. This book will mainly use the Lyapunov exponents, conditional

Lyapunov exponents, and fractal dimension for the theoretical study. Based on Matlab/Simulink simulation platform, we analyze chaotic system synchronization control mechanism and the chaotic synchronization performances. Through computer advanced language and computer network, we carry out real experimental researches, designing practical chaotic encryption software, testing it, and analyzing the security.

In the study of chaos synchronization control, both the continuous system synchronization and discrete system synchronization are studied. The relationship between the performance and the control parameters is particularly studied. In the application of chaotic systems, we will focus on the integration of chaotic dynamics, advanced control technology, modern cryptography, and modern communication technology. Combining the physical synchronization of chaotic system with computer programming technology, static data file encryption softwares are designed. The network-based self-synchronization method is investigated. Chaotic encryption software to achieve real-time voice signal transmission is designed by using the advanced computer language. Those above are the main achievements of the research group in the field of chaotic application.

Questions

1. What is chaos? What are the main characteristics of chaos?
2. What is butterfly effect? What is its philosophical significance?
3. What is the definition and significance of the Feigenbaum constant?
4. What are the main techniques of chaotic secure communication? What are the characteristics?
5. Why chaos synchronization is the key technology of secure communication?

2 Characteristic Analysis Methods for Nonlinear System

Because of the complexity of the nonlinear system, how to judge the chaotic behavior of a nonlinear system is an important question in the study of chaos theory and application. Generally, methods to determine the chaotic characteristics of a nonlinear system include phase diagram analysis, power spectral analysis, Lyapunov exponent spectrum analysis, fractal dimension analysis, dividing frequency sampling method, pseudo-phase space method, Poincaré section, 0–1 test, and so on.

2.1 Phase Diagram Analysis Method

Phase diagram analysis is one of the direct observation methods. Characteristics of the aperiodic motion of a chaotic motion can be displayed by the method of phase plane graph. Periodic motion repeats the previous movement at every cycle, so the trajectory of its movement is a closed curve. Chaotic motion is a nonperiodic motion, so its trajectory curve is a never-closed curve. The reciprocating movement reflected in the trajectory curve is confined to a bounded area. It does not diverge to infinity, and it does not converge to a stable point; thus it forms a strange attractor. When we simulate a nonlinear system, its dynamic behavior can be determined by the phase diagrams of the system directly, including limit circle, period movement, and chaos. The disadvantage of phase trajectories diagram is that when the periodic motion of the cycle is very long, it is difficult to accurately distinguish periodic motion or chaotic motion only based on the phase plane diagram.

A chaotic attractor has a complex structure by stretching and folding, which leads to the system being kept in a limited space. It is the result of the global stability and local instability of the dynamic system. The global stability makes all the movements outside the attractor get closer to the attractor, and the trajectory converges to the attractor. While the local instability makes all the arrived inside the attractor trajectory mutually reject, and the trajectory spreads in a certain direction. A small perturbation is stable in the chaotic attractor, and it will eventually reach the attractor. But in the interior of the chaotic attractor, the motion state of the system is very sensitive to the initial condition. In other words, if the location entering the chaotic attractor is slightly different, then the difference will grow exponentially and will eventually lead to different chaotic orbits. A chaotic attractor has other properties, such as infinite embedded self-similarity structure and fractal dimension.

For example, the periodic forced nonlinear system equation is [55]

$$\ddot{x} + a\dot{x} - x + b\,\mathrm{sgn}(x) = F\sin(\omega t), \tag{2.1}$$

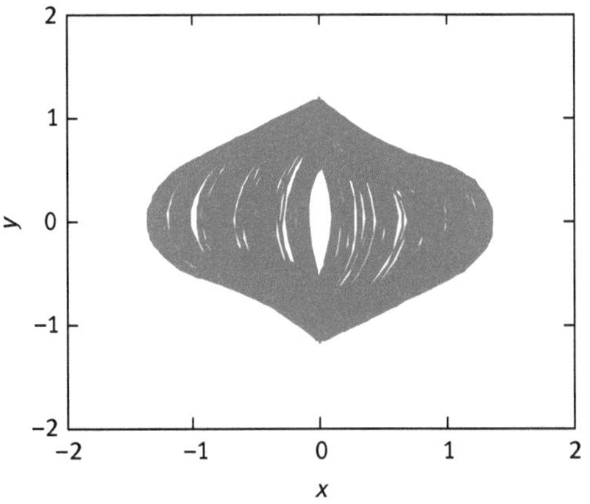

Figure 2.1: Strange attractor of system (2.1) on x–y plane.

where a, b, F, and ω are constants; sgn(x) is a symbol function. When $a = 1.05$, $b = 1$, $F = 1.1$, and $\omega = 1$, the 'Onion' chaotic attractor of the system in the x–y plane is shown in Figure 2.1.

2.2 Power Spectral Analysis Method

Power spectral density function is often used to characterize the randomness of chaotic motion. Power spectrum indicates the statistical characteristics of the frequency components in the random motion process, so it is the basic method to analyzes chaotic motion in nonlinear systems. According to the randomness of chaotic motion, we often use the spectrum analysis method to identify chaos in a nonlinear system. Chaotic motion is a nonperiodic motion that can be regarded as the superposition of periodic motion with infinite number of different frequency, and its power spectrum has the characteristics of random signal. Obviously, the power spectrum of chaotic motion is continuous spectrum. Power spectrum analysis can be used to determine whether the motion is random, but it cannot be determined whether this random motion is due to the external random disturbance or the internal randomness nature of the deterministic nonlinear system. Therefore, it is difficult for power spectrum analysis to distinguish between chaotic motion and real random motion.

The power spectrum analysis can provide the signal frequency domain information, and it can be used to observe whether a nonlinear system has chaotic characteristics. For periodic motion, the power spectrum is discrete spectrum, and for chaotic motion, the power spectrum is continuous spectrum. So we can judge whether a system is chaotic through the power spectrum diagram of the system.

According to Fourier transformation, for a function $x(t)$, $-\infty < t < \infty$, if $x(t)$ satisfies the Dirichlet condition

$$\int_{-\infty}^{+\infty} |x(t)| dt < \infty, \tag{2.2}$$

then we have

$$x(t) = \int_{-\infty}^{+\infty} F(f) e^{i2\pi ft} df, \tag{2.3}$$

where $F(f) = \int_{-\infty}^{\infty} x(t) e^{-i2\pi ft} dt$. We call $F(f)$ the Fourier transformation or the spectrum of $x(t)$. The following is the relationship between $F(f)$ and $x(t)$:

$$\int_{-\infty}^{+\infty} |x(t)|^2 dt = \int_{-\infty}^{\infty} |F(f)|^2 df, \tag{2.4}$$

where the left-side term represents the total energy of $x(t)$ over the range $(-\infty, \infty)$, while $|F(f)|^2$ in the right side represents the spectral density of $x(t)$, which means the energy or power per unit interval is on the frequency scale. Equation (2.4) is the spectral representation of the total energy of time function. The average power per unit interval on the frequency scale is

$$G(f) = \lim_{T \to \infty} \frac{1}{2T} |F(f)|^2. \tag{2.5}$$

It is also defined as the average power spectral density of $x(t)$.

According to Fourier analysis, a periodic signal $x(t)$ can be decomposed into Fourier series

$$x(t) = \sum_{n=-\infty}^{\infty} c_n e^{jn\omega_0 t}, \tag{2.6}$$

where

$$c_n = \frac{1}{T} \int_{T/2}^{T/2} x(t) e^{-jn\omega_0 t}. \tag{2.7}$$

A quasi-periodic signal also can be decomposed into a series of frequencies of the superposition of the sine vibration. Both of them are discrete spectrum.

If an aperiodic signal $x(t)$ satisfies the condition of being absolutely integrable, then it can be represented by Fourier integral

$$x(t) = \frac{1}{2\pi} \int_{-\infty}^{\infty} X(\omega) e^{j\omega t} d\omega, \tag{2.8}$$

$$X(\omega) = \int_{-\infty}^{\infty} x(t) e^{-j\omega t} dt. \tag{2.9}$$

Obviously, the frequency spectrum of nonperiodic motion signal is continuous spectrum.

To represent the frequency domain characteristics of chaotic signals, we can calculate the Fourier transform of its autocorrelation function $R_{xx}(\tau)$ and analyze its frequency domain characteristics by auto-power spectral density function $S_{xx}(f)$

$$R_{xx}(\tau) = \int_{-\infty}^{\infty} S_{xx}(f) e^{-j2\pi f\tau} df. \tag{2.10}$$

For periodic motion, the power spectrum appears at the fundamental frequency and its frequency doubling. For the quasi-periodic motion, the power spectrum is located in irreducible fundamental frequency and their superimposed frequency spikes. The power spectral can display bandwidth characteristics of the noise signal. When the period-doubling bifurcation occurs, frequency division and frequency doubling occur in power spectrum, and peaks will display at this frequency points of the power spectrum. The chaotic motion is characterized by a continuous spectrum with the noise background in the power spectrum, which contains the peak corresponding to the periodic motion. According to these characteristics, the motion of a system can be easily identified as the periodic, quasi-periodic, random, or chaotic.

In order to obtain the reliable power spectrum, the average of several sequential sampling sequence spectra is required. In addition, the transition process must be ended before the start of the sampling. If the original data contains a large amount of noise and external interference, the appropriate filtering or smoothing should be considered.

2.3 Poincaré Section Method

Poincaré section was proposed by Poincaré at the end of the nineteenth century, which was used to analyze the motion characteristics of the multivariable autonomous system.

The phase diagram of the Poincaré map well characterizes the reciprocating nonperiodic characteristics of chaos. If the Poincaré map is neither a finite set nor a closed curve, then the corresponding system motion is a chaotic motion state. More specifically, if the system does not have an external noise disturbance, and there is a certain damping, the Poincaré map will be a point set with a detailed structure. If the system disturbed by external noise or damping is small, the Poincaré map will be a set of fuzzy points. If the Poincaré is a finite set of points, then the corresponding system motion is periodic motion state. Since the resolution of Poincaré section is higher than that of the phase plane graph, the Poincaré section is a popular method for the analysis of chaotic dynamics.

A section is appropriately selected in phase space. On this cross section, if a pair of conjugate variables is fixed, it is called Poincaré section. The intersections of the

Poincaré section with the continuous trajectory in the phase space are called the cutoff points. By observing the distribution of the cutoff points of the Poincaré section, we can judge whether the chaotic phenomenon occurs. When there is only one fixed point or a small number of discrete points on the Poincaré section, the motion is periodic. When there is a closed curve on the Poincaré section, its movement is quasi-periodic. When there are pieces of dense points with a fractal structure on the Poincaré section, the motion is chaotic.

To observe the route to chaos of the system (2.1), the phase diagram and the Poincaré section of the system are shown in Figure 2.2 with $a = b = 1$ and F change. The Poincaré section is $z \bmod 2\pi = 1$. With the increase of F, the limit cycle phase diagram is observed as shown in Figure 2.2(a_1), and its corresponding Poincaré section is a point as shown in Figure 2.2(a_2). To increase the parameter F, the limit cycle phase diagram by period doubling is obtained as shown in Figure 2.2(c_1) and its corresponding Poincaré section is the two points as shown in Figure 2.2(c_2). To increase the parameters F further, the limit cycle phase diagrams by period doubling again is obtained as shown in Figure 2.2(d_1) and the corresponding Poincaré section is the four points as shown in Figure 2.2(d_2). Finally, the chaotic attractor phase diagram is shown in Figure 2.2(e_1) and the infinite points on the corresponding Poincaré section are shown in Figure 2.2(e_2). Thus the process of chaos of the system (2.1) by the period-doubling bifurcation is demonstrated by the phase diagram and the Poincaré section.

At the same time, we can observe fork bifurcation with parameter F increasing from 0.65 to 1.045 as shown in Figure 2.2(b), and then the two coexisting attractors expand as shown in Figure 2.2(c) and (d) and finally merge by an attractor-merging crisis bifurcation at the onset of chaos as shown in Figure 2.2(e).

2.4 Lyapunov Characteristic Exponent Method

To quantitatively characterize the sensitivity to initial value of chaotic system, the Lyapunov characteristic exponent is needed.

2.4.1 Definition of Lyapunov Characteristic Exponent

Lyapunov exponent refers to the average change rate of two trajectory separations or aggregates to each other with the time going on in the phase space. In the practical application, the maximum Lyapunov exponent λ_1 is of great significance, and it can be calculated by

$$\lambda_1 = \frac{1}{t_M - t_0} \sum_{k=1}^{M} \ln \frac{D(t_k)}{D(t_{k-1})}, \tag{2.11}$$

where $D(t_k)$ is the distance of most adjacent two points at t_k; M is the iterative number; λ_1 is not only the quantitative index of the chaotic attractor, but also the quantitative

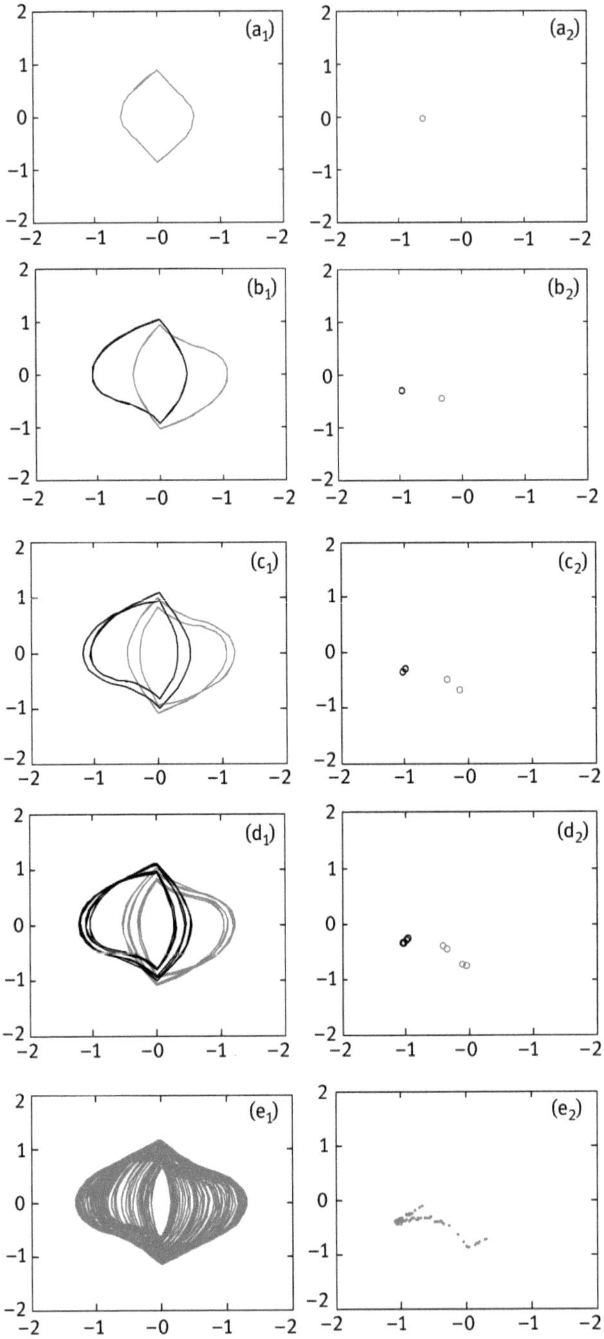

Figure 2.2: Phase diagram (left) and its corresponding Poincaré section (right) of system (2.1): (a) $F = 0.65$, $1T$ period; (b) $F = 0.95$, $1T$ period; (c) $F = 1$, $2T$ periods; (d) $F = 1.01$, $4T$ periods; and (e) $F = 1.045$, chaotic motion.

description of the initial deterministic amplification ratio of chaotic system. In addition, its reciprocal is generally considered to be the maximum predictable time scale of the metric system state.

2.4.2 Lyapunov Exponent Spectrum

For systems with dimensions greater than 1, there is a set of Lyapunov exponents, often referred to as the Lyapunov exponent spectrum. Each of the Lyapunov exponents characterizes the convergence of the trajectory in a certain direction.

Now, a definition of Lyapunov exponent spectrum is given. For a continuous dissipative dynamical system with n-dimensional phase space, it is assumed that the long-term evolution of an infinitesimal n-dimensional sphere among the initial conditions of the system can be controlled. And because of the evolution of the natural deformation, the ball will become ellipsoid. If all the spindles in the sphere center on the ellipsoid attractor are ranked by the order from the fastest to the slowest speed, then the ith Lyapunov exponent is defined according to the increase rate $p_i(t)$ of the ith axis

$$\lambda_i = \lim_{t\to\infty} \frac{1}{t} \ln\left(\frac{p_i(t)}{p_0(t)}\right). \tag{2.12}$$

Note that the linear range of the ellipse increases with $e^{\lambda_1 t}$, and the area defined by the first two principal axes increases with $e^{(\lambda_1+\lambda_2)t}$, and the volume defined by the first three principal axes increases with $e^{(\lambda_1+\lambda_2+\lambda_3)t}$, and so on. In fact, this feature is a definition of the Lyapunov exponent spectrum, i.e. the sum of the first j exponents is determined by the long-term average rate of the volume defined by the first j principal axes. If the evolution of all the points on the sphere is considered, the behavior of the phase space can be visualized.

For example, the simplified Lorenz system equation is [56]

$$\begin{cases} \dot{x} = 10(y-x) \\ \dot{y} = -xz + (24-4c)x + cy \ , \\ \dot{z} = xy - 8z/3 \end{cases} \tag{2.13}$$

where c is the system parameter, and when $c \in [-1.59, 7.75]$, the system is chaotic. Its two Lyapuov exponents are shown in Figure 2.3. When $\lambda_1 > 0$, $\lambda_2 = 0$ and when $\lambda_3 < 0$, the system is chaotic state. When $\lambda_1 = 0$, $\lambda_2 < 0$, and $\lambda_3 < 0$, the system is periodic state.

2.4.3 Physical Meaning of Lyapunov Exponent

Sensitivity of a system to the initial value can be effectively characterized by the Lyapunov exponent spectrum of the system. If the Lyapunov exponent is negative,

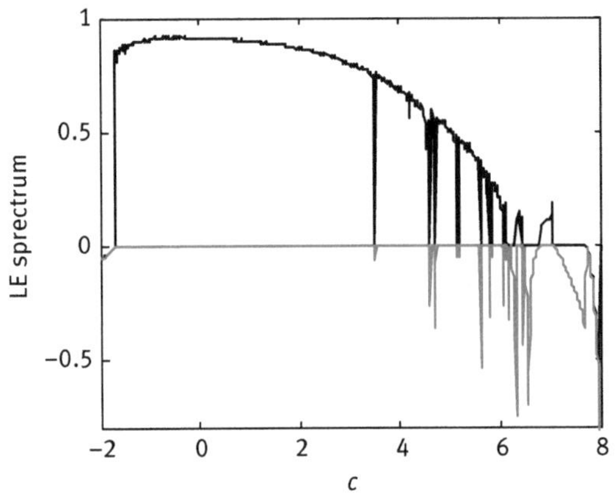

Figure 2.3: Lyapunov exponents of the simplified Lorenz system with different parameter *c*.

it indicates that the phase volume of the system shrinks in this direction, and the movement of the system is stable. For a nonlinear dissipative system, the sum of the Lyapunov exponents must be negative, and it indicates that the overall motion of the system in the phase space is stable and contraction. While a positive Lyapunov exponent indicates that the system phase volume expands and folds in one direction so that the originally adjacent trajectories in the attractor become increasingly irrelevant, and the long-term behavior of the system becomes unpredictable, namely the so-called initial value sensitivity. At this time, the motion is chaotic state, and it has a strange attractor. So, for a chaotic system, one Lyapunov exponent is greater than zero at least.

Lyapunov exponent can depict the strange attractors. The strange attractor is a phase diagram with fractal structure in the phase space. Since the chaotic attractor is often of noninteger dimension, the confirmation of strange attractors is significant for studying the chaotic motion. If a system's Lyapunov exponent spectrum is λ_1, $\lambda_2, \ldots, \lambda_n$ (from large to small), and the system has a chaotic attractor, it must satisfy the following conditions:

1. At least one positive Lyapunov exponent.
2. At least one of Lyapunov exponents $\lambda_i = 0, 1 < i < n$.
3. The sum of the Lyapunov exponents is negative, that is, $\sum_{i=1}^{n} \lambda_i < 0$.

The first condition above indicates that the adjacent track in the phase space is separated with the exponential rate, and the system is sensitive to the initial value, which is the prominent feature of the chaos. So the Lyapunov exponent spectrum must have positive component, which is an important symbol to distinguish the strange attractor. Sum of all positive Lyapunov exponents is defined as Kolmogorov entropy, which is a measure of the degree of disorder of the system, and it is a characterization of chaos. The second condition is the periodic performance of the chaotic attractor, and the period is infinite. The third condition shows that the phase volume of the system is contracted, thus ensuring the stability of the system. For a three-dimensional chaotic system, the symbols of Lyapunov exponent spectrum have only one case, namely $(+, 0, -)$, and $\lambda_1 < -\lambda_3$.

2.4.4 Condition Lyapunov Exponent

Condition Lyapunov exponent (CLE) is a generalization of the Lyapunov exponent in the chaotic synchronization system. In the research of chaos synchronization, the CLE is an important quantitative index, and it is important to guarantee synchronization. Here, the conditional Lyapunov exponent is illustrated by employing Pecora–Carroll (PC) chaos synchronization method.

According to PC chaotic synchronization method, an n-dimensional autonomous dynamical system

$$\dot{u} = f(u) \tag{2.14}$$

is decomposed into two subsystems:

$$\dot{v} = g(v, w), \dot{w} = h(v, w), \tag{2.15}$$

where the first subsystem variable $v = (u_1, u_2, \ldots, u_m)$, and the second subsystem variables $v = (u_{m+1}, u_{m+2}, \ldots, u_n)$, $g = (f_1(u), f_2(u), \ldots, f_m(u))$, $h = (f_{m+1}(u), f_{m+2}(u), \ldots, f_n(u))$. Copy a new system

$$\dot{w}' = h(v, w'). \tag{2.16}$$

It is exactly the same as the system w. Obviously, the synchronization condition between w' and w is the error $w' - w = \Delta w \to 0$ as $t \to \infty$.

If $D_w h$ is the Jacobian matrix of the subsystem, its corresponding Lyapunov exponent is called the conditional Lyapunov exponent of the subsystem w, and it is defined as

$$\lambda_j = \lim_{t\to\infty} \frac{1}{t} \ln \left\| (D_w h^t)_{x_0} v \right\| . \tag{2.17}$$

where $j = m + 1, \ldots, n$, and v is the unit matrix, and $(D_w h^t)_{x_0} = (D_w h)_{x_t} (D_w h)_{x_{t-1}}, \ldots, (D_w h)_{x_0}$.

Only when all of the CLEs are negative, system w' can synchronize with w.

2.5 Fractal Dimension Analysis Method

Dimension is an important feature of the nonlinear system. Divergence of orbits of chaotic system makes the whole phase space of the system become impossible to be filled by phase trajectories, so chaotic attractor has a noninteger dimension. In theory, the chaotic attractor has infinite-level self-similar structure. Therefore, fractal dimension is an effective parameter to distinguish chaotic motion.

There are many definitions for dimension, for example, capacity dimension, correlation dimension, information dimension, similarity dimension, and so on. The fractional dimension of the chaotic system is generally discussed in its capacity dimension. The definition of capacity dimension is the number of cubes $L(\varepsilon)$ when one uses an N-dimensional cube epsilon with the side length ε to measure a given space. It is defined by

$$D_C = \lim_{\varepsilon \to 0} \frac{\ln L(\varepsilon)}{\ln(1/\varepsilon)}. \tag{2.18}$$

The relationship between the capacity and the Lyapunov exponent spectrum is

$$D_C = j + \sum_{i=1}^{j} \frac{\lambda_i}{-\lambda_{i+1} + 1}, \tag{2.19}$$

where j satisfies $\lambda_1 \geq \lambda_2 \geq \cdots > \lambda_j \geq 0 \geq \lambda_{j+1} \geq \cdots \geq \lambda_N$. If the above formula contains $\lambda = 0$, then the number of λ should be added to j.

The fractional dimension of the simplified Lorenz system with different parameter c is shown in Figure 2.4. Obviously, when $D_C = 0$, the system converges to equilibrium point. When $D_C = 1$, the system is periodic (limit cycle). When $D_C > 2$, the system is chaotic. So the fractal dimension like the Lyapunov exponent can be used to describe the dynamics of a nonlinear system.

2.6 0–1 Test Method

Recently, Gottwald and Melbourne [57] proposed a reliable and effective binary test method for testing whether a nonlinear system is chaotic, which is called "0–1 test." The basic idea is to create a random dynamic process for the data and then study how the scale of the stochastic process changes with time. Next, we discuss this method and apply to analyze the characteristics of the fractional-order simplified Lorenz system.

2.6.1 0–1 Test Algorithm

Consider a set of continuous time series $\vec{x}(t)$ to characterize the dynamic system, and $\phi(\vec{x})$ represents the observable data set. This test method is completely independent

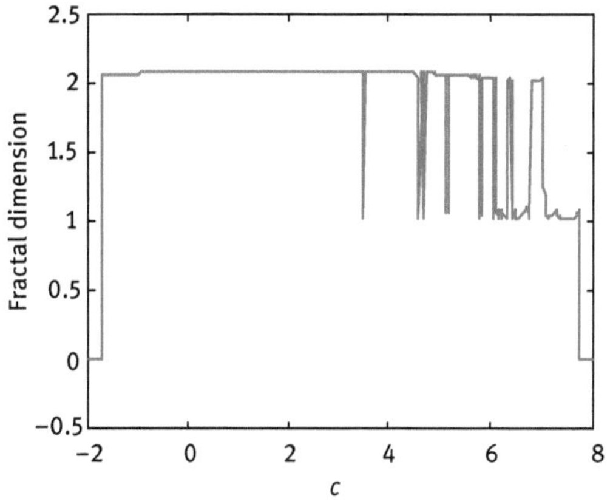

Figure 2.4: Fractional dimension of the simplified Lorenz system with different *c*.

of the specific form $\phi(\vec{x})$, that is, the condition can be satisfied for any form of ϕ. For example, if $\vec{x} = (x_1, x_2, \ldots, x_n)$, then $\phi(\vec{x}) = x_1$ satisfies the condition. Similarly $\phi(\vec{x}) = x_2$ or $\phi(\vec{x}) = x_1 + x_2$ also satisfies the condition. Choose an arbitrary constant $c > 0$ and define the following equations:

$$\theta(t) = ct + \int_0^t \phi(\vec{x}(s))ds, \tag{2.20}$$

$$p(t) = \int_0^t \phi(\vec{x}(s))\cos(\theta(s))ds. \tag{2.21}$$

Thus, $p(t)$ in eq. (2.21) is only related to the observation data set $\phi(\vec{x})$, which demonstrates the general property of the testing method. This test algorithm has no relationship with the original data of the system and the system dimension. From the above definition, we can get two conclusions as follows:

1. If the system is not chaotic, then the movement of $p(t)$ is bounded.
2. If the system is chaotic, then the movement of $p(t)$ is similar to that of the Brown movement.

To characterize the growth of $p(t)$, the root mean square displacement of $p(t)$ is defined as

$$M(t) = \lim_{T\to\infty} \frac{1}{T}\int_0^T [p(t+\tau) - p(\tau)]^2 d\tau. \tag{2.22}$$

If the motion of $p(t)$ is the Brown movement, $M(t)$ linearly increases with time. If the motion of $p(t)$ is bounded, then $M(t)$ is bounded. Define the asynchronous growth rate of $M(t)$ as

$$K = \lim_{t\to\infty} \log M(t) / \log t. \tag{2.23}$$

When the integral time t is long enough, then $K = 0$ or $K = 1$, which characterizes the nonchaotic motion or chaotic motion, respectively.

Since the integral is only suitable for continuous systems, the solution of vector integrals is not very convenient. An improved 0–1 test algorithm is introduced, which is used to replace integrating with summing so that it has wider applications.

2.6.2 Improved 0–1 Test Algorithm

Consider a set of discrete data $\phi(n)$ sampled at times $n = 1, 2, 3, \ldots$, which represents a one-dimensional observable data set. Choose a constant c at random and define [58]

$$p(n) = \sum_{j=1}^{n} \phi(j)\cos(\theta(j)), \quad n = 1, 2, 3, \ldots, \tag{2.24}$$

$$s(n) = \sum_{j=1}^{n} \phi(j)\sin(\theta(j)), \quad n = 1, 2, 3, \ldots, \tag{2.25}$$

where

$$\theta(j) = jc + \sum_{i=1}^{j} \phi(i), \quad j = 1, 2, 3, \ldots, n. \tag{2.26}$$

Then, on the basis of function $p(n)$, or $s(n)$, define the root mean square displacement as

$$M(n) = \lim_{N\to\infty} \frac{1}{N} \sum_{j=1}^{N} [p(j+n) - p(j)]^2, \quad n = 1, 2, 3, \ldots \tag{2.27}$$

Obviously, if the behavior of $p(n)$ ($s(n)$ behaves in the same way) is Brownian, then $M(n)$ grows linearly in time. If the behavior of $p(n)$ is bounded, then $M(n)$ is bounded. The asynchronous growth rate is defined by

$$K = \lim_{n\to\infty} \frac{\log M(n)}{\log n}. \tag{2.28}$$

Just as the same for continuous time system, $K = 0$ means that the system is regular (i.e. periodic or quasi-periodic), and $K = 1$ means the system is chaotic. For a finite data set $\phi(n)$, $1 \le n \le N$, we define

$$M(n) = \frac{1}{N-n} \sum_{j=1}^{N-n} [p(j+n) - p(j)]^2. \tag{2.29}$$

If $1 \leq n \leq N$, the scales of $M(n)$ is similar to that described by eq. (2.27). Therefore, by plotting $\log(M(n)+1)$ against $\log n$ for $1 \leq n \leq N/10$, the test still used to check whether K is close to 0 (regular motion) or close to 1 (chaotic motion).

Note that another interesting feature of the 0–1 test is shown that the inspection of the dynamics of the (p, s) trajectories provides a simple visual test about whether the underlying dynamics is chaotic or nonchaotic, namely bounded trajectories in the (p, s) plane imply regular dynamics, while Brownian-like (unbounded) trajectories imply chaotic dynamics.

2.6.3 Application of 0–1 Test Algorithm

The simplified Lorenz system is described by eq. (2.13). Now, we carry out the 0–1 test with $c = 7.2$ and $c = -1$, respectively, and we obtain $K = 0.0736$ at $c = 7.2$, and $K = 0.7630$ at $c = -1$, which corresponds to periodic state and chaotic state. The state space plots are shown in Figure 2.5 and the corresponding translation components (p, s) is shown in Figure 2.6, which corresponds to periodic motion and Brown motion. Obviously, the results of 0–1 test are the same with that of Lypunov exponents and phase diagram analysis [59].

2.7 Dividing Frequency Sampling Method

To avoid the confusion of complex motion in phase space, we can observe the representative points (i.e. sampling points) in the phase space at every time interval (i.e. the sampling period). Thus, the continuous trajectory in phase space is represented by a series of discrete points. Dividing frequency sampling method is a generalization of the scintillation sampling method in the experimental physics. It is the most effective method for identifying the long period of chaos. It is mainly used to study the period-doubling bifurcation and chaos phenomenon of nonlinear oscillator with the periodic force.

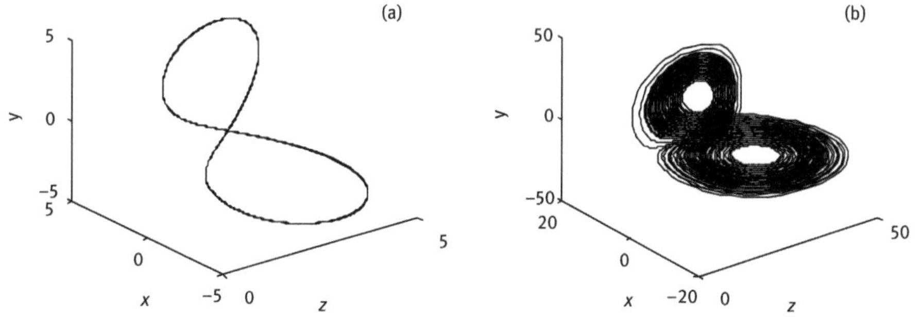

Figure 2.5: State space diagrams of the simplified Lorenz system: (a) $c = 7.2$ and (b) $c = -1$

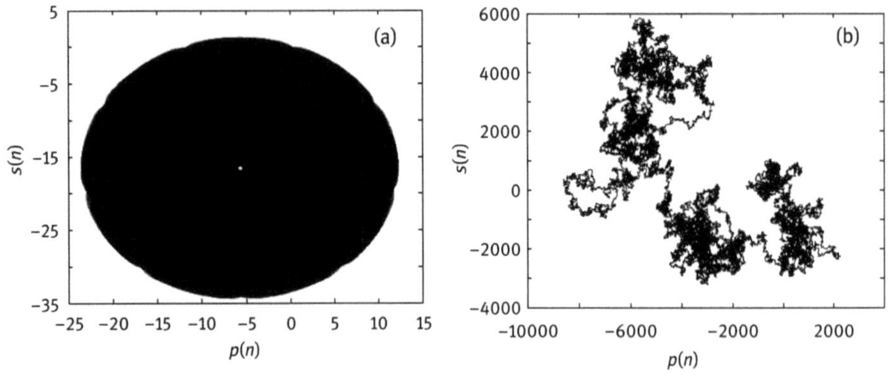

Figure 2.6: (p, s) plot of the simplified Lorenz system: (a) $c = 7.2$ and (b) $c = -1$.

For the forced vibration, the sampling period is often set to the period of the external control force. When the sampling result is a point, the system motion is a periodic motion (special case for the stable state). When the sampling results are n discrete points, the system movement is also a periodic, and its motion period is n times of the external control period. When the sampling results are infinite points, the system movement is random. When the sampling results are concentrated in a certain region and have a hierarchical structure, the system movement is chaotic. Continuously increasing resolution, we can get the fine structure of the original distribution. It is a kind of infinite-level self-similarity, and the self-similarity of the infinite level is the scaling invariance of the nonlinear system.

Dividing frequency sampling method is suitable for all the nonlinear systems driven by a periodic force for which its resolution is much higher than that of other methods. The resolution is limited by a computer word "length." But this method also has some shortcomings. One is that it is not unique for the result interpretation, and the other shortcoming is that the higher frequency over the sampling frequency cannot be distinguished.

2.8 Pseudo-Phase Space Method

When we analyze the chaotic characteristics of the unknown dynamical systems, the dividing frequency sampling method and the Poincaré section method are not applicable. Moreover, sometimes, it is easy to measure the time series of a variable in the experiments, and we can use the time series to reconstruct the phase space. The embedding theorem is used to solve the problem of establishing and describing the finite dimensional strange attractor and to reconstruct the dynamic system from a single time series. Reconstructed phase space, i.e. the dimension of the pseudo-phase space method (the embedding dimension), should satisfy $m \geq 2n + 1$, where n is the real dimension of phase space. Assuming the time series is $\{x(k), k = 1, 2, \ldots, N\}$, appropriately select the time delay τ, which is an integer multiple of the sampling period. Choosing $x(k), x(k+\tau), x(k+2\tau), \ldots, x(k+(m-1)\tau)$ as the coordinate axis, we can draw a pseudo-phase space trajectory.

The process of reconstructing the attractor is equivalent to mapping the time series $\{x(k)\}$ to the m-dimensional Euclidean space R^m, and the topological properties of the original unknown attractors may be maintained in the R^m space. Although the pseudo-phase space method is used to construct the phase space of a variable at different time, the change of one variable of the dynamical system is related to the other variables of the system. The dynamics of the whole system is implied by the change of the variable with time. So the trajectory of phase space reconstruction also reflects the evolution law of the system state. For the steady state, the results obtained by this method are still a certain point. For periodic motion, the results are limited points, and for chaotic system, the results are some discrete points with certain distribution patterns or structures.

2.9 Complexity Measure Algorithm

Complexity is a quantitative description of the complexity by employing a certain method. Obviously, different aspects of complexity are characterized by different complexity algorithms. So far, there is no unified definition about complexity. Each research field has its own understanding and interpretation about complexity. Horgan pointed out that there are at least 45 definitions about complexity at present, such as time complexity, space complexity, semantic complexity, Kolmogorov complexity, and so on [60].

Complexity of chaotic systems means the degree of the chaotic sequence close to the random sequence by use of correlation algorithm. The greater the complexity, the more the sequence is close to the random sequence, and the security of the corresponding application system is higher. In essence, the complexity of the chaotic system is a kind of chaos dynamics.

Complexity of chaotic sequences is divided into behavior complexity and structural complexity. Behavior complexity is used to measure the size of the new pattern probability of the short-time window. The larger the probability of generating a new pattern is, the more complex the sequence is. At present, there are many computational complexity algorithms for chaotic pseudo-random sequence, and most of them are based on Kolmogorov method and Shannon entropy, such as Lempl-Ziv algorithm [61], approximate entropy (ApEn) algorithm [62, 63], fuzzy entropy (FuzzyEn) algorithm [64, 65], intensity statistics complexity (ISC) algorithm [66, 67], and symbol entropy (SymEn) algorithm [68]. The calculation speed of these algorithms is fast and the result is accurate, but if the dimension is too high or the symbol space of the pseudo-random sequence is too large, the result will overflow. Structure complexity is used to measure the complexity of the sequence by the frequency characteristic and energy spectrum in the transformation domain. The more uniform distribution of the energy spectrum is, the original sequence is more close to the random signal, and the original sequence is more complex. The structural complexity focuses on analyzing the energy characteristic in the transform domain, which is based on all but not the local sequence, and thus compared with the behavior complexity, the result has a global statistical significance.

At present, structural complexity algorithms include spectral entropy [69] and wavelet entropy algorithm [70], which are based on the Fourier transform and wavelet transform and are applied to complexity analysis of discrete chaotic system and Electroencephalogram (EEG) signals. In addition, the computing speed of C_0 algorithm [71] based on fast Fourier transform (FFT) is fast, and it is used in the field of biomedicine, the brain electrical signal, brain optical images successfully for complexity analysis. The important properties of C_0 algorithm are proved in Ref. [72], which shows that C_0 algorithm can describe the random degree of the time series and can describe the complexity. The algorithm is improved in Ref. [73], and the improved C_0 algorithm is more excellent.

In fact, the complexity of chaotic systems is one of the methods to characterize the dynamics of chaotic systems. The effects of the method are similar to that of the Lyapunov exponent, the bifurcation diagram, and the phase diagram. Here, spectral entropy (SE) algorithm and C_0 algorithm are introduced here.

2.9.1 Spectral Entropy Complexity Algorithm

Spectral entropy complexity algorithm can produce energy distribution by Fourier transform and calculates the corresponding spectrum entropy according to the Shannon entropy; the algorithm is described as follows [74]:

Step 1: Remove the DC. We can use the next formula to remove the DC part of the chaotic pseudo-random sequence $\{x^N(n), n = 0, 1, 2, \ldots, N-1\}$, so that the spectrum can reflect the signal energy more effectively.

$$x(n) = x(n) - \bar{x}, \tag{2.30}$$

where $\bar{x} = \frac{1}{N}\sum_{n=0}^{N-1} x(n)$.

Step 2: Do Fourier transformation. Do discrete Fourier transformation for the sequence x^N

$$X(k) = \sum_{n=0}^{N-1} x(n)e^{-j\frac{2\pi}{N}nk} = \sum_{n=0}^{N-1} x(n)W_N^{nk}, \tag{2.31}$$

where $k = 0, 1, 2, \ldots, N-1$.

Step 3: Calculate the relative power spectrum. The first half of the sequence $X(k)$ is used to calculate. According to the Parseval theorem, the power spectrum of a certain frequency point is calculated by

$$p(k) = \frac{1}{N} |X(K)|^2 , \tag{2.32}$$

where $k = 0, 1, 2, \ldots, N/2-1$. The total power of the sequence is defined as

$$p_{\text{tot}} = \frac{1}{N} \sum_{k=0}^{N/2-1} |X(k)|^2 \tag{2.33}$$

Then, the relative power spectrum of the sequence is

$$P_k = \frac{p(k)}{p_{\text{tot}}} = \frac{\frac{1}{N}|X(k)|^2}{\frac{1}{N}\sum_{k=0}^{N/2-1} |X(k)|^2} = \frac{|X(k)|^2}{\sum_{k=0}^{N/2-1} |X(k)|^2}. \tag{2.34}$$

We know that $\sum_{k=0}^{N/2-1} P_k = 1$.

Step 4: Calculate spectral entropy. By using the relative power spectral density P_k and the Shannon entropy, the spectral entropy se is obtained by

$$se = -\sum_{k=0}^{N/2-1} P_k \ln P_k, \tag{2.35}$$

where, if $P_k = 0$, define $P_k \ln P_k = 0$. It can be proved that the value of the spectral entropy converges to $\ln(N/2)$. For comparative analysis, the spectral entropy SE is normalized as

$$SE(N) = \frac{se}{\ln(N/2)}. \tag{2.36}$$

Obviously, when the distribution of the sequence power spectrum is not uniform, the spectrum structure of the sequence is simpler and the signal has a distinct oscillation pattern. The SE measure is smaller in this case, and it indicates that the complexity is smaller. Otherwise, the complexity is larger.

The SE calculation flow chart is shown in Figure 2.7. First, the average value of the sequence is subtracted; that is, the DC part of the signal is removed. Next do FFT. The energy distribution in the transformed domain is calculated by the Shannon entropy, and the SE complexity of the sequence is obtained by the normalization.

When we calculate the SE complexity, we can do the Fourier transformation by using the MATLAB function $Y = \text{fft}(X)$. The calculation speed of this function is very fast. The obtained spectrum is symmetric, so we only choose the first $N/2$ points to calculate.

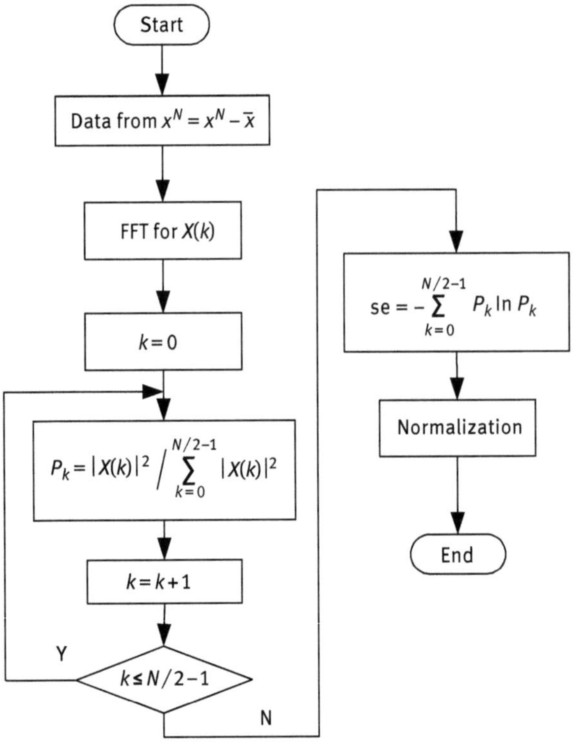

Figure 2.7: Flow chart of SE algorithm.

2.9.2 C_0 Complexity Algorithm

The main idea of C_0 complexity is to decompose the sequence into the regular and the irregular components, and the measure value is the proportion of the irregular components in the sequence. The algorithm calculation steps are as follows [75]:

Step 1: Discrete Fourier transformation. Do discrete Fourier transformation for time series $\{x^N(n),\ n = 0, 1, 2, \ldots, N-1\}$

$$X(k) = \sum_{n=0}^{N-1} x(n) e^{-j\frac{2\pi}{N}nk} = \sum_{n=0}^{N-1} x(n) W_N^{nk}, \tag{2.37}$$

where $k = 0, 1, \ldots, N-1$.

Step 2: Remove irregular parts. The mean square value of $\{X(k),\ k = 0, 1, 2, \ldots, N-1\}$ is defined by

$$G_N = \frac{1}{N} \sum_{k=0}^{N-1} |X(k)|^2. \tag{2.38}$$

To introduce a parameter r, and retain the spectrum that is more than the mean square value of r times, let the rest of the part be zero; that is

$$\tilde{X}(k) = \begin{cases} X(k) & \text{if } |X(k)|^2 > rG_N \\ 0 & \text{if } |X(k)|^2 \le rG_N \end{cases}. \tag{2.39}$$

Step 3: Fourier inverse transformation. Do inverse Fourier transformation for the sequence $\tilde{X}(k)$

$$\tilde{x}(n) = \frac{1}{N}\sum_{k=0}^{N-1}\tilde{X}(k)e^{j\frac{2\pi}{N}nk} = \frac{1}{N}\sum_{k=0}^{N-1}\tilde{X}(k)W_N^{-nk}, \tag{2.40}$$

where $n = 0, 1, \ldots, N-1$.

Step 4: Define C_0 complexity. C_0 complexity is defined as

$$C_0(r, N) = \sum_{n=0}^{N-1}|x(n) - \tilde{x}(n)|^2 / \sum_{n=0}^{N-1}|x(n)|^2. \tag{2.41}$$

C_0 algorithm can also be implemented in MATLAB, and the flow chart is shown in Figure 2.8. FFT and Inverse Fast Fourier Transform (IFFT) are, respectively, implemented by function $Y = \text{fft}(X)$ and $Y = \text{ifft}(X)$.

In addition, there are some other chaotic dynamical characteristics analysis methods, such as measure entropy method. It is often combined with qualitative analysis methods (phase diagram, dividing frequency sampling, or Poincaré section) and quantitative analysis method (Lypunov exponents, power spectral density, or 0–1 test).

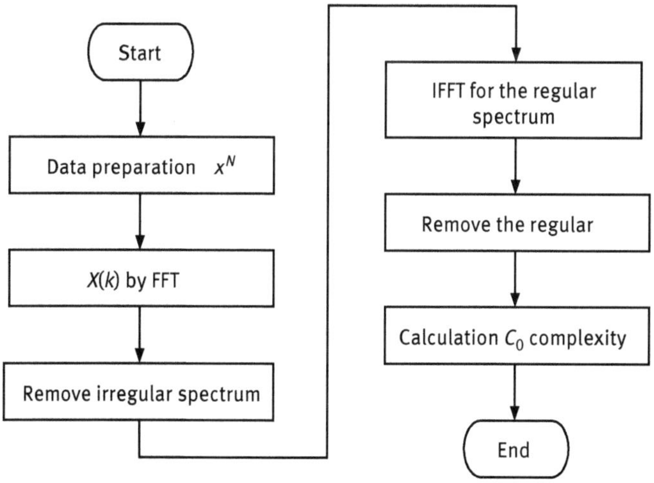

Figure 2.8: Flow chart of the C_0 complexity algorithm.

Questions

1. What is the relationship between the Lyapunov exponent spectrum and the capacity dimension?
2. What is the conditional Lyapunov exponent?
3. What are the characteristics and deficiencies of the 0–1 test method for analyzing chaotic dynamics?
4. What is the complexity of the chaotic system? How to measure it?

3 Typical Chaotic Systems

Chaos exists widely in nature. During the research and development of chaos theory, researchers found a lot of chaotic dynamics models, and some of them are typical for the chaos theory and application research, including discrete-time chaotic maps, continuous-time chaotic systems, and hyperchaotic systems. In this chapter, we will discuss the mathematical model and its main characteristics of some typical chaotic systems.

3.1 Discrete-Time Chaotic Map

First, the definition of discrete-time chaotic map is presented. Consider a nonlinear system

$$x(k+1) = G_{u(k)} \circ F(x(k)), \quad x(0) = x_0, \tag{3.1}$$

where "$\circ$" represents a compound operation between two functions. $x(k) \in M \subset R$, $u(k) \in R^P$, $\{G_u\}_{u \in R^P}$ is the diffeomorphism group with p parameters on M. $F: M \to M$ is a diffeomorphism. If $x(k)$ is chaotic, and $x_0 \in M$, then we call $x(k)$, which all pass through x_0 on smooth n-dimensional manifold M, is discrete chaos. The physical meaning is that chaos is generated by a discrete map described by a nonlinear difference equation, which can usually be achieved by a software program or sampler.

3.1.1 Logistic Map

Logistic map is the most widely used discrete chaotic map model, and its equation is [4]

$$x_{n+1} = \mu x_n(1 - x_n), \quad n = 1, 2, 3, \ldots, \tag{3.2}$$

where the parameter $\mu \in (0, 4]$ and $x_n \in (0, 1)$. When $3.5699\ldots < \mu \le 4$, it is chaotic. The dynamical state of the system changes with the system parameter, which includes the following aspects:

1. When $\mu \in [0, 1]$, the system has a fixed point $x_n \to 0$.
2. When $\mu \in [1, 3]$, the system has fixed points $x_n \to 0$ and $x_n \to 1 - 1/\mu$. This is called period 1 solution.
3. When $\mu \in [3, \mu')$, $\mu' = 3.569945672\ldots$, where orbital period changes such as $1 \to 2 \to 4 \to 8\ldots$ occur. This is called period-doubling bifurcation phenomenon.
4. When $\mu \in [\mu', 4)$, the system shows a chaotic phenomenon.

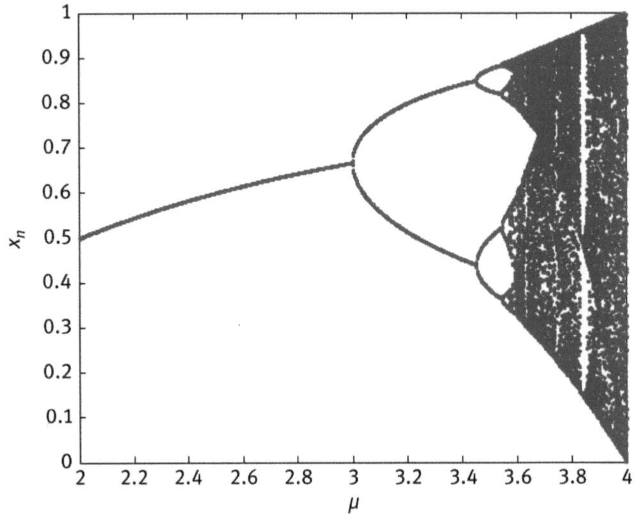

Figure 3.1: Bifurcation diagram of Logistic map.

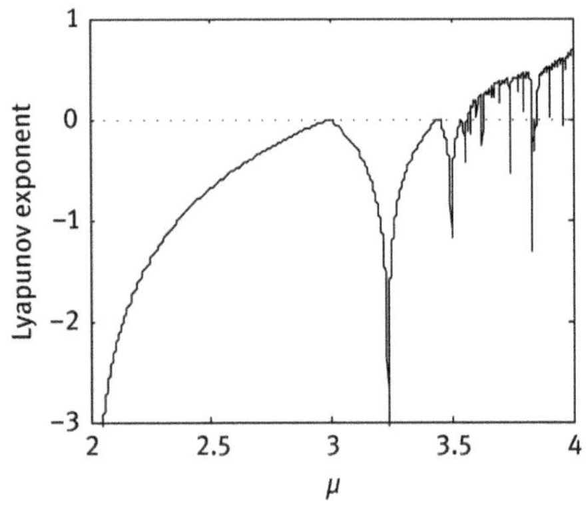

Figure 3.2: Lyapunov exponent of Logistic map.

The bifurcation diagram of the system is shown in Figure 3.1. The iterative initial value of x_0 is 0.6, and the iteration number N is 2,000. The step size of μ is 0.01. As shown in Figure 3.1, when time series experiences four stages of evolution, from stable fixed point, unstable fixed points, period to chaos, complexity of the system increases. If μ = 4.0, the complexity is largest, which is shown in Figure 3.2 by the Lyapunov exponent. When μ = 4.0, Lyapunov exponent is equal to 0.69, which is the maximum.

Next, we will discuss the statistical properties of Logistic map.

The probability density function of Logistic mapping system is

$$\rho(x) = \begin{cases} \frac{1}{\pi\sqrt{x(1-x)}}, & 0 < x < 1 \\ 0, & \text{others} \end{cases}. \tag{3.3}$$

The average value is

$$\bar{x} = \lim_{N\to\infty} \sum_{i=1}^{N} x_i = \frac{1}{2}. \tag{3.4}$$

The autocorrelation function is

$$R_{aa}(\tau) = \lim_{N\to\infty} \frac{1}{N} \sum_{n=0}^{N-1} a_n a_{n+\tau} = \begin{cases} 0.125, & \tau = 0 \\ 0, & \tau \neq 0 \end{cases}. \tag{3.5}$$

The cross-correlation function is

$$R_{ab}(\tau) = \lim_{N\to\infty} \frac{1}{N} \sum_{n=0}^{N-1} a_n b_{n+\tau} = 0. \tag{3.6}$$

The advanced Logistic map, which is also called improved logistic map, is

$$x_{n+1} = 1 - 2(x_n)^2, \quad x_n \in (-1, 1). \tag{3.7}$$

The probability density function is

$$\rho(x) = \begin{cases} \frac{1}{\pi\sqrt{1-x^2}}, & x \in (-1, 1) \\ 0, & \text{others} \end{cases}. \tag{3.8}$$

The average value is

$$\bar{x} = \lim_{N\to\infty} \sum_{i=1}^{N} x_i = 0. \tag{3.9}$$

The autocorrelation function is

$$R_{aa}(\tau) = \lim_{N\to\infty} \frac{1}{N} \sum_{n=0}^{N-1} a_n a_{n+\tau} = \begin{cases} 0.5, & \tau = 0 \\ 0, & \tau \neq 0 \end{cases}. \tag{3.10}$$

The cross-correlation function is

$$R_{ab}(\tau) = \lim_{N\to\infty} \frac{1}{N} \sum_{n=0}^{N-1} a_n b_{n+\tau} = 0. \tag{3.11}$$

3.1.2 Tent Map

Tent map is another commonly used discrete chaotic map. The chaotic sequence generated by the tent map has been widely applied in the field of chaotic spread spectrum communication, chaotic encryption system, chaotic optimum algorithm, and so on. Its equation is

$$x_{n+1} = \begin{cases} x_n/q, & 0 < x_n \le q \\ (1 - x_n)/(1 - q), & q < x_n < 1 \end{cases}, \tag{3.12}$$

where the system parameter $q \in (0, 1)$ and the variable $x_n \in (0, 1)$. Tent map and logistic map are topological conjugates. When q varies as (0, 1), the system is chaotic. In particular, when $q = 0.5$, the system is in a short-cycle state, such as (0.2, 0.4, 0.8, 0.4). In this case, the system is simple and its complexity is small. Therefore, for the application of tent map system, we should choose reasonable system parameter and initial value. Tent map system has the following statistical properties.

The average value is

$$\bar{x} = \lim_{N\to\infty} \sum_{i=1}^{N} x_i = \frac{1}{2}. \tag{3.13}$$

The autocorrelation function is

$$R_{aa}(\tau) = \lim_{N\to\infty} \frac{1}{N} \sum_{n=0}^{N-1} a_n a_{n+\tau} = \begin{cases} 1/12, & \tau = 0 \\ 0, & \tau \ne 0 \end{cases}. \tag{3.14}$$

The cross-correlation function is

$$R_{ab}(\tau) = \lim_{N\to\infty} \frac{1}{N} \sum_{n=0}^{N-1} a_n b_{n+\tau} = 0. \tag{3.15}$$

3.1.3 Hénon Map

Hénon map is a two-dimensional discrete chaotic system, which was proposed by Hénon in 1976. The map equation is [76]

$$\begin{cases} x_{n+1} = 1 - a x_n^2 + y_n \\ y_{n+1} = b x_n \end{cases}. \tag{3.16}$$

When $1.07 \le a \le 1.4$ and $b = 0.3$, it is chaotic. The variable $x_n \in (-1.5, 1.5)$. When $a = 1.4$, the complexity of the system is largest. Generally, in the study of the chaotic synchronization and encryption, we choose $a = 1.4$ and $b = 0.3$. The

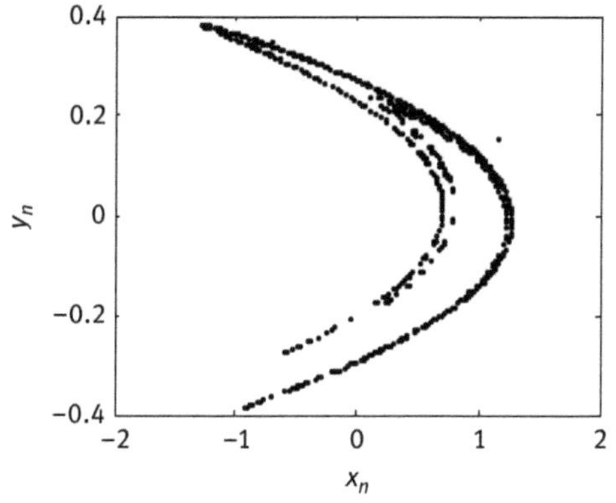

Figure 3.3: Strange attractor of Hénon map.

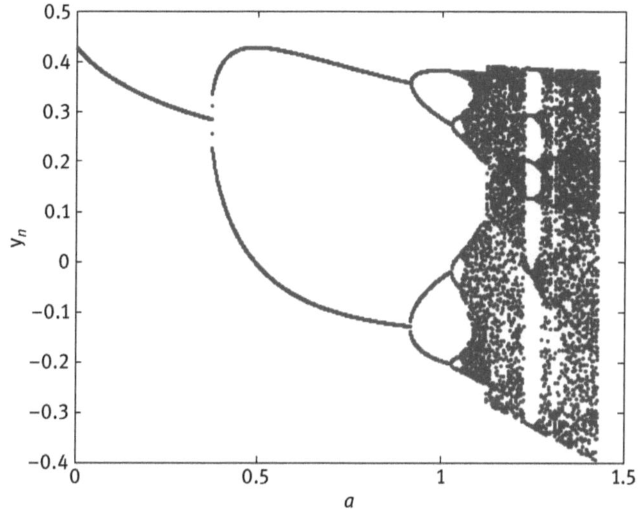

Figure 3.4: Bifurcation diagram of Hénon map.

strange attractor of Hénon map system is shown in Figure 3.3, which looks like the galaxy in the sky, where the initial value $x_n(0) = y_n(0) = 0.5$, and the iteration times are 1,000. When $b = 0.3$, and $a \in [0, 1.5]$, the bifurcation diagram of Hénon map system is shown in Figure 3.4. Obviously, it enters into chaos by period-doubling bifurcation.

3.1.4 Tangent Delay-Ellipse Reflecting Cavity System Map

In 2004, Sheng et al. proposed a discrete chaotic system based on tangent delay (TD), which is called as TD-ellipse reflecting cavity system (TD-ERCS) map [77]. The map equation is

$$\begin{cases} x_n = -\dfrac{2k_{n-1}y_{n-1} + x_{n-1}(\mu^2 - k_{n-1}^2)}{\mu^2 - k_{n-1}^2} \\ k_n = \dfrac{2k'_n - k_{n-1} + k_{n-1}k'^2_{n-1}}{1 + 2k_{n-1}k'_n - k'^2_n}, n = 1, 2, 3, \ldots, \\ k'_{n-m} = -\dfrac{x_{n-m}}{y_{n-m}}\mu^2 \qquad (m \le n) \\ y_n = k_{n-1}(x_n - x_{n-1}) + y_{n-1} \end{cases} \tag{3.17}$$

where the system parameter $\mu \in (0, 1]$, and the variables $|x_n| \le 1$ and $|y_n| \le 1$. m is an integer, which represents the number of TD. k'_{n-m} is the ellipse tangent slope after delaying m. k_0 is determined by incident angle α, and the method is as follows.

First, we calculate y_0 and the slope of k'_0 through the initial value x_0 and system parameter μ

$$y_0 = \mu\sqrt{1 - x_0^2}, \tag{3.18}$$

$$k'_0 = -\frac{x_0}{y_0}\mu^2. \tag{3.19}$$

Then according to trigonometric function, we can obtain:

$$k_0 = \frac{\tan\alpha + k'_0}{1 - k'_0 \tan\alpha}. \tag{3.20}$$

So, if we know the system parameters μ and m, and the initial values x_0 and α, we can get the chaotic sequence $\{x_n, k_n\}$. When the TD m= 0, the system becomes ERCS map. When $m \ge 1$, the TD-ERCS is chaotic.

With the change in system parameter μ, the largest Lyapunov exponent of the system is shown in Figure 3.5. Because the curve chart looks like a flying goose, it is called "Goose." The iteration number n is 400,000. Obviously, when $\mu = 1$, the maximum Lyapunov exponent LE = 0.92. When $\mu = 0.5$, the maximum Lyapunov exponent LE = 0.33. Thus, the maximum Lyapunov exponent is greater than zero in the range of μ, which shows that TD-ERCS is chaotic in the whole range.

When $m = 1$, there exists a square attractor for TD-ERCS as shown in Figure 3.6, where we choose $\mu = 0.8123$, $x_0 = 0.7654$, $\alpha = 0.9876$, and the iteration number N is 1,500. At the beginning, the phase points are not on the attractor, but with the increase of iteration, phase points will fall on the attractor. The experiment also shows that if $0 < \mu < 1$, the position of square attractor is unchangeable and has no relation with

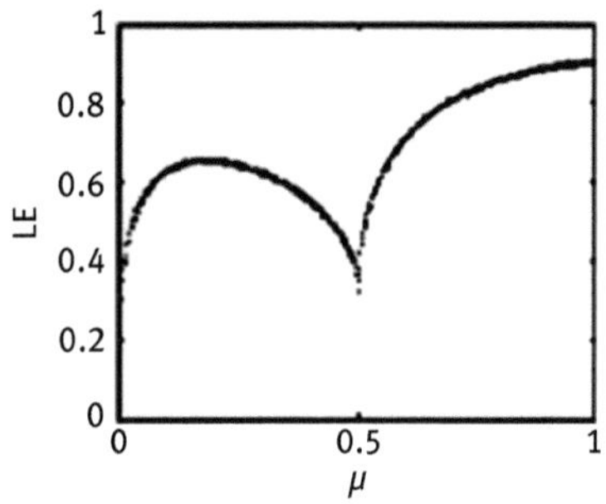

Figure 3.5: Maximum Lyapunov exponent of TD-ERCS system (x_0 = 0.7654, α = 0.9876).

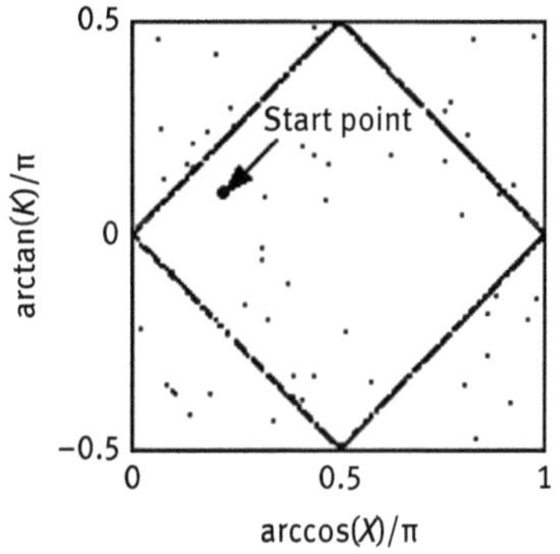

Figure 3.6: Square attractor of TD-ERCS with m = 1.

the initial value. When $\mu = 1$, the position of attractor is related to the initial values, but all phase points are on the attractor.

3.2 Continuous-Time Chaotic System

First, we present the definition of continuous-time chaotic system. Consider the nonlinear system

$$\dot{x} = F(x, \mu, t), \quad x(t_0) = x_0, \tag{3.21}$$

where $x(t) \in M \subset R^n$; it is an integral curve through x_0 in the vector field of F. $x_0 \in M$ and $\mu \subset R^P$ are parameter vectors. If $x(t)$ is chaotic, all $x(t)$ going through x_0 on the smooth n-dimensional manifold M are continuous chaos. From physical meaning, a continuous chaotic system is a nonlinear dynamical system and can produce chaotic phenomenon, which is described by the differential equation $\dot{y} = F(x)$ and usually can be realized by analog circuit.

There are two types of continuous chaotic systems. One is autonomous system, such as the Lorenz system, Rössler system, and Chua system. The other is nonautonomous system, such as Duffing oscillator and van der Pol oscillator.

3.2.1 Duffing Oscillator

Duffing oscillator is a typical forced vibration nonlinear dynamic system, and it belongs to nonlinear vibration problem. The Duffing system equation is [78]

$$\ddot{x} + k\dot{x} - f(x) = e(t), \tag{3.22}$$

where k is damping coefficient, $f(x)$ is a nonlinear function, and $e(t)$ is the drive term with period T. When $f(x)$ and $e(t)$ take different forms, we will get Duffing equation with different forms. Here, consider the following Duffing equation:

$$\ddot{x} + k\dot{x} - x + x^3 = F\cos\omega t, \tag{3.23}$$

where k is the dissipative coefficient, and F and ω are amplitude and frequency of the periodic force, respectively. It is an important type that can generate chaos in nonlinear vibration.

Let $\dot{x} = y$, then eq. (3.23) is translated into

$$\begin{cases} \dot{x} = y \\ \dot{y} = -ky + x - x^3 + F\cos\omega t \end{cases}. \tag{3.24}$$

When $k = 0$, $F = 1.5$, and $\omega = 1$, the attractor of the system is shown in Figure 3.7.

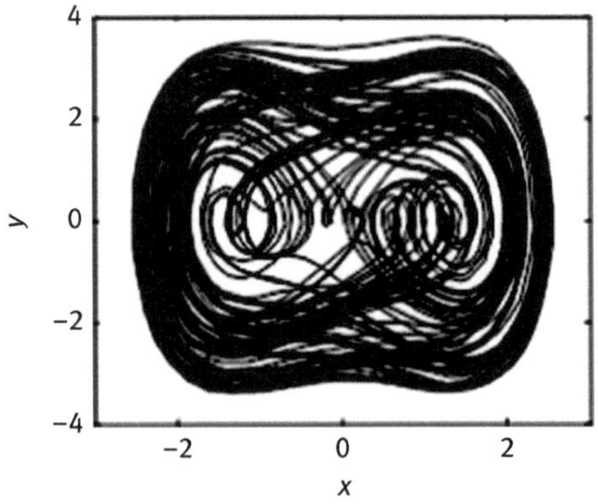

Figure 3.7: Chaotic attractor of Duffing system in x–y plane.

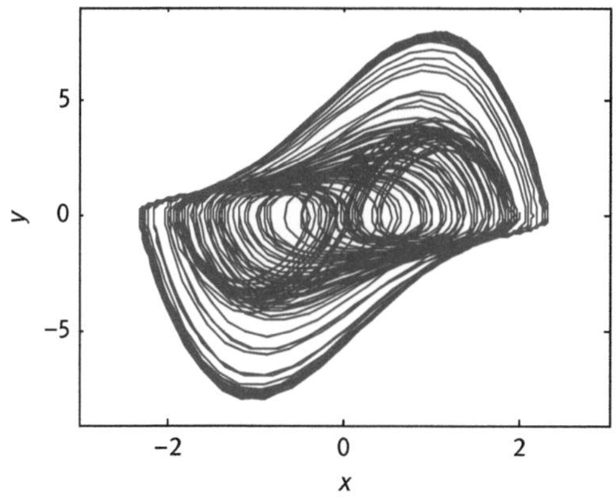

Figure 3.8: Chaotic attractor of van der Pol oscillator in x–y plane.

3.2.2 van der Pol Oscillator

In 1926, van der Pol proposed a nonlinear nonautonomous system that is driven by external force, and the equation is [79]

$$\begin{cases} \dot{x} = y \\ \dot{y} = -x + b(1 - x^2)y + A\sin\Omega t \end{cases}. \tag{3.25}$$

When parameters $b = 3$, $A = 5$, and $\Omega = 1.788$, the chaotic attractor of the system is shown in Figure 3.8.

3.2.3 Lorenz System

When American meteorologist Lorenz investigated atmospheric convection model, he summarized a three-dimensional equation and found "butterfly effect," which lays a foundation for chaos research. For Lorenz system, the mathematical model is a three-dimensional one-order nonlinear differential equation with three parameters, and it becomes one of the most typical examples to study chaos theory and its applications. Lorenz system has abundant bifurcation and chaotic phenomenon.

Lorenz system equation is [2]

$$\begin{cases} \dot{x} = \sigma(y - x) \\ \dot{y} = rx - y - xz \\ \dot{z} = xy - bz \end{cases}, \tag{3.26}$$

where σ, r, and b are the system parameters. The third-order ordinary differential equation is based on the simplified model of heat fluid convection movement between the infinite flats. Its variables do not contain time, so it is called autonomous equation. x represents convection strength; y represents the temperature difference between

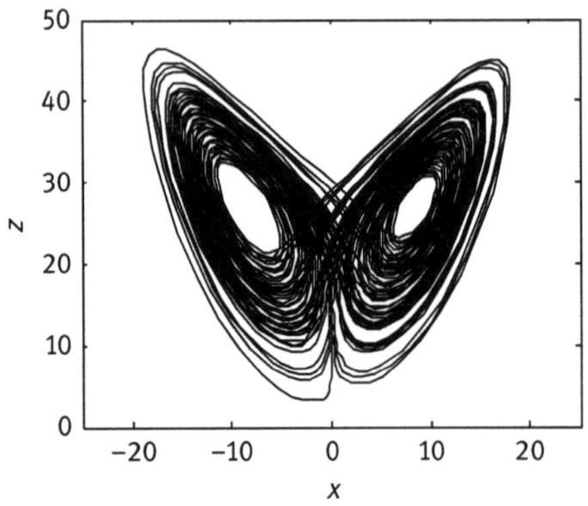

Figure 3.9: Chaotic attractor of Lorenz system in x–z plane.

upflow and downflow between the units; z represents the nonlinear intensity in the vertical temperature distribution; r represents the ratio of driving factors and inhibited factor that causes convection and turbulent, which are the main control parameters of the system. $\sigma = v/k$ is Prandtl value (where v and k are molecular viscosity coefficient and heat transfer coefficient, respectively); b represents the ratio of shape that is related to convection in aspect ratio; and σ and b are dimensionless constant values.

Numerical results show that when $\sigma = 10$, $b = 8/3$, and $28 \leq r < \infty$, except some values near b (such as $b = 100, 160, 260$, and so on), the Lorenz system is in a chaotic state. Let $\sigma = 10$, $r = 28$, and $b = 8/3$ and take numerical simulation for the Lorenz system, we get the famous attractor – "butterfly," as shown in Figure 3.9. The butterfly attractor has two fixed points, and the orbit line transfers randomly around the two fixed points and forms the spiral shape, just like two wings of a butterfly.

3.2.4 Rössler System

Rössler equation looks like the Lorenz system equation, but only a nonlinear term, which is also one of the typical systems in chaotic application research.

Rössler system model is described as [80]

$$\begin{cases} \dot{x} = -y - z \\ \dot{y} = x + ay \\ \dot{z} = b + z(x - c) \end{cases}, \tag{3.27}$$

where a, b, and c are system parameters. When $a = b = 0.2$, $c = 5.7$, the system is in a chaotic state. The Lyapunov exponent spectra are (0.0714, 0, −5.3943). The movement of the attractor describes the complexity of the system. The attractor of Rössler system is shown in Figure 3.10.

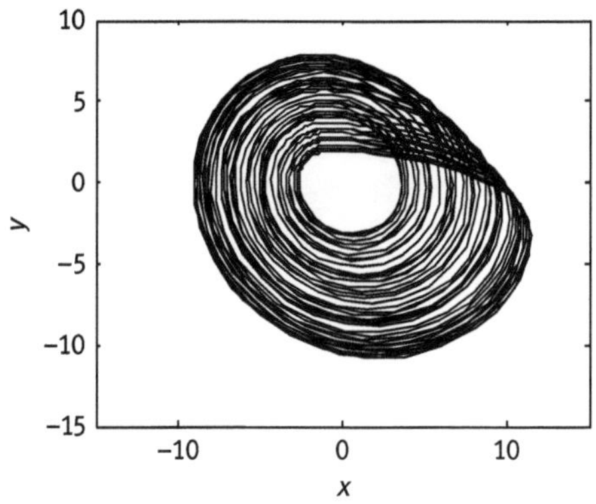

Figure 3.10: Chaotic attractor of Rössler system in x–y plane.

3.2.5 Chua's Circuit

In 1983, Professor Chua designed the famous nonlinear chaotic circuit at the first time, which is called as Chua's circuit. Not only it can generate the rich and complex dynamic behaviors, but also the structure of the circuit is relatively simple. It has become a classical chaotic system for application of chaos control and chaos secure communication. It can produce period-doubling bifurcation, single-scroll, period 3, double-scroll attractor, and other complex nonlinear dynamic behaviors by adjusting the circuit parameters. On the basis of Chua's circuit, the study of multi-scroll Chua's chaotic attractors has become a research hotspot at present.

The normalized state equation of Chua's circuit is [15]

$$\begin{cases} \dot{x} = \alpha[y - h(x)] \\ \dot{y} = x - y + z \\ \dot{z} = -\beta y \end{cases}, \tag{3.28}$$

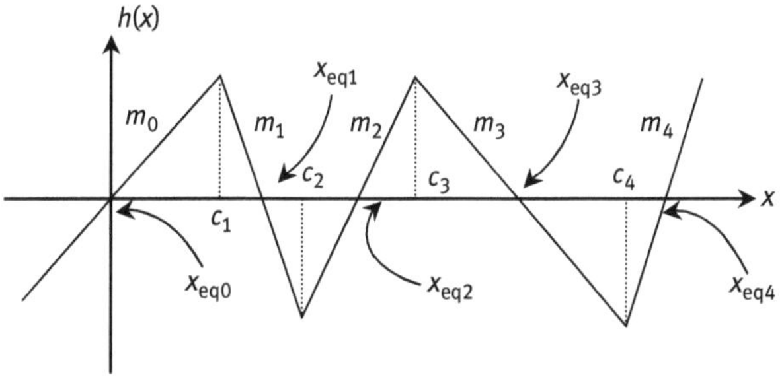

Figure 3.11: Schematic diagram of piecewise linear function $h(x)$.

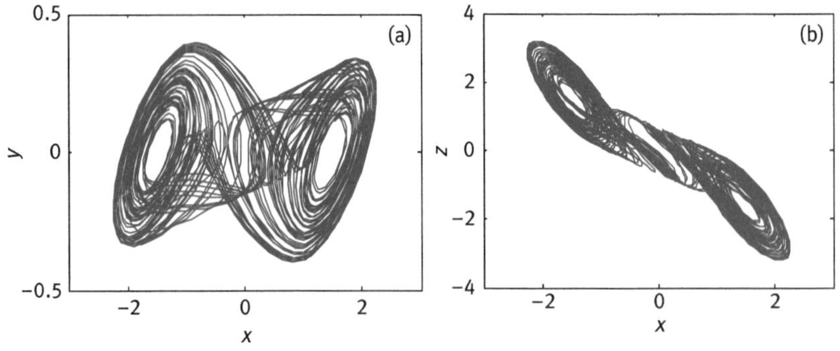

Figure 3.12: Chaotic attractors of the double scroll: (a) x–y plane and (b) x–z plane.

where $h(x)$ is a piecewise linear function, and its waveform is shown in Figure 3.11. When $h(x)$ is a different piecewise linear function, the system produces double-scroll and multi-scroll chaotic attractors. For chaos secure communication, the more the scroll is, the more secure the chaotic secure communication system is. Here only the double-scroll, three-scroll, five-scroll, and seven-scroll chaotic systems are simulated.

3.2.5.1 Double-Scroll Chaotic Attractor

For double-scroll attractor, $h(x)$ and other parameters are defined by

$$\begin{cases} h(x) = m_{2q-1}x + \frac{1}{2}\sum\limits_{i=1}^{2q-1}(m_{i-1} - m_i)\cdot(|x + c_i| - |x - c_i|) \\ q = 1,\ \alpha = 9,\ \beta = 100/7 \\ m = [m_0; m_1] = [-1/7; 2/7] \\ c = [c_1] = [1] \end{cases} . \tag{3.29}$$

By Matlab simulation, we obtain the chaotic attractor as shown in Figure 3.12.

3.2.5.2 Three-Scroll Chaotic Attractor

For three-scroll chaotic attractor, $h(x)$ and other parameters are defined by

$$\begin{cases} h(x) = m_{2q-1}x + \frac{1}{2}\sum\limits_{i=1}^{2q-1}(m_{i-1} - m_i)\cdot(|x + c_i| - |x - c_i|) \\ q = 2,\ \alpha = 9,\ \beta = 100/7 \\ m = [m_0, m_1, m_2, m_3] = [0.9/7, -3/7, 3.5/7, -2.4/7] \\ c = [c_1, c_2, c_3] = [1, 2.15, 4] \end{cases} \tag{3.30a}$$

$$\text{Or} \begin{cases} m = [m_0, m_1, m_2, m_3] = [-1/7, 2/7, -4/7, 2/7] \\ c = [1, 2.15, 3.6] \end{cases} \tag{3.30b}$$

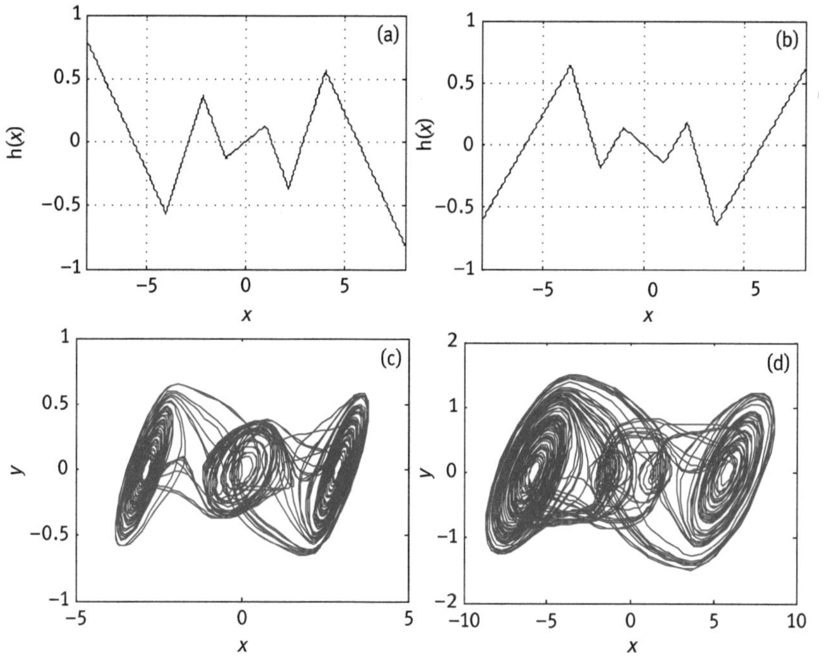

Figure 3.13: Nonlinear curve and the three-scroll chaotic attractor (a), (b) nonlinear curve; (c), (d) The three-scroll chaotic attractor in x-y plane.

Through Matlab simulation, the nonlinear function and three-scroll chaotic attractor are shown in Figure 3.13.

3.2.5.3 Five-Scroll Chaotic Attractor

For five-scroll chaotic attractor, $h(x)$ and other parameters are given by

$$\begin{cases} h(x) = m_{2q-1}x + \frac{1}{2}\sum\limits_{i=1}^{2q-1}(m_{i-1} - m_i)\cdot(|x + c_i| - |x - c_i|) \\ q = 3,\ \alpha = 9,\ \beta = 100/7 \\ m = [m_0, m_1, m_2, m_3, m_4, m_5] \\ \quad = \left[0.9/7, -3/7, 3.5/7, -2.7/7, 4/7, -2.4/7\right] \\ c = [c_1, c_2, c_3, c_4, c_5] = [1, 2.15, 3.6, 6.2, 9] \end{cases} \tag{3.31a}$$

$$\text{Or} \begin{cases} m = [m_0, m_1, m_2, m_3, m_4, m_5] = \left[-1/7, 2/7, -4/7, 2/7, -4/7, 2/7\right] \\ c = [c_1, c_2, c_3, c_4, c_5] = [1, 2.15, 3.6, 6.2, 9] \end{cases}. \tag{3.31b}$$

By Matlab simulation, the five-scroll chaotic attractor is shown in Figure 3.14

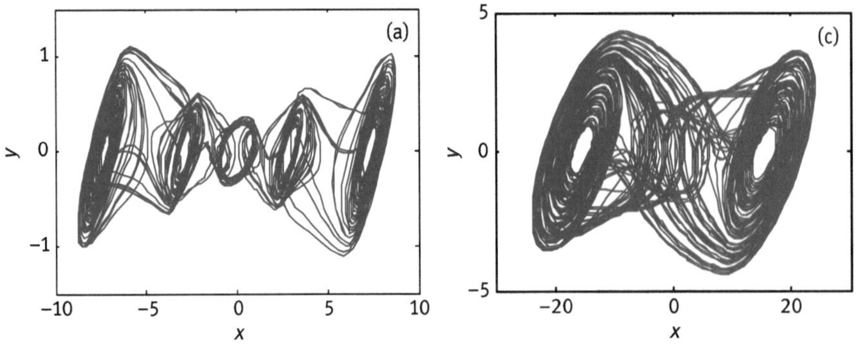

Figure 3.14: Five-scroll chaotic attractor: (a) Parameter 1 (x–y plane) and (b) parameter 2 (x–y plane).

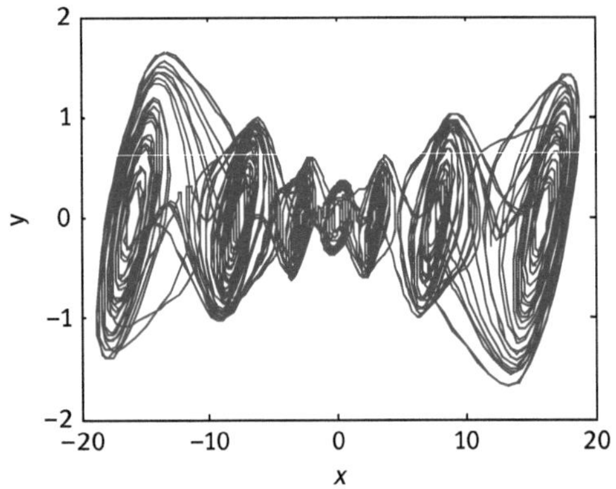

Figure 3.15: Seven-scroll chaotic attractor (x–y plane).

3.2.5.4 Seven-Scroll Chaotic Attractor

For seven-scroll chaotic attractor, $h(x)$ and other parameters are described as

$$\begin{cases} h(x) = m_{2q-1}x + \frac{1}{2}\sum\limits_{i=1}^{2q-1}(m_{i-1} - m_i)\cdot(|x + c_i| - |x - c_i|) \\ q = 4,\ \alpha = 9,\ \beta = 100/7 \\ m = [m_0, m_1, m_2, m_3, m_4, m_5] \\ \quad = \left[0.9/7, -3/7, 3.5/7, -2.4/7, 2.52/7, -1.68/7, 2.52/7, -1.68/7\right] \\ c = [c_1, c_2, c_3, c_4, c_5] = [1, 2.15, 3.6, 6.2, 9, 14, 25] \end{cases}. \tag{3.32}$$

Through Matlab simulation, the seven-scroll chaotic attractor is shown in Figure 3.15.

3.2.6 Chen System

In 1999, in the research of anti-control of chaos, Professor Chen discovered a new chaotic attractor. From the phase diagram of the system, compared with the "butterfly" attractor, it is more complicated on the phase space.

The mathematical model of Chen attractor is [16]

$$\begin{cases} \dot{x} = a\,(y - x) \\ \dot{y} = (c - a)\,x - xz + cy\,, \\ \dot{z} = xy - bz \end{cases} \tag{3.33}$$

where $X = (x, y, z)^{\mathrm{T}} \in R^3$ is the state of the system; $a > 0$, $b > 0$, and $c > 0$ are the system parameters. Although Chen system is similar to the other two famous chaotic systems – Lorenz system and Rössler system in the form, Professor Chen proved that it isn't topological equivalence with the two chaotic systems. The attractor of this system is different from any previous chaotic attractors, and qualitatively analysis the features of the attractor, such as bifurcation, the route to chaos. If $a = 35$, $b = 3$, and $c = 28$, the strange attractor of Chen chaotic system is shown in Figure 3.16.

Hardware circuit of Chen system is designed and tested in Ref. [81], and the circuit diagram and experimental results are shown in Figures 3.17 and 3.18, respectively. Obviously, the experimental results are consistent with the simulation results.

3.2.7 Lü System

In 2002, Lü and Chen discovered a new chaotic attractor, which connects the famous Lorenz attractor and Chen attractor. The mathematical model of Lü system is [82]

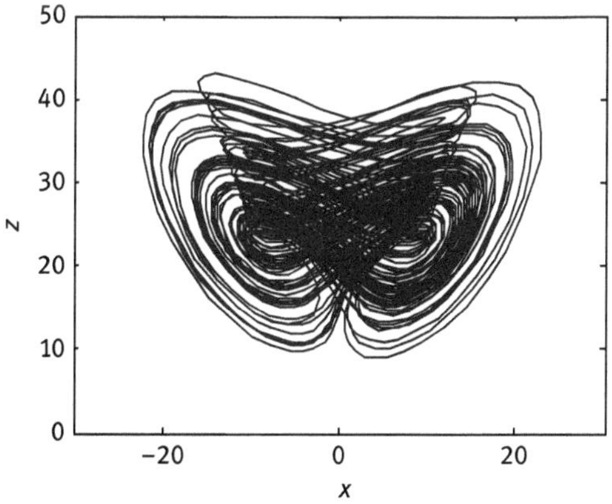

Figure 3.16: Phase diagram of Chen's chaotic attractor in x–z plane.

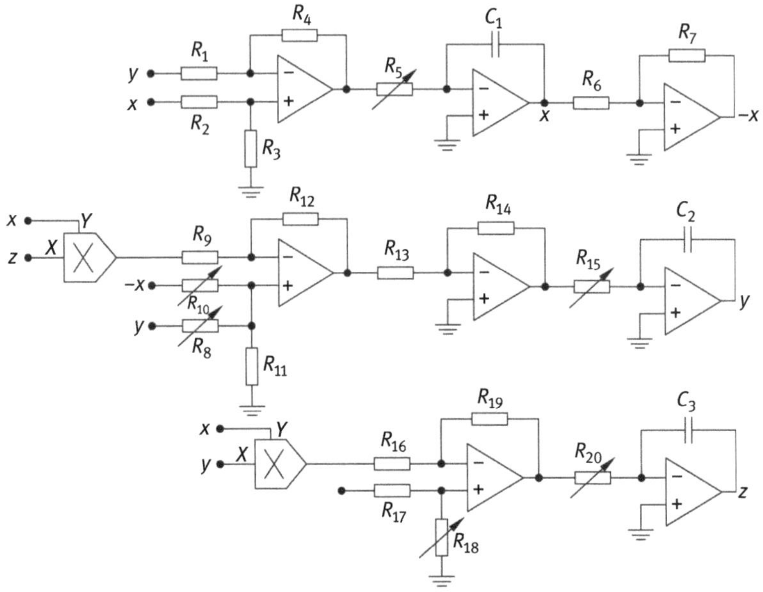

Figure 3.17: Circuit diagram of Chen system.

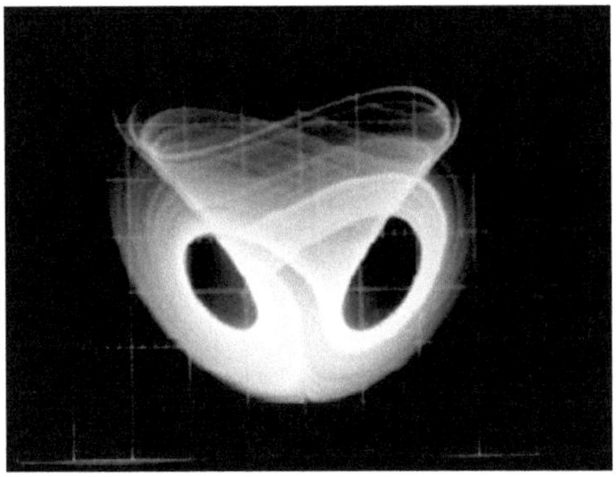

Figure 3.18: Chaotic attractor of Chen system from oscilloscope.

$$\begin{cases} \dot{x} = a(y - x) \\ \dot{y} = -xz + cy \ , \\ \dot{z} = xy - bz \end{cases} \tag{3.34}$$

where a, b, and c are system parameters. When $a = 36$, $b = 3$, and $c = 20$, the system is in a chaotic state. The chaotic attractor of the system is shown in Figure 3.19.

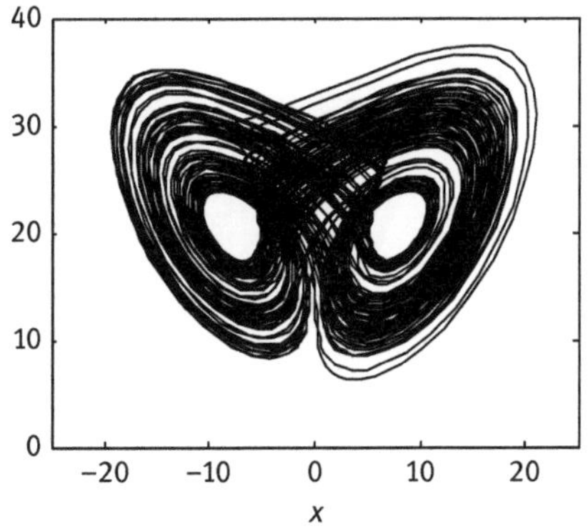

Figure 3.19: Attractor of Lü system in x–z plane.

3.2.8 Unified Chaotic System

In 2002, Lü and Chen proposed a new chaotic system, and the system connects the Lorenz attractor and Chen attractor. Lü system is just a special one of the new chaotic system, so it is called the unified chaotic system. According to the definition of Vaněček and Čelikovský [83], Lorenz system, Chen system, and Lü system belong to different topology types. Mathematical model of the unified chaotic system is [84]

$$\begin{cases} \dot{x} = (25\alpha + 10)(y - x) \\ \dot{y} = (28 - 35\alpha)x - xz + (29\alpha - 1)y \\ \dot{z} = xy - (8 + \alpha)z/3 \end{cases}, \tag{3.35}$$

where the system parameter α varies in the interval [0, 1]. At this range, the unified chaotic system has full chaotic characteristics. According to the definition of Vaněček and Čelikovský [83], when α varies in the interval [0, 0.8), the unified chaotic system belongs to the generalized Lorenz system. When α varies in the interval (0.8, 1], the unified chaotic system belongs to the generalized Chen system. When $\alpha = 0.8$, the unified chaotic system belongs to Lü system. So the unified chaotic system plays an important role in connecting Lorenz system with Chen system, and it is a one-parameter continuous chaotic system. Namely, using only one parameter α, we can control the entire system. When α gradually increases from 0 to 1, the system also gradually transits from the generalized Lorenz system to the generalized Chen system [85]. Under the different parameters of unified chaotic system, the chaotic attractors in the x–z plane are shown in Figure 3.20. The unified chaotic system parameter α has the largest Lyapunov exponent in the range of [0, 1] as shown in Figure 3.21, except the three obvious periodic windows, $W_1 = [0.369, 0.371]$, $W_2 = [0.468, 0.470]$, $W_3 = [0.575, 0.597]$. The largest Lyapunov exponent is greater than zero in the range

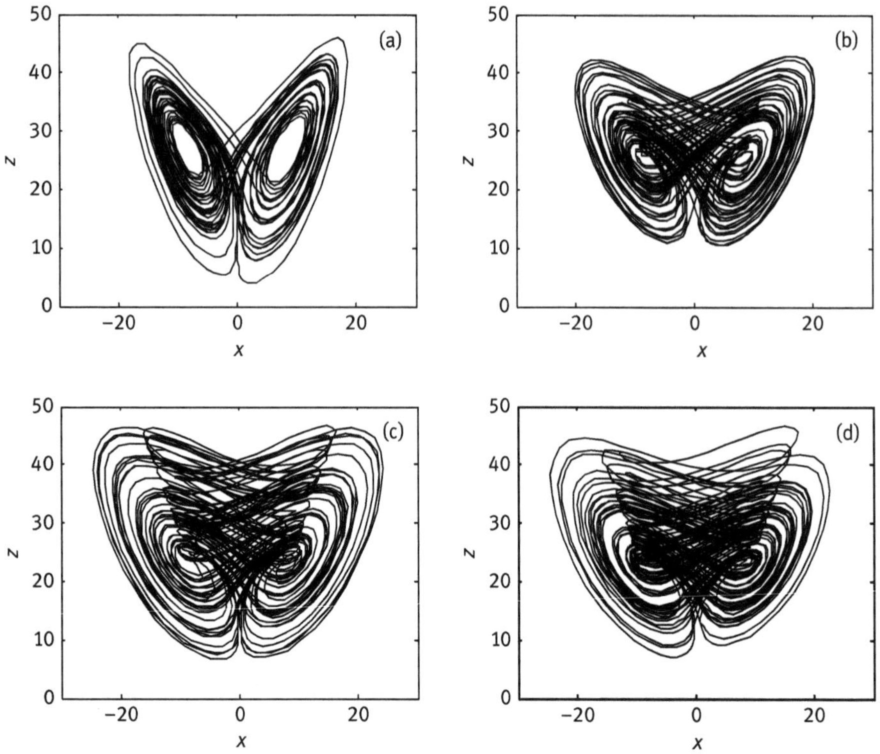

Figure 3.20: Attractors of unified chaotic system with different parameters in x–z plane. (a) $\alpha = 0$; (b) $\alpha = 0.5$; (c) $\alpha = 0.8$; and (d) $\alpha = 1$.

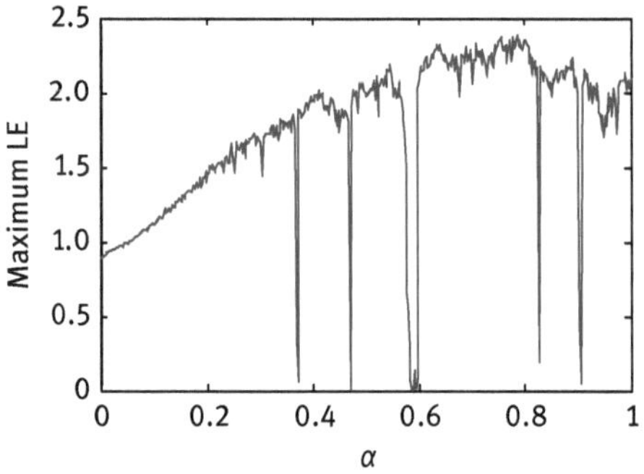

Figure 3.21: Maximum Lyapunov exponent of unified system with different α.

of [0, 1], so the unified chaotic system has full chaos characteristics. Due to its excellent performances, the control and synchronization of the system become a hot topic for the chaotic research [86], and it has extensive application prospect in the field of secure communications [87].

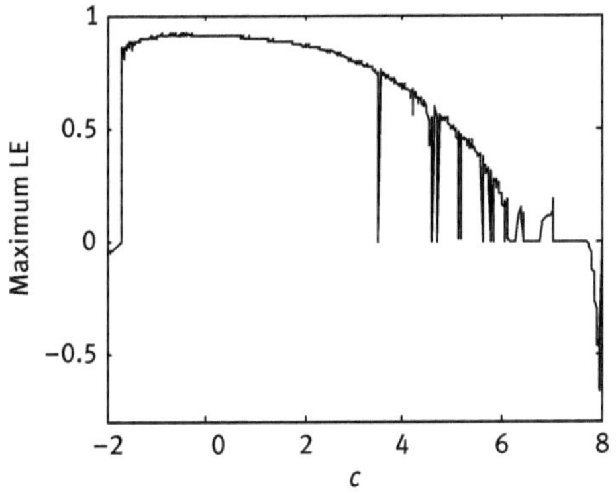

Figure 3.22: Maximum Lyapunov exponent of simplified Lorenz system with different *c*.

3.2.9 Simplified Lorenz System

Dynamic equation of simplified Lorenz system is [56]

$$\begin{cases} \dot{x} = 10(y - x) \\ \dot{y} = -xz + (24 - 4c)x + cy \ , \\ \dot{z} = xy - 8z/3 \end{cases} \tag{3.36}$$

where c is a bifurcation parameter of the system. The maximum Lyapunov exponent of simplified Lorenz system is shown in Figure 3.22. When $c \in [-1.59, 7.75]$, the system is in a chaotic state obviously. But there exists nine period windows in this range at least, namely, $W_1 = [3.507, 3.509]$, $W_2 = [4.581, 6.612]$, $W_3 = [4.6911, 4.722]$, $W_4 = [5.122, 5.127]$, $W_5 = [5.167, 5.169]$, $W_6 = [5.599, 5.600]$, $W_7 = [5.7601, 5.770]$, $W_8 = [5.820, 5.830]$, and $W_9 = [6.415, 6.425]$.

With the change of the system parameter c, the bifurcation diagram of the simplified Lorenz system is shown in Figure 3.23. The chaotic attractor of the simplified Lorenz system is shown in Figure 3.24.

Simplified Lorenz system has the following properties:

1. It is chaotic over most of the range $c \in [-1.59, 7.75]$.
2. For $c = -1$, it is the usual Lorenz system with the standard parameters.
3. For $c = 0$, the variable y is removed from the second equation.
4. For $c = 6$, the variable x is removed from the second equation.
5. There is a rich set of bifurcation as c is varied over the range.
6. According to the topological definition by Vaněček and Čelikovský [83], the linearization of system (3.36) about the origin produces a 3×3 constant matrix of partial derivatives, $A = [a_{ij}]_{3\times3}$, in which the sign of $a_{12}a_{21}$ distinguishes nonequivalent topologies. According to this criterion, when $c < 6$, $a_{12}a_{21} > 0$; when $c = 6$, $a_{12}a_{21} = 0$; and when $c > 6$, $a_{12}a_{21} < 0$.

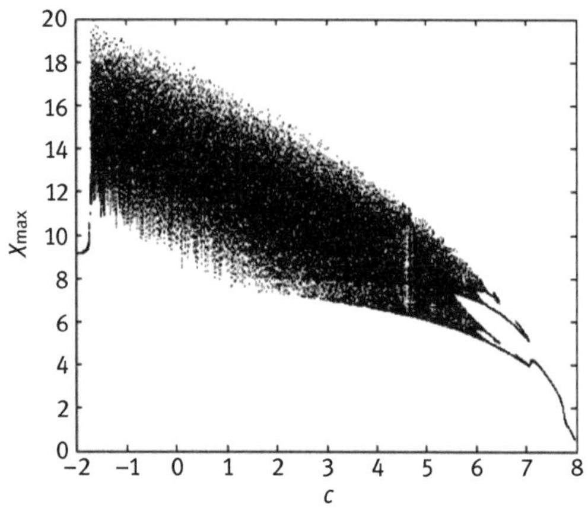

Figure 3.23: Bifurcation diagram of the simplified Lorenz system.

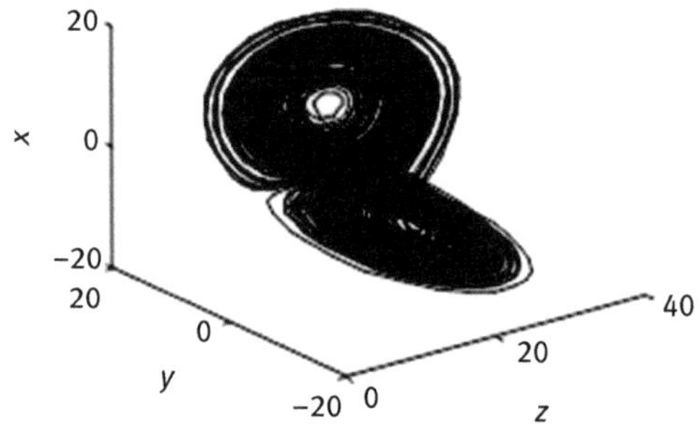

Figure 3.24: Chaotic attractor of the simplified Lorenz system (c = 2).

Simplified Lorenz system can be implemented by analog circuit. The nonlinear part uses operational amplifier LM741 and multiplier AD633 to implement. According to the simulation, the variable amplitude exceeds the voltage range that operational amplifier and multiplier provide, so the system state variables cannot be directly as the voltage variables. To realize the circuit systems, the state variables must be scaled for transformation appropriately. Simplified Lorenz system circuit is shown in Figure 3.25 [88], and c is the system parameter of simplified Lorenz system; thus, the change of R_{14} is equal to the change of system parameter c value. Figure 3.26 is an attractor phase diagram captured by an oscilloscope.

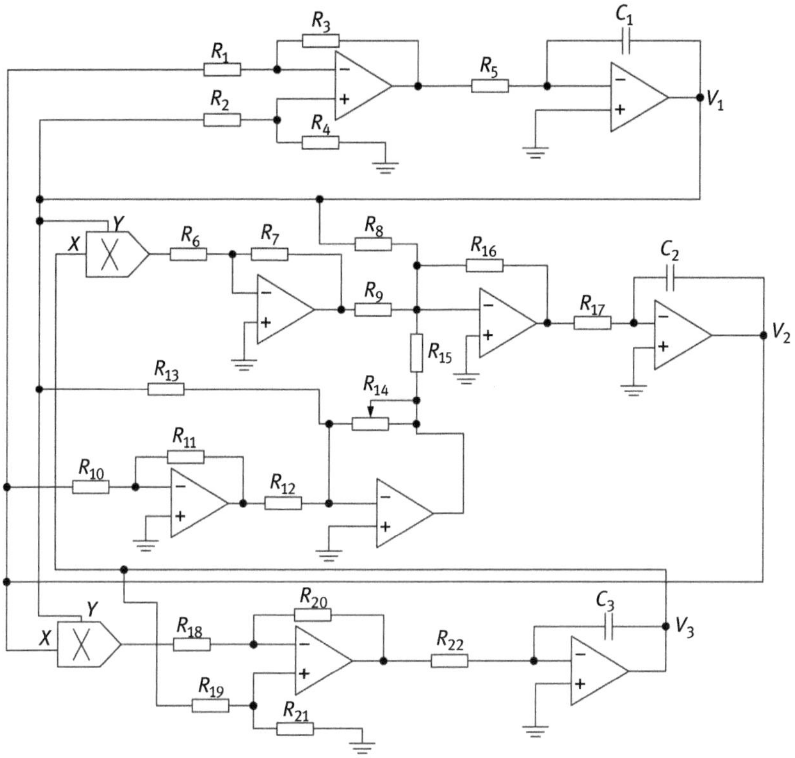

Figure 3.25: Circuit diagram of the simplified Lorenz system.

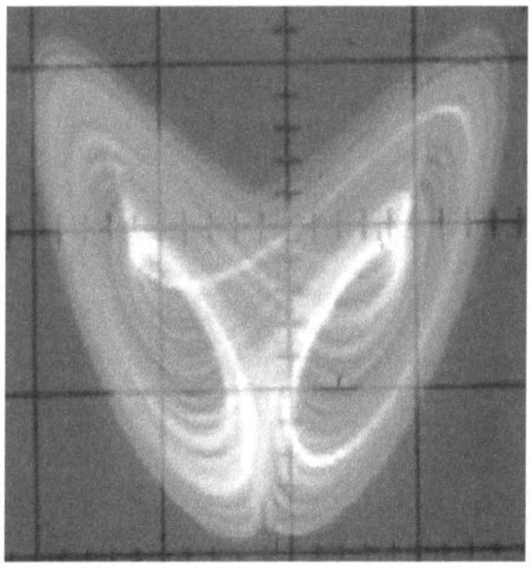

Figure 3.26: Attractor of simplified Lorenz system in x–z plane from oscilloscope.

3.2.10 Standard for a New Chaotic System

At present, many mathematical models of low-dimensional discrete time and continuous time have been put forward, and their solutions are chaotic, which are proved by a positive Lyapunov exponent. Researchers are continuously finding new chaotic systems and publishing them in a pile of journals about nonlinear dynamics. These papers just reported some characteristics about nonperiodic orbit in the state space and bifurcation diagram, Lyapunov exponents, and other dynamics and topological properties. Obviously, these papers almost are not wrong, but they usually only provide another example that is known and understood fully to all of us. So they can't be called as a new chaotic system.

For this phenomenon, Professor Sprott thought if you want to say the proposed system is a new chaotic system, the system must satisfy at least one of the following criteria [89]:

1. The system should credibly model some important unsolved problems in nature and shed insight on that problem.
2. The system should exhibit some behaviors previously unobserved.
3. The system should be simpler than all other known examples exhibiting the observed behaviors.

The above standards provide a sufficient condition for the publication of new chaotic system. The famous Lorenz system meets the three standards when first published since it deals with atmospheric turbulence, exhibited sensitive dependence on initial conditions, and is the simplest system in its time known to have these properties. Lorenz devoted considerable effort to simplifying the system from the original seven-dimensional system with 13 quadratic nonlinearities that was suggested by Saltzman [90].

3.3 Hyperchaotic System

For the chaotic movement of high-dimensional partial differential equations, in many cases, there is more than one direction of instability, which is called as hyperchaotic system. Hyperchaotic system has two or more positive Lyapunov characteristic index, which has more complex dynamical behaviors than general chaotic system.

3.3.1 Rössler Hyperchaotic System

Rössler hyperchaotic system is a simple four-dimensional oscillator model that was found by Rössler in 1979. It can generate hyperchaotic attractor in more than two directions of hyperbolic unstable. Rössler equation is [91]

$$\begin{cases} \dot{x}_1 = -x_2 - x_3 \\ \dot{x}_2 = x_1 + ax_2 + x_4 \\ \dot{x}_3 = b + x_1x_3 \\ \dot{x}_4 = -cx_3 + dx_4 \end{cases} . \tag{3.37}$$

When the system parameters $a = 0.25$, $b = 3.0$, $c = 0.5$, and $d = 0.05$, the system is in a chaotic state, and the Lyapunov exponent spectrum is (0.112, 0.019, 0, −25.188). Its attractor is shown in Figure 3.27.

3.3.2 Chen Hyperchaotic System

In 2005, Li et al. designed the hyperchaotic Chen system by employing state feedback control, and the equation is [92]

$$\begin{cases} \dot{x} = a(y - x) + w \\ \dot{y} = dx - xz + cy \\ \dot{z} = xy - bz \\ \dot{w} = yz - rw \end{cases} , \tag{3.38}$$

where x, y, z, and w are state variables of the system, and a, b, c, d, and r are control parameters of the system. When $a = 35$, $b = 3$, $c = 12$, and $d = 7$, r changes in the interval [0, 0.1085], (0.1085, 0.1798], (0.1798, 0.1900], respectively, then the system shows chaotic motion, hyperchaotic motion, and periodic motion, respectively. When $a = 35$, $b = 3$, $c = 12$, $d = 7$, and $r = 0.16$, the hyperchaotic system has two positive Lyapunov exponents, 0.1567 and 0.1126. The attractor of the system is shown in Figure 3.28.

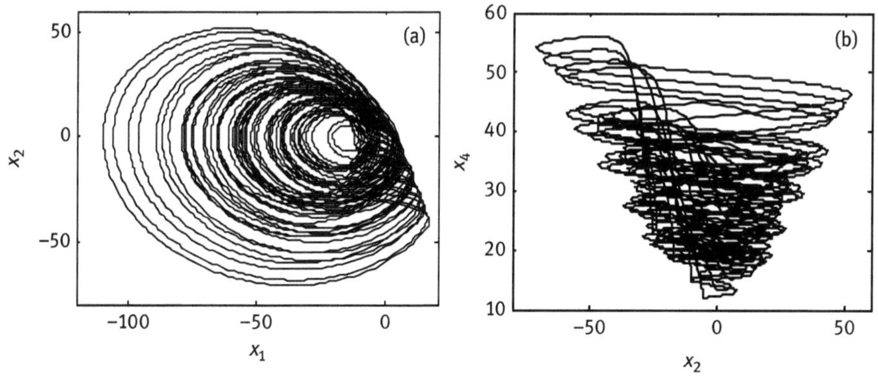

Figure 3.27: Attractors of Rössler hyperchaotic system: (a) x_1–x_2 plane and (b) x_2–x_4 plane.

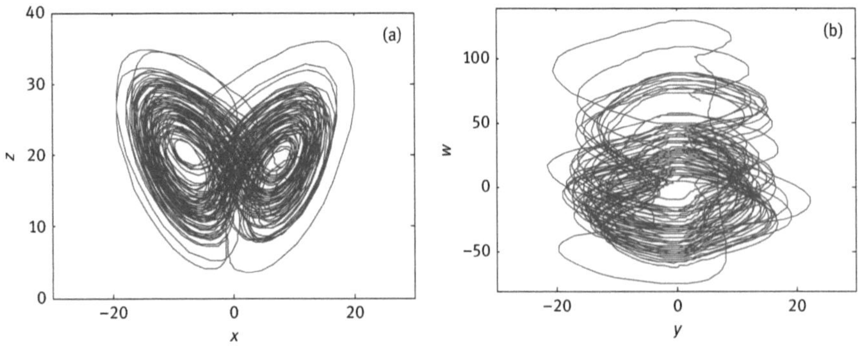

Figure 3.28: Attractors of Chen chaotic system: (a) x–z plane and (b) y–w plane.

3.3.3 Folded Towel Hyperchaotic Map

Folded towel map is presented by Rössler in 1979, which is described as [91]

$$\begin{cases} x_{n+1} = ax_n(1 - x_n) - 0.05(y_n - 0.35)(1 - 2z_n) \\ y_{n+1} = 0.1((y_n + 0.35)(1 + 2z_n) - 1)(1 - 1.9x_n) \\ z_{n+1} = 3.78z_n(1 - z_n) + by_n \end{cases}, \tag{3.39}$$

where a and b are control parameters, and x, y, z are system variables. If $a = 3.8$, $b = 0.2$, the system is hyperchaotic, and the Lyapunov exponents are (0.418, 0.373, −2.502) [93]. The phase diagram of the folded towel map function for $a = 3.8$, $b = 0.2$, $x_0 = 0.1$, $y_0 = 0.2$, and $z_0 = 0.3$ is shown in Figure 3.29.

The chaotic system models are much more than those models discussed earlier. In this chapter, we mainly introduce typical chaotic systems, and they are common models in chaotic theory and applied research. These models not only have the actual physical meaning, but also have unique characteristics. Researchers often use them

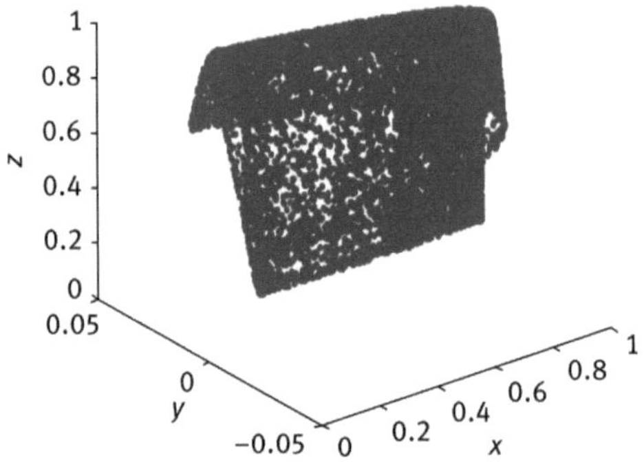

Figure 3.29: Chaotic attractor of folded towel map.

to research chaotic theory and application. With the study of chaos, researchers will continuously find and reveal chaotic system model and its dynamics in nature.

Questions

1. Try to build a new chaos model.
2. What kind of characteristics a new chaotic system should have?
3. If $x_{n+1} = x_n^2 + c$ and $-2 < c < 0.3$, calculate its bifurcation diagram, and approximately prove Feigenbaum constant.
4. What is hyperchaotic system? What are its features?

4 Chaotic Synchronization Principle and Method

Due to the extremely sensitive initial value, researchers once thought it is impossible to control chaos. Until 1990, American scientists, Ott, Grebogi, and Yorke, proposed the OGY method to realize chaos control [12]. In the same year, Pecora and Carroll proposed a self-synchronizing method for chaos [10], and the synchronization between two chaotic systems is realized by using the drive–response method for the first time [94]. The breakthrough research has broken the traditional concept that chaos movement is not controllable and dangerous. It is proved that chaos can not only be controlled and synchronized, but also it is the basis of information transmission and processing. So it can be used in the field of information security communication, and thus it opened the prelude to the application of chaotic secure communication.

Chaos synchronization, in general, belongs to chaos control. The problem of chaos synchronization can be considered as a kind of control problem that the chaotic orbit of the controlled system evolves as the orbit of the target system. The traditional chaos control is generally the stable system on the unstable periodic orbits, while the synchronization of the chaotic system is the complete reconstruction of the chaotic state of the two systems.

In this chapter, based on the definition of chaos synchronization and the performance index of chaotic synchronization system, the principle and performance of various chaotic synchronization control methods are studied.

4.1 Definition of Chaos Synchronization

Definition 4.1 ([95]): Consider two nonlinear systems as follows:

$$\dot{x} = F(t, x), \tag{4.1}$$

$$\dot{y} = F'(t, y) + U(t, x, y), \tag{4.2}$$

where state variables $x, y \in R^n$, and $F, F' : [R_+ \times R^n] \to R^n$ is the nonlinear map. $U : [R_+ \times R^n \times R^n] \to R^n$ is the synchronization control variable of chaotic system. R_+ is the nonnegative real number set. If we have $D(t_0) \subseteq R^n, \forall x_0, y_0 \in D(t_0)$, and $\lim \left\| x(t; t_0, x_0) - y(t; t_0, y_0) \right\| = 0$ when $t \to \infty$, then systems (4.2) and (4.1) are synchronous. We call system (4.1) as the drive system, and system (4.2) as the response system, and $D(t_0)$ is the synchronization region. If $D(t_0)$ belongs to the whole space R^n, then the synchronization is defined as global (complete) synchronization. If $D(t_0)$ is a subset of R^n, then the synchronization is defined as local (partial) synchronization. If $F = F'$, then it is called self-synchronization.

If we set error vector $e = x-y$, then the synchronization error system of systems (4.1) and (4.2) is

$$\dot{e} = F(t, x) - F'(t, y) - U(t, x, y) = G(t, e, x). \tag{4.3}$$

Similarly, based on the stability of error system (4.3) at the origin of the system, the stability conditions of the system are derived.

In the study of chaos synchronization, the stability of the synchronization system is determined mainly by the Lyapunov stability principle or by the condition Lyapunov exponent (CLE) criterion proposed by Pecora and Carroll. The former is the main criterion, and the latter is a necessary condition for synchronization. But the linear stability theorem shows that the necessary condition is also a sufficient condition for most of the dynamic systems, and it is used as a criterion during the synchronization research [96].

The above synchronization definition is suitable for synchronization of not only chaotic systems, but also nonchaotic systems. It is suitable for both of the synchronization of autonomous systems and nonautonomous systems, and it can be extended to the synchronization of hyperchaotic systems and spatiotemporal chaotic systems.

4.2 Performance Index of Chaotic Synchronization System

The study of the performance of the synchronization system is directly related to the practical application of the chaotic system, and it is of great significance to study the mechanism of the synchronization performance. Control parameters have a great influence on the performance of the synchronization system, and how to choose the control parameters to make the optimal synchronization performance is an important part of the study of chaos synchronization. The synchronization performance of the system can be improved by choosing the initial value and controlling parameters of the system, such as shortening the setup time of synchronization, improving the stability and convergence speed, and increasing the anti-disturbance ability for the synchronization system. Here, we put forward the performance index of chaotic synchronization system first, including synchronization stability, synchronization robustness, synchronization steady-state error, synchronization accuracy, synchronization setup time, synchronization region, synchronization dynamic characteristics, and so on [97].

Definition 4.2: Synchronization stability refers to that the chaos synchronization system with different initial states satisfies the Lyapunov asymptotic stability conditions, namely the solution of system equation is convergent. It is the primary characteristic of the chaotic synchronization system, and it is the necessary condition for the synchronization system to work. It is the key to realize the application of the secure communication. Stability is divided into large-scale (global) asymptotic stability and small-scale (local) asymptotic stability.

Definition 4.3: Synchronization robustness refers to the drive–response system that maintains stability in the case of the system parameters mismatch. From the practical view, if the system's robustness is better, then the system is easier to implement. But from the view of communication security, if the system's robustness is better, then the security is worse. This is a contradiction, which must be considered in the practical application.

Definition 4.4: Synchronization steady-state error is the accurate degree of synchronization of the two chaotic systems to achieve synchronization. In general, it is denoted by $\Delta\varphi$.

Definition 4.5: Synchronization accuracy refers to the consistent degree of the chaotic signal generated by the controlled system and the target system in the amplitude and phase, and it is denoted by δ. The relationship between the steady-state error and the synchronization accuracy is $\delta = 10^{[\lg \Delta\varphi]}$, where $[\Delta\varphi]$ is the smallest integer that is not more than $\Delta\varphi$.

Definition 4.6: Synchronization setup time is an important index for the application of chaotic secure communication, and it is the transition time experienced by the response signal for the given synchronization accuracy, and it is denoted by T_S. To describe the synchronization of chaotic systems, the synchronization accuracy can be calculated by

$$p = \left[\sum_{i=1}^{n} (y_i - x_i)^2 \right]^{\frac{1}{2}} \tag{4.4}$$

or

$$p = \max\{|y_i - x_i|, \quad i = 1, 2, \ldots, n\} \tag{4.5}$$

where n is the number of system variables. When $p \leq \delta$, synchronization is achieved.

Definition 4.7: Synchronization region is the initial value range for the synchronization of the chaotic system. He and Vaidya proved the solution of the system is asymptotically stable when the initial value of chaotic system is in a certain region (synchronization region) [98]. That is, the response system will eventually be synchronized with the drive system. Obviously, a successful or practical synchronization method should have sufficiently large synchronous region.

Table 4.1: Comparison of the maximum CLE with the synchronization setup time.

Chaotic system	Drive variable	Maximum CLE	Theory value [99]	Setup time (s)
Lorenz system 1	x	−1.81	−1.86	2.8
($\sigma = 10, b = 8/3, r = 28$)	y	−2.67	−9.99	1.8
Lorenz system 2	x	−2.50	−2.50	2.0
($\sigma = 16, b = 4, r = 45.92$)	y	−3.95	−16.00	1.4

Definition 4.8: Synchronization dynamic characteristics are also known as the transient characteristics of the synchronization establishment. It shows as monotonic or nonmonotonic decay. Nonmonotonic decay shows that there is a trend of nonsynchronization in the synchronization process, which should be avoided.

Research shows that the control parameters are closely related to the performance index of the synchronization. Reasonable choice of control parameters can make the CLE of the response system from positive to negative, so that the response system can be controlled. That all CLEs of the response system are negative is a sufficient condition for the realization of chaos synchronization control, and it is a necessary condition in most cases. The smaller the CLE is, the faster the error convergence is, and the shorter the synchronization setup time is. The maximum CLE in Table 4.1 is calculated by employing the algorithm presented in Chapter 2. When the maximum CLE of the response system is negative, and it is significantly less than zero, then it can guarantee a large enough synchronization region. So changing the control parameters means changing the system's CLE, and it can improve the performance of the synchronous system.

4.3 Principle and Performance of Feedback Control Synchronization

4.3.1 Multivariable drive feedback synchronization for chaotic system

Consider two continuous chaotic systems with the same structure and different initial values as follows:

$$\text{Drive system:} \begin{cases} \dot{X} = AX + Bf(X), \\ X(0) \in R^n, \end{cases} \tag{4.6}$$

$$\text{Response system:} \begin{cases} \dot{Y} = AY + Bf(Y) + U(X, Y), \\ Y(0) \in R^n, \end{cases} \tag{4.7}$$

where drive variables $X \in R^n$ and $Y \in R^n$, and matrices $A \in R^{n\times m}$ and $B \in R^{n\times m}$; F: $R^n \times R \to R^m$ is the nonlinear function. AX is the linear part, and $Bf(X)$ is the nonlinear part. Control variable $U \in R^m$, and (A, B) is controllable. If the state variable error $E = X-Y$, then the synchronization problem between response system (4.7) and drive system (4.6) is used to find a suitable control law $U(X, Y)$ to satisfy

$$\lim_{t\to\infty} \|X - Y\| = \lim_{t\to\infty} \|E\| = 0. \tag{4.8}$$

Obviously, the synchronization control problem of two continuous chaotic systems is transformed into the control problem of the error system.

According to Lyapunov stability theorem, we design a positive definite quadratic function $V(E) = E^T PE$ as a Lyapunov function. If $\dot{V}(E) = -E^T QE$, and Q is positive definite (that is, $\dot{V}(E)$ is negative definite), then the system is asymptotically stable. The synchronous error system equation for eqs (4.6) and (4.7) is

$$\dot{E} = AE + B(f(X) - f(Y)) - U(X, Y). \tag{4.9}$$

So the problem of synchronization of two systems is transformed into the problem of the control law and the choice of the control parameters. In addition, how to select a set of control parameters from a certain range of parameters must be based on the principle of synchronization performance optimization. It is very important whether it is from the study of the synchronization control itself, or the practical application of the synchronization system.

4.3.1.1 Control Law of the Feedback Controller

The multivariable drive feedback synchronization principle for continuous chaotic system is shown in Figure 4.1. The response system is driven by all the variables of the drive system, and the error feedback controller is composed of a linear feedback part and a nonlinear feedback part.

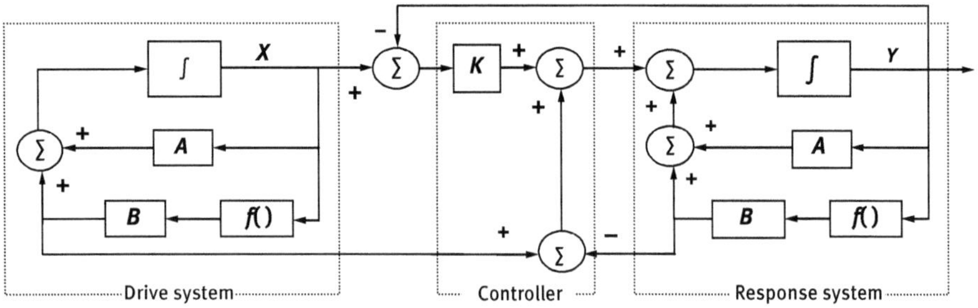

Figure 4.1: Synchronization principle block diagram for continuous chaotic system.

Theorem 4.1: For the n-dimensional continuous chaotic system and its replication system, the synchronous control method based on multivariable drive error feedback control is employed. The drive system is $\dot{X} = AX + Bf(X)$, and the response system is $\dot{Y} = AY + Bf(Y) + U$, where $X, Y \in R^n$ and $K \in R^{n\times m}$. The control variable $U = K(X-Y) + B(f(X)-f(Y))$. If the elements in the parameter control matrix K satisfy: when $i = j$, then $k_{ij} > a_{ij}$; or when $i \neq j$, then $k_{ij} = a_{ij}$, and then the asymptotically stable synchronization is achieved between the response system and the drive system.

Proof. Let $E = X-Y$, then the error system equation $\dot{E} = AE + B(f(X) - f(Y)) - U(X, Y)$. If the control law of the controller is $U = KE + B(f(X) - f(Y))$, then $\dot{E} = (A - K)E$.

Define $V(E)$ as the Lyapunov function

$$V(E) = E^{\mathrm{T}}PE \geq 0, \tag{4.10}$$

where P is a positive definite matrix. Its differential is

$$\begin{aligned} \dot{V}(E) &= \dot{E}^{\mathrm{T}}PE + E^{\mathrm{T}}P\dot{E} = [(A-K)E]^{\mathrm{T}}PE + E^{\mathrm{T}}P[(A-K)E] \\ &= E^{\mathrm{T}}[(A-K)^{\mathrm{T}}P + P(A-K)]E = -E^{\mathrm{T}}QE \end{aligned} \tag{4.11}$$

Obviously, as long as the feedback controller parameter k_{ij} meet: when $i = j$, $k_{ij} > a_{ij}$; when $i \neq j$, $k_{ij} = a_{ij}$, then $Q = (A - K)^{\mathrm{T}}P + P(A - K)$ is positive definite. That is, $\dot{V}(E)$ is negative definite. If and only if $E = 0$, the equality is right. That is, the drive–response system is asymptotically stable. ■

Inference 4.1 When $i = j$, then $k_{ij} > a_{ij}$; When $i \neq j$, and $a_{ij} = a_{ji}$, then $k_{ij} = k_{ji} = 0$, the asymptotic stability synchronization is realized and the structure of the controller is simplified, which can be proved by Lyapunov stability theorem (omitted).

Because the response system is driven by all drive system variables, and the synchronization of two systems is achieved by linear and nonlinear feedback controls, it is called multivariable driving feedback synchronization control method. The method does not need to decompose the system, which is not only suitable for simple chaotic systems, such as the Lorenz chaotic system, but also be applied to hyperchaotic system, such as the Rössler hyperchaotic system. In this method, the selection of feedback parameters is presented, and the performance of the synchronization system can be effectively improved by choosing proper feedback control parameters.

4.3.1.2 Synchronization Simulation and Discussion

Example 1: Multivariable drive feedback synchronization control for Rössler hyperchaotic system.

As shown in Section 3.3.1, the Rössler hyperchaotic system equation is

$$\begin{cases} \dot{x}_1 = x_2 - x_3 \\ \dot{x}_2 = x_1 + ax_2 + x_4 \\ \dot{x}_3 = b + x_1x_3 \\ \dot{x}_4 = -dx_3 + cx_4 \end{cases} . \tag{4.12}$$

When system parameters $a = 0.25$, $b = 3.0$, $c = 0.05$, and $d = 0.5$, it is hyperchaotic.

According to Theorem 4.1 and Inference 4.1, the drive system is

$$\dot{X} = \begin{bmatrix} 0 & 1 & -1 & 0 \\ 1 & a & 0 & 1 \\ 0 & 0 & 0 & 0 \\ 0 & 0 & -d & c \end{bmatrix} X + \begin{bmatrix} 0 \\ 0 \\ 1 \\ 0 \end{bmatrix} (b + x_1x_3), \tag{4.13}$$

and the response system is

$$\dot{Y} = \begin{bmatrix} 0 & 1 & -1 & 0 \\ 1 & a & 0 & 1 \\ 0 & 0 & 0 & 0 \\ 0 & 0 & -d & c \end{bmatrix} Y + \begin{bmatrix} 0 \\ 0 \\ 1 \\ 0 \end{bmatrix} (b + y_1y_3) + U. \tag{4.14}$$

The synchronization control law is

$$U = \begin{bmatrix} k_{11} & 0 & -1 & 0 \\ 0 & k_{22} & 0 & 1 \\ 0 & 0 & k_{33} & 0 \\ 0 & 0 & -d & k_{44} \end{bmatrix} E + \begin{bmatrix} 0 \\ 0 \\ 1 \\ 0 \end{bmatrix} (x_1x_3 - y_1y_3). \tag{4.15}$$

When feedback coefficient $k_{11} > 0$, $k_{22} > a$, $k_{33} > 0$, and $k_{44} > c$, then the synchronization is achieved for the drive system and the response system. Set the initial conditions of the drive system and the response system as (−20, −30, 20, 45) and (−40, 25, 40, 15), respectively. The control parameter $k_{ii} = 1$, and the simulation time is 20 s. The synchronization phase portrait between x_1 and y_1 is shown in Figure 4.2.

The error curves are obtained with the same initial values and the different control parameters as shown in Figure 4.3, where $|E| = [(x_1 - y_1)^2 + (x_2 - y_2)^2 + (x_3 - y_3)^2 + (x_4 - y_4)^2]^{1/2}$. The corresponding feedback control parameters of curves a, b, and c are (0.6, 0.6, 0.6, 0.6), (1, 1, 1, 1), and (2, 2, 2, 2). Obviously, in a certain range, the greater the control parameters, i.e., the greater the control energy, the shorter the synchronization setup time, and the better the dynamic performance in a certain convergence error range. The simulation results show that the error feedback control can realize the synchronization of two hyperchaotic systems. To realize the synchronization control, the feedback control parameters must be selected correctly.

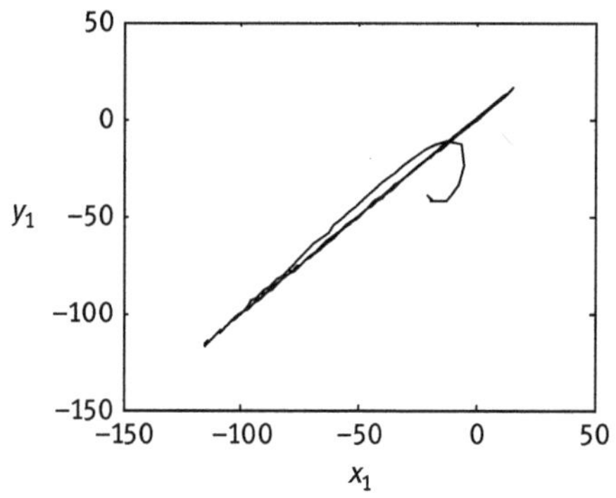

Figure 4.2: Synchronization phase portrait for Rössler hyperchaotic system.

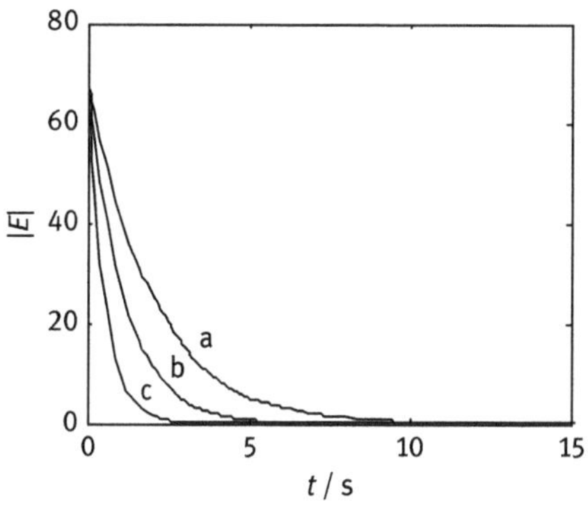

Figure 4.3: Error curves with different control parameters.

The synchronization performance of the hyperchaotic synchronization system can be improved by choosing appropriate parameters.

4.3.2 Linear and Nonlinear Feedback Synchronization for Discrete Chaotic System

Consider an m-dimensional discrete chaotic system (drive system)

$$X_{n+1} = F(X_n) = AX_n + Bf(X_n), \tag{4.16}$$

then the response system is

$$Y_{n+1} = F(Y_n) + U(X_n, Y_n) = AY_n + Bf(Y_n) + U(X_n, Y_n), \tag{4.17}$$

and the control law is

$$U(X_n, Y_n) = K(X_n - Y_n) + P(f(X_n) - f(Y_n)) \tag{4.18}$$

where $X_n, Y_n \in R^m$. f is the nonlinear function and U is the feedback control law. The feedback parameters $K = \text{diag}[k_1, k_2, \ldots, k_m]$, $P = \text{diag}[p_1, p_2, \ldots, p_m]$, where $k_i \geq 0$ and $p_i \geq 0$ $(i = 1, 2, \ldots, m)$. The synchronization of the response system and the drive system is actually the problem of the stability of the synchronous manifold $S = \{(X_n, Y_n) : Y_n = X_n\}$.

Here the problem is analyzed by using local linearization. Set

$$Y_n = X_n + \delta_n, \tag{4.19}$$

where δ_n is a small synchronization error. From eqs (4.16) to (4.19), we obtain the linear evolution equation of the synchronization error as

$$\delta_{n+1} = \left[(A - K) + (B - P)\frac{\partial f}{\partial Y}|_{Y=X_n}\right]\delta_n = D_n\delta_n. \tag{4.20}$$

Obviously, if D_n is the constant matrix, and all the characteristic roots are within the unit circle, then $\delta_n \to 0$ with $n \to \infty$, that is, $Y_n \to X_n$. So the synchronization manifold is stable, and the two systems will be synchronized. But if D_n in eq. (4.20) is a time-varying matrix, then the above method does not work.

Pecora–Carroll Chaos Synchronization Theorem For all the time t, the zero solution of the nonlinear unstable system $\dot{x} = A(t)x + O(x, t)$ with $O(x, t)$ will be asymptotically stable, as long as the system meets the following conditions.

1. For time t, $\lim\limits_{\|x\| \to 0} \|O(x, t)\| / \|x\| = 0$.
2. For all the time t, $A(t)$ are all bounded.
3. The zero solution of the nonlinear system $\dot{x} = A(t)x$ is asymptotically stable.

That is to say, the Pecora–Carroll chaos synchronization theorem can be concluded as follows. Only when all the Lyapunov exponents of the response system are negative, we can achieve the chaos synchronization between the response system and drive system. This theorem is a sufficient condition for synchronization, but the linear stability theorem proves that the sufficient condition is also a necessary condition for many dynamical systems, so it is used as a criterion in the study of chaos synchronization.

Define the maximum conditional Lyapunov exponent of response system (4.17) as

$$\lambda = \lim_{n \to \infty} \frac{1}{n} \ln \frac{|\delta_n|}{|\delta_0|}. \tag{4.21}$$

Obviously, λ is an average statistic. It reflects the global characteristics of the system and determines the final behavior of the system. When $\lambda > 0$, the error δ_n will increase by exponential law. When $\lambda < 0$, the error δ_n will decrease by exponential law and leads to $\delta_n = 0$ finally. That is,

$$\lim_{n\to\infty} |Y_n - X_n| = 0. \tag{4.22}$$

It means that the synchronization is achieved between the response system and drive system. So, we can determine whether the synchronization is realized and obtain the range of the feedback parameters by calculating the maximum CLE of the response system.

The control law $U(X_n, Y_n)$ consists of a linear feedback and nonlinear feedback, so it is called the linear and nonlinear feedback synchronization method. This method has many advantages, such as simple structure, optional parameters, and fast synchronous speed.

4.3.2.1 Control Law of the Synchronization Controller

The synchronization principle of discrete chaotic system is shown in Figure 4.4.

According to the stability criterion, the parameter selection theorem for the synchronization of two discrete chaotic systems is given.

Theorem 4.2: For an m-dimensional discrete-time chaotic system $X_{n+1} = AX_n + Bf(X_n)$, its response system is $Y_{n+1} = AY_n + Bf(Y_n) + U(X_n, Y_n)$. By employing the linear and nonlinear feedback control synchronization method, the control law is $U(X_n, Y_n) = K(X_n - Y_n) + P(f(X_n) - f(Y_n))$, where $X_n, Y_n \in R^m$, $K = \mathrm{diag}[k_1, k_2, \ldots, k_m]$. When $P = B$, the necessary and sufficient condition for the synchronization between the response system and the drive system is that all poles of the characteristic polynomial $T(z) = |Iz - (A - K)|$ are located in the unit circle.

Proof. For the drive and response systems, the error system equation is

$$\delta_{n+1} = \left[(A - K) + (B - P)\frac{\partial f}{\partial Y}|_{Y=X_n}\right]\delta_n. \tag{4.23}$$

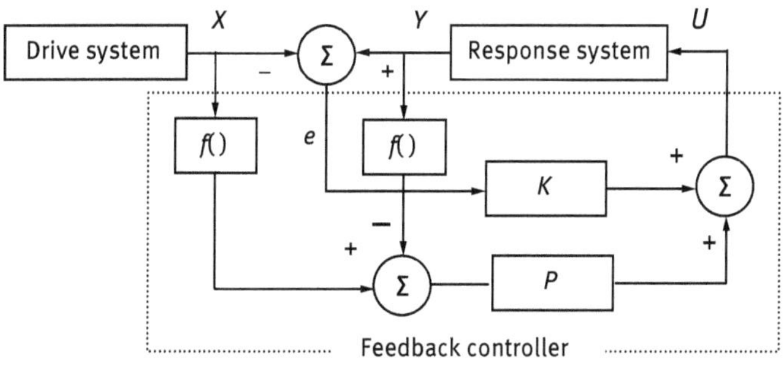

Figure 4.4: Synchronization principle block diagram of discrete chaotic system.

If $P = B$, then we have

$$\delta_{n+1} = (A - K)\delta_n. \tag{4.24}$$

Obviously, the coefficient of the characteristic polynomial of the error system is a function of the linear feedback coefficient K, which is independent of the initial value. If $T(z) = |Iz - (A - K)|$ is a Hurwitz polynomial, then all the eigenvalues of the matrix $A-K$ are located in the unit circle. $\delta_n \to 0$ with $n \to \infty$, that is, $Y_n \to X_n$. Synchronization between the drive system and the response system is achieved. ■

Theorem 4.3: For a one-dimensional discrete chaotic system $x_{n+1} = ax_n + bf(x_n)$, its response system is $y_{n+1} = ay_n + bf(y_n) + u(x_n, y_n)$. By employing the linear and nonlinear feedback synchronization control method, the control law is $u(x_n, y_n) = k(x_n - y_n) + p(f(x_n) - f(y_n))$, where $x_n, y_n \in R$. When $k \geq 0$, and $p \geq b$, then the sufficient condition for the asymptotic stability synchronization between the response system and the drive system is $|a{-}k| < 1$.

Proof. For a one-dimensional discrete chaotic system, the error system equation is

$$\delta_{n+1} = \left[(a - k) + (b - p)\frac{\partial f}{\partial y}\Big|_{y=x_n}\right]\delta_n. \tag{4.25}$$

So we have

$$\delta_n = \prod_{i=1}^{n}\left[(a - k) + (b - p)\frac{\partial f}{\partial y}\Big|_{y=f^{(i)}_{(x_0)}}\right]\delta_0, \tag{4.26}$$

where $f^{(i)}_{(x_0)}$ represents the ith iteration from x_0. According to the definition in eq. (4.21), the maximum CLE is

$$\lambda = \lim_{n\to\infty}\frac{1}{n}\ln\frac{|\delta_n|}{|\delta_0|} = \lim_{n\to\infty}\frac{1}{n}\ln\left|\prod_{i=1}^{n}\left[(a - k) + (b - p)\frac{\partial f}{\partial y}\Big|_{y=f^{(i)}_{(x_0)}}\right]\right|. \tag{4.27}$$

If $p = b$, we have

$$\lambda = \lim_{n\to\infty}\frac{1}{n}\ln\frac{|\delta_n|}{|\delta_0|} = \lim_{n\to\infty}\frac{1}{n}\ln\left|[(a - k)]^n\right| = \ln|a - k|. \tag{4.28}$$

According to eq. (4.28), as long as k meets $|a - k| < 1$, then we have $\lambda < 0$. ■

Theorem 4.4: For an m-dimensional discrete chaotic system $X_{n+1} = AX_n + Bf(X_n)$, its response system is $Y_{n+1} = AY_n + Bf(Y_n) + U(X_n, Y_n)$. By employing the linear and nonlinear feedback control synchronization method, the control law of the controller is $U(X_n, Y_n) = K(X_n - Y_n) + P(f(X_n) - f(Y_n))$, where $X_n, Y_n \in R^m$. When $K = dA$, and $P = dB$,

$d > 0$, then the sufficient condition for the asymptotic stability between the response system and the drive system is $1 - e^{-\lambda'} < d < 1 + e^{-\lambda'}$, where λ' is the maximum Lyapunov exponent of the drive system. Obviously, $\lambda' > 0$. (Proof omitted.)

For one-dimensional discrete chaotic system, we have the following theorem.

Theorem 4.5: For a one-dimensional discrete chaotic system $x_{n+1} = ax_n + bf(x_n)$, its response system is $y_{n+1} = ay_n + bf(y_n) + u(x_n, y_n)$. By employing the linear and nonlinear feedback control synchronization method, the control law of the controller is $u(x_n, y_n) = k(x_n - y_n) + p(f(x_n) - f(y_n))$, where $x_n, y_n \in R$. When $k = ma$, $p = mb$, $m > 0$. The sufficient conditions for the asymptotic stability of the response system and the drive system are $a - ae^{-\lambda'} < k < a + ae^{-\lambda'}$, $b - be^{-\lambda'} < p < b + be^{-\lambda'}$, that is, $1 - e^{-\lambda'} < m < 1 + e^{-\lambda'}$, where λ' is the maximum Lyapunov exponent of the drive system, and $\lambda' > 0$.

According to the above theorems, the linear and nonlinear feedback control can be used to realize the synchronization of two discrete chaotic systems by choosing appropriate feedback coefficient. Although Theorems 4.4 and 4.5 are sufficient conditions to guarantee the synchronization of two chaotic systems, it is also a necessary condition for most of the dynamical systems. It is worth noting that the synchronization speed is very fast. When $d = 1$ or $m = 1$, the synchronization can be achieved as D-B synchronization.

4.3.2.2 Simulation and Discussion

Example 2: Linear and nonlinear feedback synchronization for logistic map.

Logistic map is

$$x_{n+1} = \mu x_n(1 - x_n), \tag{4.29}$$

where $x \in (0, 1)$ and $\mu \in (0, 4]$. When $\mu = 4$, the system is chaotic. The experimental conditions are as follows. The initial value is 0.2. The step size of parameter p and k is 0.05. The iteration number is 40,000.

The drive system is

$$x_{n+1} = F(x_n) = 4x_n - 4x_n^2. \tag{4.30}$$

The response system is

$$y_{n+1} = F(y_n) = 4y_n - 4y_n^2 + u(x_n, y_n). \tag{4.31}$$

The control law is

$$u(x_n, y_n) = k(x_n - y_n) - p(x_n^2 - y_n^2). \tag{4.32}$$

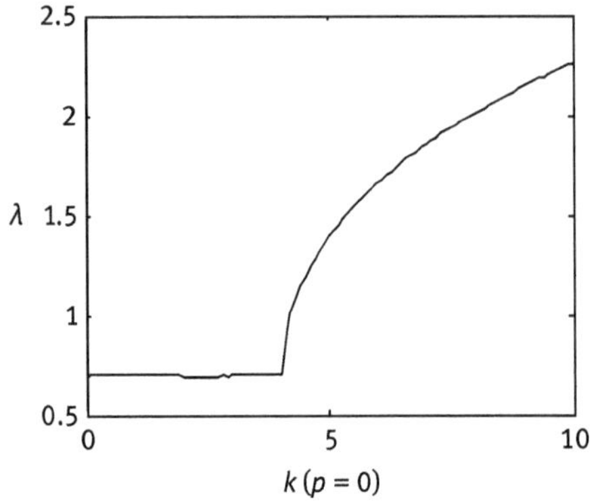

Figure 4.5: CLE versus feedback coefficient *k*.

1. When $p = 0$, the linear feedback control method is adopted. The maximum Lyapunov exponent λ of the response system varies with the linear feedback coefficient k as shown in Figure 4.5. Obviously, the value of λ is always greater than zero, so the two discrete systems cannot be synchronized.
2. When $k = 0$, the nonlinear feedback control strategy is adopted. The maximum Lyapunov exponent λ of the response system is shown in Figure 4.6 with the linear feedback coefficient p. The value of λ cannot meet the requirements of less than zero, so the synchronization between two discrete systems cannot be achieved.
3. When $k \neq 0, p \neq 0$, it means using linear feedback and nonlinear feedback strategy at the same time. According to Theorem 4.3, when $p = 4$, and $3 < k < 5$, then the CLE $\lambda < 0$, and the two systems can realize synchronization. The maximum Lyapunov exponent of the response system varies with the linear feedback coefficient k as shown in Figure 4.7.

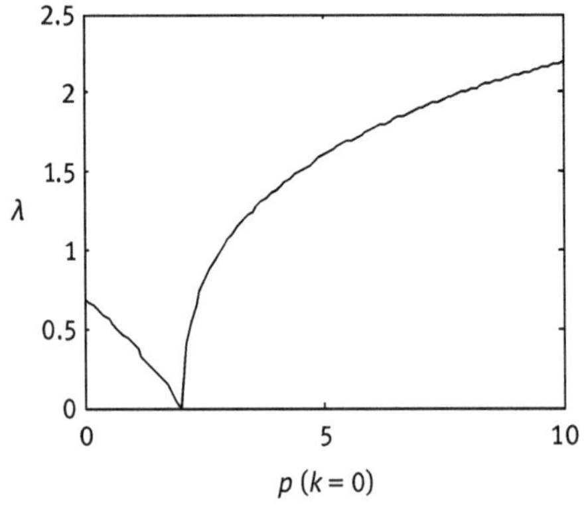

Figure 4.6: CLE versus feedback coefficient *p*.

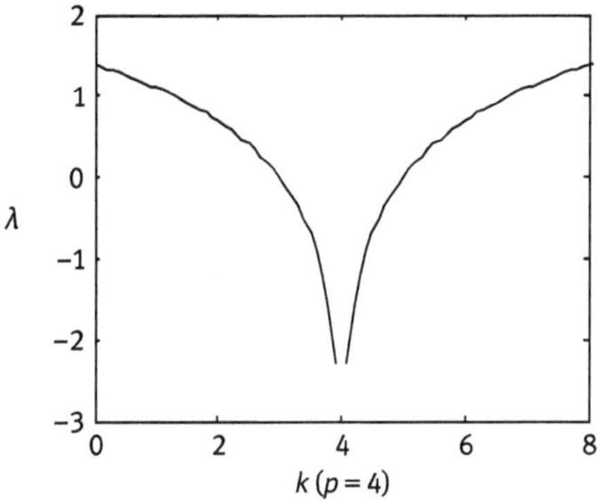

Figure 4.7: CLE versus feedback coefficient k.

4. When $k = ma, p = mb$, according to Theorem 4.5, and $4 - 4e^{-\lambda'} < k = p < 4 + 4e^{-\lambda'}$, the two systems can realize synchronization. Due to $\lambda' = 0.6931$ for logistic map, $2 < k = p < 6$. The maximum CLE of the response system varies with the linear feedback coefficient as shown in Figure 4.8.
5. The synchronization performance of the system is related to the selection of the control parameters. Experimental results show that the more negative the maximum CLE is by choosing the control parameters, the faster the error converges, and the shorter the synchronization setup time is. The error convergence with different feedback coefficients is shown in Figure 4.9. When the initial conditions of the drive and response systems are 0.1 and 0.8, respectively, and $k = p = 3.2$, then $\lambda = -1$. The synchronization time is 4 s as shown in Figure 4.9(a). When $k = p = 4$, then $\lambda = -3$. The synchronization time is 1 s as shown in Figure 4.9(b). So the synchronous speed is fast, and the D-B synchronization is realized.

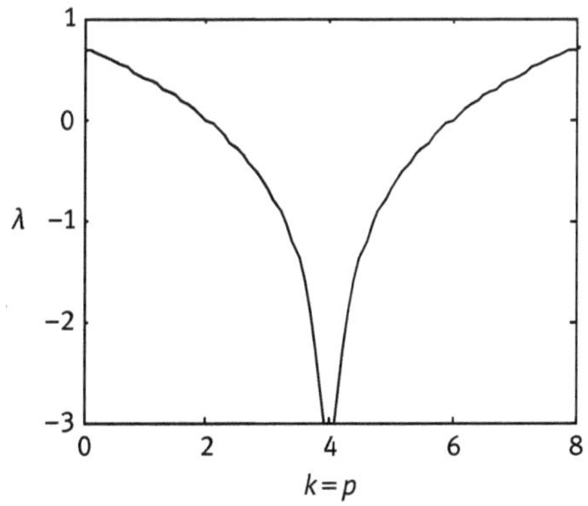

Figure 4.8: CLE versus feedback coefficient.

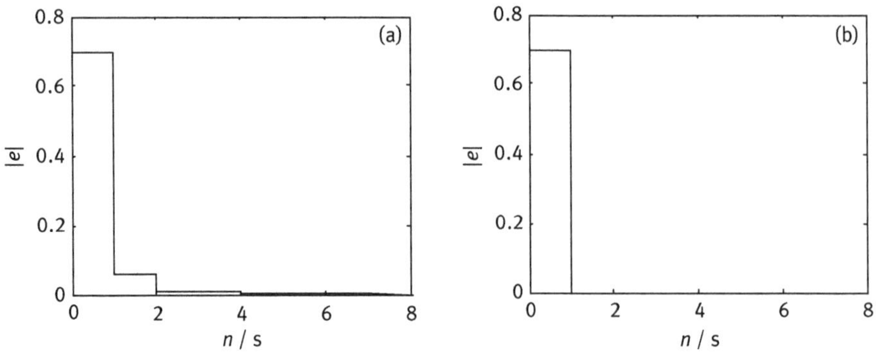

Figure 4.9: Synchronization error curves with different feedback coefficients: (a) $k = p = 3.2$ and (b) $k = p = 4$.

Example 3: Linear and nonlinear feedback synchronization for Hénon map.

Hénon map is

$$\begin{cases} x_{n+1} = 1 - ax_n^2 + y_n \\ y_{n+1} = bx_n \end{cases}. \tag{4.33}$$

When $a = 1.4$ and $b = 0.3$, the system is chaotic. Next, we apply Theorem 4.2 to Hénon map.

The drive system is

$$X_{n+1} = \begin{pmatrix} 0 & 1 \\ 0.3 & 0 \end{pmatrix} X_n + \begin{pmatrix} 1 \\ 0 \end{pmatrix} (1 - 1.4x_n^2). \tag{4.34}$$

The response system is

$$Y_{n+1} = \begin{pmatrix} 0 & 1 \\ 0.3 & 0 \end{pmatrix} Y_n + \begin{pmatrix} 1 \\ 0 \end{pmatrix} (1 - 1.4y_n^2) + U(X_n, Y_n). \tag{4.35}$$

The control law is

$$U(X_n, Y_n) = \begin{pmatrix} k_1 & 0 \\ 0 & k_2 \end{pmatrix} (X_n - Y_n) + \begin{pmatrix} 1.4 \\ 0 \end{pmatrix} (y_{1n}^2 - x_{1n}^2). \tag{4.36}$$

Then the error system is

$$\delta_{n+1} = \begin{pmatrix} -k_1 & 1 \\ 0.3 & -k_2 \end{pmatrix} \delta_n = D\delta_n. \tag{4.37}$$

Obviously, the characteristic equation of matrix D is

$$T(z) = |Iz - D| = \begin{vmatrix} z + k_1 & -1 \\ -0.3 & z + k_2 \end{vmatrix} = z^2 + (k_1 + k_2)z + k_1k_2 - 0.3\,. \tag{4.38}$$

According to Jury stability criteria, the necessary and sufficient condition of all the characteristic roots of $T(z)$ to locate in the unit circle is

$$\begin{cases} T(1) > 0 \\ T(-1) > 0 \\ 1 > |k_1k_2 - 0.3| \end{cases}, \text{ that is, } \begin{cases} k_1k_2 + k_1 + k_2 > -0.7 \\ k_1k_2 - (k_1 + k_2) > -0.7 \\ -0.7 < k_1k_2 < 1.3 \end{cases}. \tag{4.39}$$

According to eq. (4.39), when $k_1 = 0$ and $|k_2| < 0.7$, or when $k_2 = 0$ and $|k_1| < 0.7$, or when $k_1 = k_2 = k$ and $|k| < 0.4522$, the error system is asymptotically stable. That is to say, the response system can be synchronized with the drive system. When we set the initial values as $[0.1\ 1.0]^{\mathrm{T}}$ and $[0\ 0.1]^{\mathrm{T}}$, respectively, then the control parameters are $k_1 = 0$, $k_2 = 0.2$, and $P = B$. The error convergence curve of the two systems is shown in Figure 4.10(a). It shows that there exist nonsynchronization trends in the synchronization process. When the steady-state error is $\Delta\varphi = 10^{-4}$, the synchronization setup time is about 22 s. The synchronization performance of the synchronization system can be improved by choosing the appropriate parameters. For example, when $k_1 = k_2 = 0$ and $P = B$, the error convergence curve is shown in Figure 4.10(b). It can be seen that the synchronous speed is accelerated, and the synchronization time is shortened. The simulation synchronization time is about 17 s under the same steady-state error, and it realizes the monotonous synchronization.

By applying Theorem 4.4 to Hénon map and setting the simulation conditions, the initial value is $[0.1, 0.5]^{\mathrm{T}}$, the step size of parameter is 0.05, and the iteration number is 40,000.

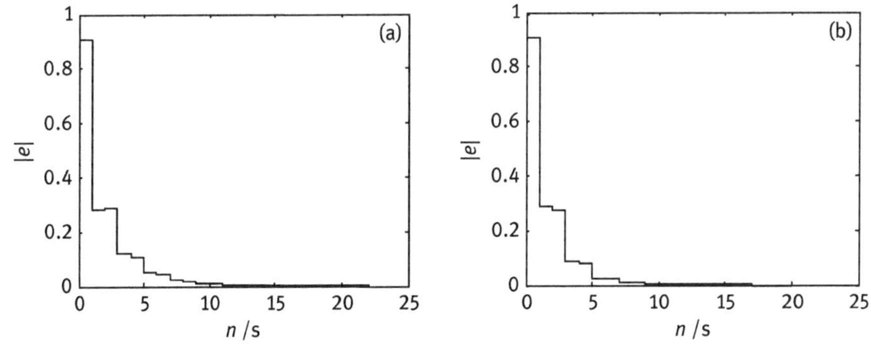

Figure 4.10: Synchronization error curves with different feedback parameters: (a) $k_1 = 0$, $k_2 = 0.2$ and (b) $k_1 = k_2 = 0$.

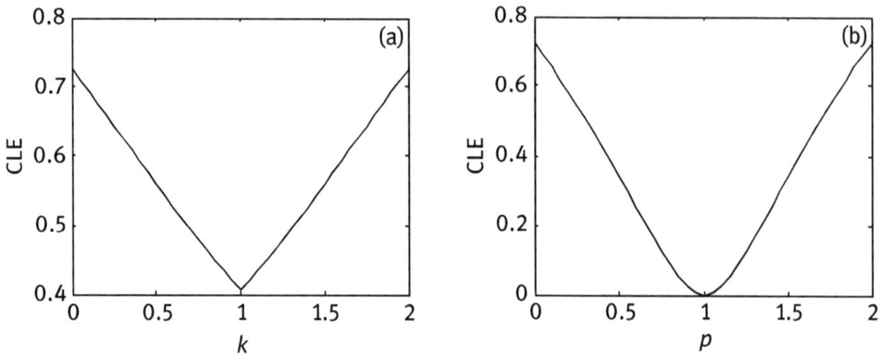

Figure 4.11: Maximum CLE curves with linear or nonlinear feedback: (a) $p = 0$ and (b) $k = 0$.

From drive system (4.34) and response system (4.35), the control law is chosen as

$$U(X_n, Y_n) = kA(X_n - Y_n) + pB(f(X_n) - f(Y_n)) \tag{4.40}$$

1. When $p = 0$, only the linear control method is employed. The maximum conditional Lyapunov exponent varies with the linear feedback coefficient k as shown in Figure 4.11(a). Obviously, regardless of the value of k, λ is always greater than zero, and the synchronization between the two discrete chaotic systems cannot be realized.
2. When $k=0$, only the nonlinear control method is employed. The maximum conditional Lyapunov exponent varies with the nonlinear feedback coefficient p as shown in Figure 4.11(b). Regardless of the value of p, λ cannot meet the requirements of less than zero, so the two systems cannot achieve synchronization.
3. When $p \neq 0$ and $k \neq 0$, that both the linear feedback and nonlinear feedback strategies are used at the same time. The conditional Lyapunov exponent λ is greater than zero by rational selection of control parameters. The system can realize synchronization. According to Theorem 4.4, when $p = 1$, the maximum CLE varies with the linear feedback coefficient k as shown in Figure 4.12(a). It can be seen that when $0 < k < 2$, then $\lambda < 0$.
4. When $k = dA$, $p = dB$, according to Theorem 4.4, if we want $\lambda < 0$, then $1 - e^{-\lambda'} < d < 1 + e^{-\lambda'}$ should be satisfied. As the maximum Lyapunov exponent of Hénon system is $\lambda' = 0.7258$, we have $0.52 < d < 1.48$. The maximum CLE of the response system varies with the feedback control parameters as shown in Figure 4.12(b).
5. The synchronization performance of the system is related to the control parameters. Experimental results show that the more negative the maximum CLE is, the faster the synchronization error converges, and the shorter the

synchronization setup time is. When $d = 1$, then the CLE is the smallest (infinitesimal). At this time, only one iteration step is needed to realize the chaos synchronization, as shown in Figure 4.13. It is equivalent to the D-B synchronization.

Both the stability criterion of discrete system and the Pecora–Carroll asymptotic synchronization theorem can synchronize two discrete chaotic systems, but they are different. The selection of the control parameters obtained by the stability criterion of the discrete system is a necessary and sufficient condition for the synchronization of the systems. While the selection of the control parameters according to the Pecora–Carroll chaos synchronization theorem is a sufficient condition for the synchronization of the systems. That is, it is possible to choose the control parameters to synchronize the drive system and the response system, but the maximum CLE is not negative. For example, for Hénon map, when $K = \mathrm{diag}[k, k]$ and $P = B$, the maximum CLE is greater

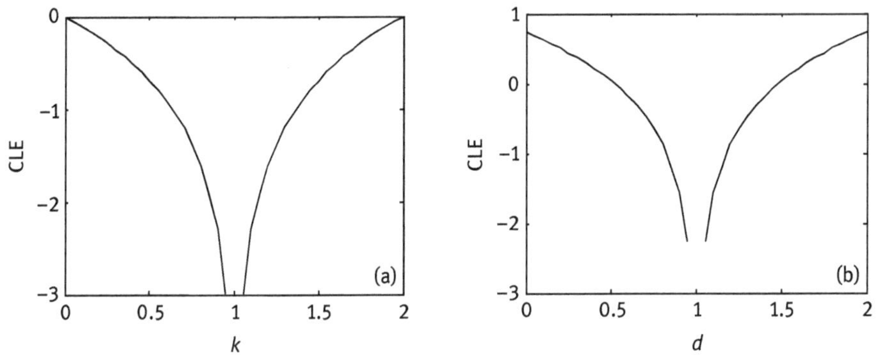

Figure 4.12: Maximum CLE curves with linear or nonlinear feedback: (a) $p = 1$, $k \neq 0$ and (b) $k = dA$, $p = dB$.

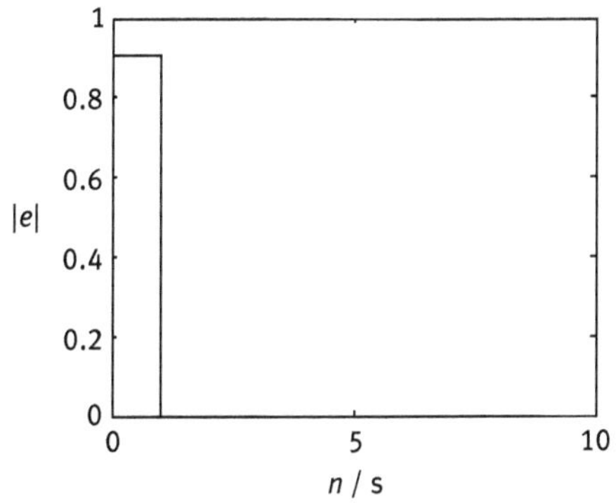

Figure 4.13: Synchronization error curve with $k = A$, $p = B$.

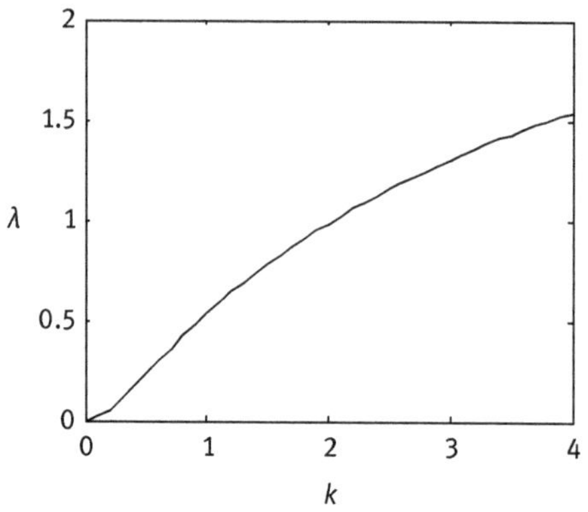

Figure 4.14: CLE versus k.

than zero as shown in Figure 4.14. But according to Theorem 4.2, when $|k| < 0.4522$, the response system and drive system can be synchronized.

4.4 Parameter Adaptive Synchronization Based on Pecora–Carroll Synchronization Criterion

4.4.1 Principle for Adaptive Synchronization

Consider an autonomous system

$$\dot{x} = f(x, \alpha), \tag{4.41}$$

where the state variable $x \in R^n$, the system parameter $\alpha \in R^m$, and nonlinear function $f: R^{n\times m} \to R^n$.

The given system that generates the desired orbit is

$$\dot{y} = f(y, \alpha), \tag{4.42}$$

where $y \in R^n$, $\alpha \in R^m$. The variable y starts from the initial and evolves with the known parameter α^*, and one obit $O(y)$ is obtained, which is called as a desired orbit or reference orbit. System (4.42) that produces the orbit is called a target system or a reference system. The adaptive control synchronization method is to introduce a control mechanism for the parameter α in system (4.41), so that the orbit $O(x)$ of system (4.41) from any initial value follows the orbit $O(y)$ of the target system. System (4.41) is called a controlled system or response system. If $\lim \|y - x\| = 0$, then the controlled system and the target system are synchronized.

The parameter α_j in system (4.41) can be obtained by introducing a certain amount of perturbation to modify the evolution of its orbit, which can make the $O(x)$ and $O(y)$ realize the synchronization. For parameter α_j, the control law is

$$\dot{\alpha}_j = -\varepsilon \sum h\left[(x_i - y_i), \operatorname{sgn}\left[\frac{df_i}{d\alpha_j}\right]\right] - \delta g(\alpha_j - \alpha_j^*), \tag{4.43}$$

where $i = 1, 2, \ldots, L$, $L \leq n$, $j = 1, 2, \ldots, k$, $k \leq m$. α_j^* is the parameter value to generate the desired obit. $h(\cdot,\cdot)$ and $g(\cdot)$ are control laws, respectively, and ε, δ are positive control constants. When the control law is selected, by changing ε and δ, the combined system from systems (4.43) and (4.41) is

$$\begin{cases} \dot{x} = f(x, \alpha) \\ \dot{\alpha}_j = -\varepsilon \sum h\left[(x_i - y_i), \operatorname{sgn}\left[\frac{df_i}{d\alpha_j}\right]\right] - \delta g(\alpha_j - \alpha_j^*) \end{cases} . \tag{4.44}$$

If all the CLEs are less than zero, the synchronization between the two autonomous chaotic systems is realized.

4.4.2 Choose Control Law and Control Parameters

Formula (4.43) only gives a general parameter control law, and it does not make a specific description for the choice of h and g. On the other hand, that all the Lyapunov exponents of the combined system are less than zero is only the sufficient conditions for the synchronization of chaotic systems. Therefore, it is necessary to make a further discussion about the control law and control parameter for the chaos synchronization.

In the following simulation, the control law is chosen as

$$h\left[(x_i - y_i), \operatorname{sgn}\left(\frac{df_i}{d\alpha_i}\right)\right] = (x_i - y_i)\operatorname{sgn}\left(\frac{df_i}{d\alpha_j}\right), \tag{4.45}$$

$$g(\alpha_j - \alpha_j^*) = (\alpha_j - \alpha_j^*). \tag{4.46}$$

Therefore, the adaptive control law of the controllable parameter is

$$\dot{\alpha}_j = -\varepsilon \sum (x_i - y_i)\operatorname{sgn}\left(\frac{df_i}{d\alpha_j}\right) - \delta(\alpha_j - \alpha_j^*). \tag{4.47}$$

So the change of controlled parameters depends on two factors. One is the difference $(x_i - y_i)$ between the state variable x in eq. (4.41) and the corresponding variable y of the desired orbit. The second is the difference $(\alpha_j - \alpha_j^*)$ between the values of the controlled parameters and the corresponding parameter values of the desired orbit.

Here, we take the unified chaotic system (3.35) as an example to illustrate the synchronization principle. To realize the synchronization control of two unified chaotic

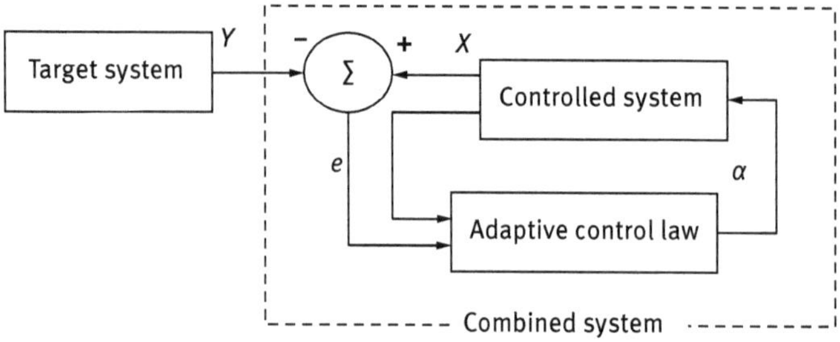

Figure 4.15: Principle block diagram of the adaptive synchronization.

systems, the adaptive control law for parameter α is given according to eqs (4.45) and (4.46), that is,

$$\begin{aligned}\dot{\alpha} = &-\varepsilon[(x_1 - y_1)\mathrm{sgn}(25(x_2 - x_1)) + (x_2 - y_2)\mathrm{sgn}(-35x_1 + 29x_2) \\ &+ (x_3 - y_3)\mathrm{sgn}(-x_3/3)] - \delta(\alpha - \alpha^*).\end{aligned} \tag{4.48}$$

According to the Pecora–Carroll synchronization stability criterion, the sufficient condition for the synchronization of two unified chaotic systems with different parameters is determined by the fact that the maximum CLE of the combined system is less than zero. The principle block diagram is shown in Figure 4.15.

4.4.3 Determine the Boundary and Range of the Control Constant

According to the analysis above, once adaptive control law (4.48) of the parameter is determined, the size of ε and δ becomes the key to achieve synchronization. The ranges can be determined by two methods, such as numerical calculation and theoretical analysis.

The numerical calculation method is used mainly to determine the range of ε and δ by calculating the Lyapunov exponent of the composite system which is composed of the unified chaotic system and formula (4.48). The boundary of the constants ε and δ is determined by the fact that the maximum CLE of the composite system is equal to zero.

The theoretical analysis method is described as follows. Assuming the reference system is at equilibrium, set $y = y^*$ as the equilibrium point of the reference system. α^* is the parameter corresponding to y^*. So α^* and y^* satisfy $f(y^*, \alpha^*)$. Obviously, the most basic requirement for the synchronization control is that the composite system is stable in the neighborhood of y^*. That is, all the eigenvalues of the Jacobian matrix of the composite system have negative real part, or the characteristic polynomial of Jacobian matrix is Hurwitz polynomial. So we can obtain the range of control constant ε and δ. Here the Jacobian matrix of the composite system is

$$J = \begin{bmatrix} -a & a & 0 & b \\ 28 - 35\alpha - x_3 & 29\alpha - 1 & -x_1 & c \\ x_2 & x_1 & d & e \\ -\varepsilon\,\mathrm{sgn}(b) & -\varepsilon\,\mathrm{sgn}(c) & -\varepsilon\,\mathrm{sgn}(e) & -\delta \end{bmatrix}, \tag{4.49}$$

where $a = 25\alpha + 10$, $b = 25(x_2 - x_1)$, $c = -35x_1 + 29x_2$, $d = -(8+\alpha)/3$, and $e = -x_3/3$. Obviously, the Jacobian matrix contains not only variables but also the symbol function, so the characteristic polynomial is very complex, and it is difficult to determine the range of ε and δ. So here the numerical calculation method of the maximum CLE presented in the second chapter is used to determine the range of ε and δ.

4.4.4 Simulation Results and Discussion

When $\alpha \in [0, 1]$, the reference system is chaotic. The initial values of the controlled system (response system) are (10, 10, 10) and $\alpha(0) = 0.5$. The initial values of the reference system (drive system) are (−5, −5, −5) and $\alpha^* = 0.3$. Let $|e| = [(x_1 - y_1)^2 + (x_2 - y_2)^2 + (x_3 - y_3)^2]^{1/2}$, and we have the following conclusions.

1. Average evolution time $\bar{\tau}$ from initial deviation to synchronization is related with the constant ε, δ and the expected parameters α^*, that is, $\bar{\tau} = \bar{\tau}(\varepsilon, \delta, \alpha^*)$.
2. When ε is determined, there is a range for constant $\delta < \delta_{\min}$ or $\delta > \delta_{\max}$, and synchronization control of two chaotic systems is not achieved. When δ is determined, there is a range for constant $\varepsilon < \varepsilon_{\min}$ or $\varepsilon > \varepsilon_{\max}$, and synchronization control of two chaotic systems is not achieved either, which as shown in Table 4.2.
3. When ε is determined, the larger δ is, the better the synchronization performance is, as shown in Figure 4.16.
4. When ε increases, $\delta_{\min}$ and $\delta_{\max}$ become larger, and the range of δ increases. Similarly, when δ increases, $\varepsilon_{\min}$ and $\varepsilon_{\max}$ become larger, and the range of ε also increases, but the increased magnitude is small as shown in Table 4.3. It shows that the control effect of the control constant ε is better than that of control constant δ.
5. When δ is determined, the larger ε is, the better the synchronization performance is, as shown in Figure 4.17.

Table 4.2: Values of ε and δ with different desired parameters for synchronization.

α^*	ε	δ	Maximum CLE
0	50	161–297	−1.7894
	39–95	230	−1.5513
0.3	50	145–328	−2.0530
	35–126	230	−4.3644
0.6	50	170–267	−4.1972
	14–49	230	−2.2715
0.9	25	195–340	−1.3862
	18–33	230	−1.5510

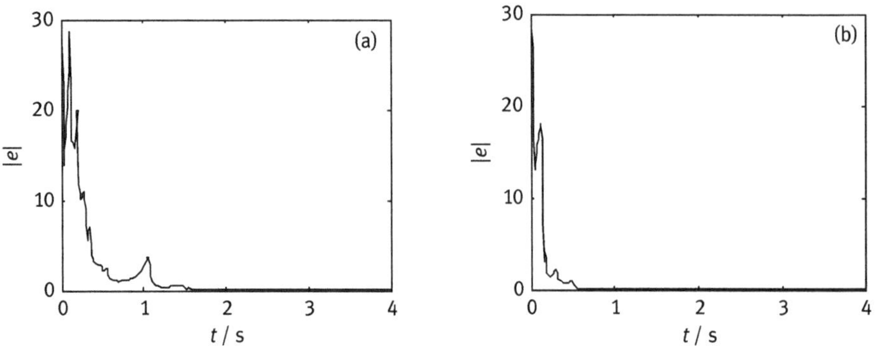

Figure 4.16: Error curves with $\varepsilon = 40$ and varying δ: (a) $\delta = 140$ and (b) $\delta = 200$.

Table 4.3: Change relation between ε and δ with $\alpha^* = 0.9$.

ε	δ	$\triangle_\delta$	Δ	ε	$\triangle_\varepsilon$
10	80–110	30	200	16–26	10
20	155–255	100	210	17–27	10
30	228–380	150	220	17–28	11
40	300–530	230	230	18–30	12
50	380–700	320	240	19–31	12

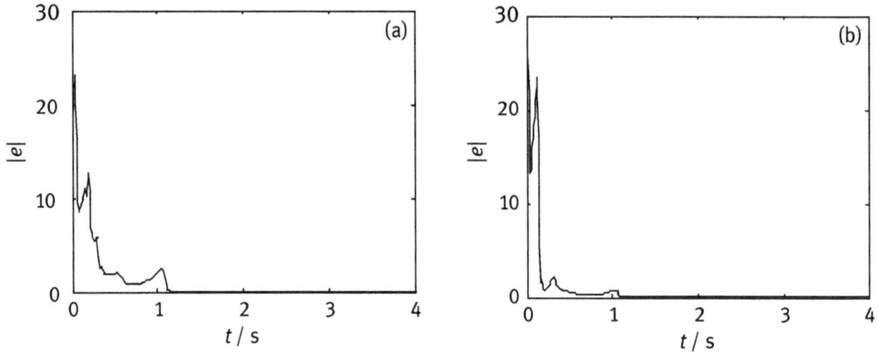

Figure 4.17: Error curves with $\delta = 220$ and varying ε: (a) $\varepsilon = 35$ and (b) $\varepsilon = 50$.

4.5 Adaptive Synchronization Control Based on Lyapunov Stable Theory

4.5.1 Adaptive Synchronization Control with Certain Parameter

To study synchronization between two unified chaotic systems with different initial values and different certain parameters, the parameters of the drive system and response system are set to be α_1 and α_2, respectively, and the drive system equation is

$$\begin{cases} \dot{x}_1 = (25\alpha_1 + 10)(y_1 - x_1) \\ \dot{y}_1 = (28 - 35\alpha_1)x_1 - x_1z_1 + (29\alpha_1 - 1)y_1 \\ \dot{z}_1 = x_1y_1 - (8 + \alpha_1)z_1/3 \end{cases} . \tag{4.50}$$

The response system equation is

$$\begin{cases} \dot{x}_2 = (25\alpha_2 + 10)(y_2 - x_2) \\ \dot{y}_2 = (28 - 35\alpha_2)x_2 - x_2z_2 + (29\alpha_2 - 1)y_2 \\ \dot{z}_2 = x_2y_2 - (8 + \alpha_2)z_2/3 \end{cases} . \tag{4.51}$$

To realize the synchronization, the control functions $u_1(t)$, $u_2(t)$, and $u_3(t)$ are added to the right-hand side of each equation, and the response system becomes

$$\begin{cases} \dot{x}_2 = (25\alpha_2 + 10)(y_2 - x_2) + u_1(t) \\ \dot{y}_2 = (28 - 35\alpha_2)x_2 - x_2z_2 + (29\alpha_2 - 1)y_2 + u_2(t) \\ \dot{z}_2 = x_2y_2 - (8 + \alpha_2)z_2/3 + u_3(t) \end{cases} . \tag{4.52}$$

The error system is obtained by subtracting eq. (4.52) from eq. (4.50)

$$\begin{cases} \dot{e}_1 = 25\alpha_1(e_2 - e_1) + 25(\alpha_2 - \alpha_1)(y_2 - x_2) + 10(e_2 - e_1) + u_1(t) \\ \dot{e}_2 = (28 - 35\alpha_1)e_1 + 29y_2 + (29\alpha_1 - 1)e_2 - 35(\alpha_2 - \alpha_1)x_2 \\ \quad -x_2z_2 + x_1z_1 + u_2(t) \\ \dot{e}_3 = x_2y_2 - x_1y_1 - (8 + \alpha_1)e_3/3 - (\alpha_2 - \alpha_1)z_2/3 + u_3(t) \end{cases} , \tag{4.53}$$

where synchronization error $e_1 = x_2 - x_1$, $e_2 = y_2 - y_1$, and $e_3 = z_2 - z_1$. Obviously, the problem of synchronization between two chaotic systems with different parameter is transformed into the stability problem of error system (4.53) at the origin $e_1 = 0$, $e_2 = 0$, and $e_3 = 0$. If the appropriate control variables $\boldsymbol{u} = [u_1(t), u_2(t), u_3(t)]^{\mathrm{T}}$ is chosen to make error system (4.53) stable, then response system (4.51) can be synchronized with drive system (4.50).

Theorem 4.6: For drive system (4.50) and response system (4.51), if we choose the control function

$$\begin{cases} u_1(t) = -25\hat{\alpha}(e_2 - e_1) - 10e_2 - 25(\alpha_2 - \alpha_1)(y_2 - x_2) \\ u_2(t) = (35\hat{\alpha} - 28)e_1 + x_2z_2 - x_1z_1 + 35(\alpha_2 - \alpha_1)x_2 \\ \quad -29(\alpha_2 - \alpha_1)y_2 - 29\hat{\alpha}e_2 \\ u_3(t) = -x_2y_2 + x_1y_1 + (\hat{\alpha} + 5)e_3/3 + (\alpha_2 - \alpha_1)z_2/3 \end{cases} , \tag{4.54}$$

where the parameter $\hat{\alpha}$ is an estimate of the parameter α_1, and the adaptive law of the estimate parameter is $\dot{\hat{\alpha}} = -10e_1e_2 - 25e_1^2 + 29e_2^2 - e_3^2/3 + \lambda(\alpha_1 - \hat{\alpha})$. If the control parameter satisfy $\lambda \geq 0$, then the equilibrium point $e = 0$ and $\alpha_1 = \hat{\alpha}$ exist in the error system, and the asymptotic synchronization is realized between the response system and drive system. That is, $\lim_{t\to\infty} \|e(t)\| = 0$ for any initial values.

Proof. Substitute control function (4.54) into error system (4.53), then we have

$$\begin{cases} \dot{e}_1 = 25(\alpha_1 - \hat{\alpha})(e_2 - e_1) - 10e_1 \\ \dot{e}_2 = -35(\alpha_1 - \hat{\alpha})e_1 + 29(\alpha_1 - \hat{\alpha})e_2 - e_2 \\ \dot{e}_3 = -(\alpha_1 - \hat{\alpha})e_3/3 - e_3 \end{cases} . \tag{4.55}$$

So error dynamics system (4.55) has no relationship with the response system parameter α_2. As long as the error system is stable, it can realize the synchronization of two chaotic systems with different parameters. According to the Lyapunov stable theory, the Lyapunov function is constructed as

$$E(e, \tilde{\alpha}) = \frac{1}{2}e^T e + \frac{1}{2}\tilde{\alpha}^2, \tag{4.56}$$

where $e = [e_1, e_2, e_3]^T$, $\tilde{\alpha} = \alpha_1 - \hat{\alpha}$. If the adaptive law of the estimate parameter is

$$\dot{\hat{\alpha}} = -10e_1e_2 - 25e_1^2 + 29e_2^2 - e_3^2/3 + \lambda(\alpha_1 - \hat{\alpha}), \tag{4.57}$$

then the derivative of eq. (4.56) is

$$\dot{E} = e_1\dot{e}_1 + e_2\dot{e}_2 + e_3\dot{e}_3 - (\alpha_1 - \hat{\alpha})\dot{\hat{\alpha}}. \tag{4.58}$$

Substituting eq. (4.55) and eq. (4.57) into eq. (4.58), we have

$$\dot{E} = -[10e_1^2 + e_2^2 + e_3^2 + \lambda(\alpha_1 - \hat{\alpha})^2]. \tag{4.59}$$

According to the Lyapunov stable theory, when $\lambda \geq 0$, $\dot{E} \leq 0$, and equilibrium point $e = 0$, $\alpha_1 = \hat{\alpha}$. The error system is asymptotically stable, so the two systems are synchronized. ■

We use Matlab/Simulink for numerical simulation. Set $\alpha_1 = 0.8$, $\alpha_2 = 0.3$, and initial values $x_1(0) = 10$, $y_1(0) = 10$, $z_1(0) = 10$, $x_2(0) = -5$, $y_2(0) = -5$, $z_2(0) = -5$ and $\hat{\alpha}(0) = 0.5$, and do numerical simulation for control parameter $\lambda = -0.1$, $\lambda = 1$, $\lambda = 2$, and $\lambda = 5$. The synchronization error curves between the two chaotic systems with different control parameters are shown in Figure 4.18.

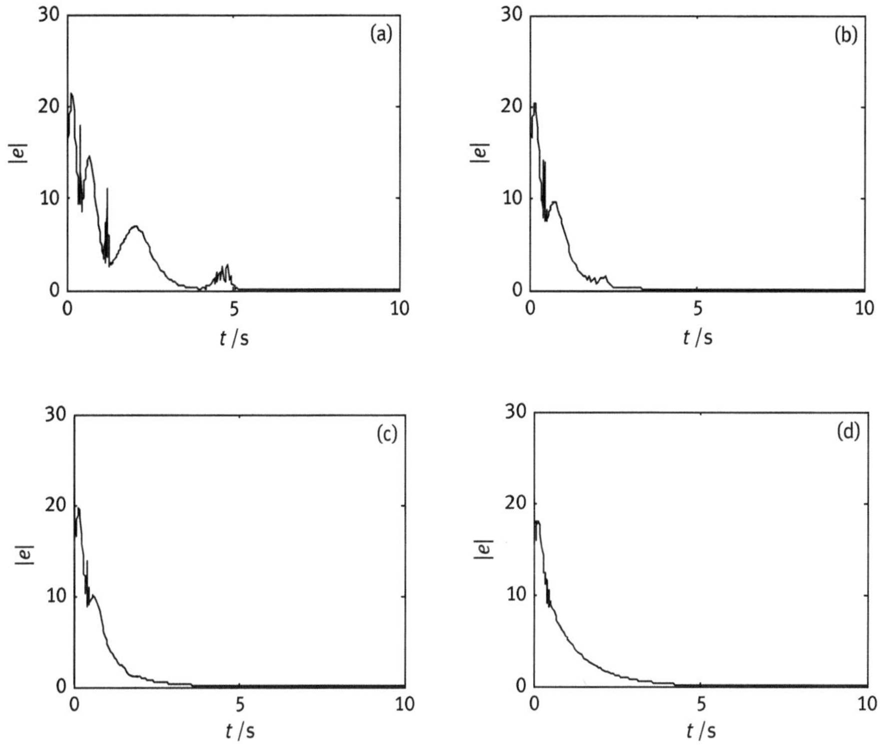

Figure 4.18: Synchronization error curve with different control parameters: (a) $\lambda = -0.1$; (b) $\lambda = 1$; (c) $\lambda = 2$; and (d) $\lambda = 5$.

The following conclusions are obtained by theoretical analysis and simulations. (1) The performance of the synchronization system can be improved by increasing the control parameter, and the monotonous synchronization of two chaotic systems is realized as shown in Figure 4.18(b)–(d). (2) The selection of the initial value of the adaptive parameters has no effect on the results. For example, the synchronization error curve with $\alpha_1 = 0.3$, $\hat{\alpha}(0) = 2$, and $\lambda = 5$ is the same with that with $\alpha_1 = 0.8$, $\hat{\alpha}(0) = 0.5$, and $\lambda = 5$ as shown in Figure 4.18(d). (3) There is parameter λ that does not meet the synchronization conditions, but it can make the synchronization of two systems [99], as shown in Figure 4.18(a), while the performance is poor at this time.

4.5.2 Adaptive Synchronization Control with Uncertain Parameters

Taking the unified chaotic system as an example, the synchronization of two chaotic systems with unknown parameters is discussed.

When the two system parameters are unknown, the drive and response systems based on unified chaotic system are

$$\begin{cases} \dot{x}_1 = (25\alpha + 10)(y_1 - x_1) \\ \dot{y}_1 = (28 - 35\alpha)x_1 - x_1z_1 + (29\alpha - 1)y_1 \\ \dot{z}_1 = x_1y_1 - (8 + \alpha)z_1/3 \end{cases}, \tag{4.60}$$

$$\begin{cases} \dot{x}_2 = (25\alpha + 10)(y_2 - x_2) + v_1(t) \\ \dot{y}_2 = (28 - 35\alpha)x_2 - x_2z_2 + (29\alpha - 1)y_2 + v_2(t) \\ \dot{z}_2 = x_2y_2 - (8 + \alpha)z_2/3 + v_3(t) \end{cases}, \tag{4.61}$$

where $v = [v_1, v_2, v_3]^T$ is the control vector and α is the unknown parameter. Set the error variables as $e_1 = x_2 - x_1$, $e_2 = y_2 - y_1$, and $e_3 = z_2 - z_1$, then the error system of systems (4.60) and (4.61) is

$$\begin{cases} \dot{e}_1 = (25\alpha + 10)(e_2 - e_1) + v_1(t) \\ \dot{e}_2 = (28 - 35\alpha)e_1 + x_1z_1 - x_2z_2 + (29\alpha - 1)e_2 + v_2(t) \\ \dot{e}_3 = x_2y_2 - x_1y_1 - (8 + \alpha)e_3/3 + v_3(t) \end{cases}. \tag{4.62}$$

To make the error system stable and realize the synchronization between the two systems, the following theorem is proposed for the control function and the estimated parameters.

Theorem 4.7: For drive system (4.60) and response system (4.61), if the control function is defined as

$$\begin{cases} v_1(t) = -(38 - 10\alpha)e_2 \\ v_2(t) = -29\hat{\alpha}e_2 + e_1z_2 + e_3x_1 \\ v_3(t) = e_1y_2 + e_2x_1 \end{cases}, \tag{4.63}$$

the parameter $\hat{\alpha}$ is the estimate of the parameter α. If the adaptive law of the estimate parameter is $\dot{\hat{\alpha}} = 29e_2^2$, then there exist the equilibrium points $e = 0$ and $\alpha = \hat{\alpha}$ in the error system, and the asymptotic synchronization is realized between the response system and drive system. That is, $\lim_{t\to\infty} \|e(t)\| = 0$ for any initial values.

Proof. Substituting control function (4.63) into error system (4.62), we have

$$\begin{cases} \dot{e}_1 = (35\alpha - 28)e_2 - (25\alpha + 10)e_1 \\ \dot{e}_2 = (28 - 35\alpha)e_1 + (29\alpha - 1)e_2 - 29\hat{\alpha}e_2 \\ \dot{e}_3 = -(\alpha_1 - \hat{\alpha})e_3/3 \end{cases}. \tag{4.64}$$

As long as the error system is stable, the synchronization of two chaotic systems with uncertain parameters is realized. According to the Lyapunov stable theory, the Lyapunov function V is constructed as

$$V(e, \tilde{\alpha}) = \frac{1}{2}e^T e + \frac{1}{2}\tilde{\alpha}^2, \tag{4.65}$$

where $\tilde{\alpha} = \alpha - \hat{\alpha}$, and $\hat{\alpha}$ is the estimate of uncertain α. If we choose the adaptive law of the estimate parameter $\dot{\hat{\alpha}} = 29e_2^2$, then the derivative of eq. (4.65) is

$$\dot{V}(e, \tilde{\alpha}) = -(25\alpha + 10)e_1^2 - e_2^2 - (8 + \alpha)e_3^2/3. \tag{4.66}$$

Because V is positive, and $\alpha \geq 0$, $\dot{V}$ is negative. According to the Lyapunov stable theory, there are equilibrium points $e = 0$ and $\alpha = \hat{\alpha}$, that is,

$$\lim_{t\to\infty} \left\| e(t) \right\| = 0, \quad \alpha \in [0, 1]. \tag{4.67}$$

The response system and the drive system are synchronized. ■

We use Matlab/Simulink for numerical simulation. When $\alpha \in [0, 1]$, both drive system (4.60) and response system (4.61) are chaotic. Let the initial values $x_1(0) = 10$, $y_1(0) = 10$, $z_1(0) = 10$, $x_2(0) = -5$, $y_2(0) = -5$, $z_2(0) = -5$, and $\alpha \in [0, 1]$, $\hat{\alpha}(0) = 0.5$. The synchronization error curve of two chaotic systems with unknown parameters is shown in Figure 4.19. The phase diagram of variables x_1 and x_2 is displayed in Figure 4.20.

The following conclusions are obtained by theoretical analysis and simulations. (1) The performance of the synchronization is stable as shown in Figure 4.19. When steady-state error is $\Delta\varphi = 10^{-4}$, the setup time of synchronization system is short ($t < 2$ s), and the monotonous synchronization of two systems is achieved. (2) The selection of the initial value of the adaptive parameters has no effect on the results. For example, the synchronization error curve with $\hat{\alpha}(0) = 2$ is the same with that with $\hat{\alpha}(0) = 0.5$. (3) By the Lyapunov stability theory, the synchronization condition is only a sufficient condition.

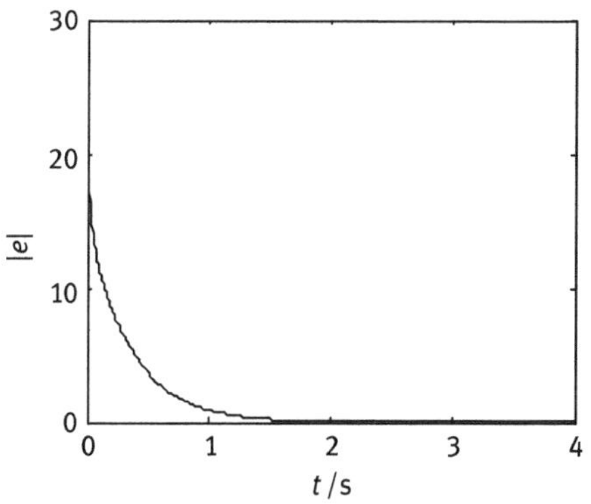

Figure 4.19: Error curve of chaotic synchronization system.

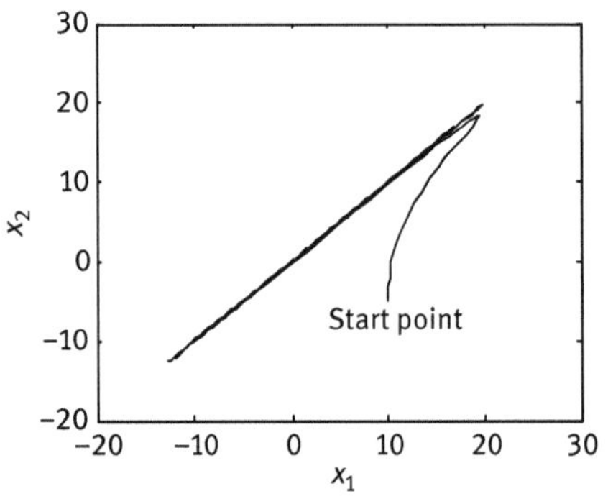

Figure 4.20: Phase portrait of x_1 and x_2.

4.6 Intermittent Feedback Synchronization Control

Among the synchronization control methods, the intermittent feedback control method is used to obtain the desired synchronization at low cost, and it is simple and easy to realize. In this section, the intermittent feedback synchronization between the unified chaotic system and its deformation system is investigated.

4.6.1 Intermittent Synchronization Control between Different Systems

Here, we discuss the problem of intermittent synchronization between two different chaotic systems with topological similarity. For the unified chaotic system

$$\begin{cases} \dot{x}_1 = (25\alpha + 10)(x_2 - x_1) \\ \dot{x}_2 = (28 - 35\alpha)x_1 - x_1x_3 + (29\alpha - 1)x_2 \\ \dot{x}_3 = x_1x_2 - (8 + \alpha)x_3/3 \end{cases} . \tag{4.68}$$

When the parameter $\alpha \in [0, 1]$, the system is chaotic.

The deformation system of the unified chaotic system is [100]

$$\begin{cases} \dot{y}_1 = (25\alpha + 10)(y_2 - y_1) \\ \dot{y}_2 = (28 - 35\alpha)y_1 - y_1y_3 + (29\alpha - 1)y_2 \\ \dot{y}_3 = y_1^2 - (8 + \alpha)y_3/3 \end{cases} . \tag{4.69}$$

Correspondingly, when $\alpha = 0.8$, systems (4.68) and (4.69) are Lü system and deformation Lü system, respectively. The phase portraits with the initial value [−1, 1, 0] are shown in Figure 4.21. Obviously, the phase track curves of Lü system and deformation Lü system are similar, but the evolution of the deformation system in the phase space is more divergent and the system is more complex.

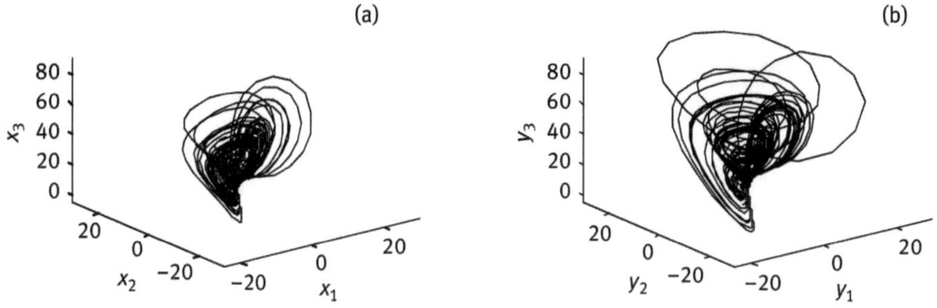

Figure 4.21: Strange attractors with $\alpha = 0.8$: (a) Lü system and (b) deformation Lü system.

To realize the synchronization between chaotic system (4.68) and deformation system (4.69), the synchronization control signal $u_i(t)$ is added to eq. (4.69), and then eq. (4.69) becomes

$$\begin{cases} \dot{y}_1 = (25\alpha + 10)(y_2 - y_1) + u_1(t) \\ \dot{y}_2 = (28 - 35\alpha)y_1 - y_1y_3 + (29\alpha - 1)y_2 + u_2(t) \\ \dot{y}_3 = y_1^2 - (8 + \alpha)y_3/3 + u_3(t) \end{cases} . \tag{4.70}$$

Set $e_i = x_i - y_i$, then the error system equation is

$$\begin{cases} \dot{e}_1 = (25\alpha + 10)(e_2 - e_1) - u_1(t) \\ \dot{e}_2 = (28 - 35\alpha - x_3)e_1 + (29\alpha - 1)e_2 - y_1e_3 - u_2(t) \\ \dot{e}_3 = (x_1x_2 - x_1^2) + 2x_1e_1 - e_1^2 - (8 + \alpha)e_3/3 - u_3(t) \end{cases} . \tag{4.71}$$

Set the control variables to

$$\begin{cases} u_1(t) = 0 \\ u_2(t) = -x_3e_1p(t) + ke_2p(t) - y_1e_3p(t) \\ u_3(t) = (x_1x_2 - x_1^2) + 2x_1e_1p(t) + ke_3p(t) \end{cases} , \tag{4.72}$$

where $p(t)$ is a pulse signal with period T and the ratio D of pulse length and pulse period, and it is defined by

$$p(t) = \begin{cases} 1 & iT \le t \le iT + TD \\ 0 & iT + TD < t < (i + 1)\mathrm{T} \end{cases} . \tag{4.73}$$

It can be seen that eq. (4.72) is a composite controller composed of nonlinear continuous feedback and linear intermittent feedback. So, eq. (4.71) becomes

$$\begin{cases} \dot{e}_1 = (25\alpha + 10)(e_2 - e_1) \\ \dot{e}_2 = (28 - 35\alpha - x_3 + x_3 p(t))e_1 + (29\alpha - 1 - kp(t))e_2 - y_1(1 - p(t))e_3 \\ \dot{e}_3 = 2x_1 e_1(1 - p(t)) - e_1^2 - (8 + \alpha)e_3/3 - ke_3 p(t) \end{cases} . \tag{4.74}$$

If it is asymptotically stable at the origin, the synchronization between systems (4.68) and (4.70) is achieved.

Theorem 4.8: For systems (4.68) and (4.69), if it is controlled by eq. (4.72), and the feedback parameter satisfies $k > 27{-}6\alpha$, and $D \geq 0.4$, then the asymptotically stable synchronization will be realized between systems (4.70) and (4.68).

Proof. Decompose the error system eq. (4.74), and when $iT < t < iT + TD$, we have

$$\begin{bmatrix} \dot{e}_1 \\ \dot{e}_2 \\ \dot{e}_3 \end{bmatrix} = A_1 \begin{bmatrix} e_1 \\ e_2 \\ e_3 \end{bmatrix} + \begin{bmatrix} 0 \\ 0 \\ -e_1^2 \end{bmatrix}, \tag{4.75}$$

where A_1 is the Jacobian matrix of the error system when the system has a control pulse

$$A_1 = \begin{bmatrix} -25\alpha - 10 & 25\alpha + 10 & 0 \\ 28 - 35\alpha & 29\alpha - 1 - k & 0 \\ 0 & 0 & -(8 + \alpha)/3 - k \end{bmatrix}. \tag{4.76}$$

When $iT + TD < t < (i + 1)T$, we have

$$\begin{bmatrix} \dot{e}_1 \\ \dot{e}_2 \\ \dot{e}_3 \end{bmatrix} = A_2 \begin{bmatrix} e_1 \\ e_2 \\ e_3 \end{bmatrix} + \begin{bmatrix} 0 \\ 0 \\ -e_1^2 \end{bmatrix}, \tag{4.77}$$

where A_2 is the Jacobian matrix of the error system when the system has no control pulse

$$A_2 = \begin{bmatrix} -25\alpha - 10 & 25\alpha + 10 & 0 \\ 28 - 35\alpha - x_3 & 29\alpha - 1 & -y_1 \\ 2x_1 & 0 & (8 + \alpha)/3 \end{bmatrix} \tag{4.78}$$

and $-e_1^2$ is the nonlinear term of the error system. For all t, when $\|e\| \to 0$, $\|-e_1^2\| / \|e\| \to 0$, so the evolution of errors depends on A_1 and A_2. During $iT < t < iT + TD$, the error is controlled by A_1. For $\alpha \in [0, 1]$, according to the stability criterion of continuous system, if k satisfies $k > 27{-}6\alpha$, then the eigenvalues of the matrix A_1 have negative real part. The zero point of the error system is asymptotically stable at this time. During $iT + TD < t < (i+1)T$, the controlled systems lose some additional feedback, and the error is controlled by A_2. Because A_2 does not satisfy the stability condition, the trajectory of the two systems will diverge, and the synchronization of the two systems will be lost. As long as the error is still small enough, and the exert control pulse is applied again, it can ensure that the error will not deviate too far, so that the synchronization of the two systems will be achieved gradually. After the continuous system is discretized, the error equation is written as

$$\begin{bmatrix} e_1(i) \\ e_2(i) \\ e_3(i) \end{bmatrix} = (A_1)^{m_1}(A_2)^{m_2} \begin{bmatrix} e_1(i-1) \\ e_2(i-1) \\ e_3(i-1) \end{bmatrix}, \tag{4.79}$$

where $e(i)$ is the error vector of the ith pulse period. $DT = m_1 h$, $(1-D)T = m_2 h$, where h is the integral step size, and m_1, m_2 are the integral iteration numbers for a pulse and no pulse, respectively. Obviously, in the process of synchronization, the system will generate oscillation due to $(A_2)^{m_2}$, and the stable synchronization process will depend on $|(A_1)^{m_1}(A_2)^{m_2}|$. If $|(A_1)^{m_1}(A_2)^{m_2}| = |J| < 1$, then the system will become stable [101]. The exact range of D can be obtained by numerical experiments. Simulations show that when $D \geq 0.4$, the exact synchronization can be achieved within 10 s. ∎

4.6.2 Synchronization Simulation and Performance Analysis

The synchronization control and performance improvement for the unified chaotic system and its deformation system are studied based on Matlab simulation. Here, we observe the changes of the time performance of the synchronization from the aspects of the initial value, the feedback coefficient k, the proportion factor D, the step size h, and the system parameter α.

4.6.2.1 Effect of Initial Value on Synchronization Performance

Set $\alpha = 0.8$, $k = 40$, $T = 0.1$ s, $D = 0.6$, $h = 0.001$, and the initial values are (1) $X(0) = [20, 0, 0]$, $Y(0) = [-2, 1, 0]$; (2) $X(0) = [10, 0, 0]$, $Y(0) = [-2, 1, 0]$; and (3) $X(0) = [5, 0, 0]$, $Y(0) = [-2, 1, 0]$. The synchronization error curve is shown in Figure 4.22. So, the synchronization between the two systems is achieved within 1.5 s. The oscillation phenomenon occurs in the process of establishing the synchronization, which is related to the characteristic value of A_2. Because at the beginning of $iT + TD < t < (i+1)T$, the error has not reached the steady state. In addition, the oscillation for initial value (3) is minimum, while the oscillation for initial value (1) is maximum. It can be seen that the initial value of the two systems is more; the larger the oscillation, the more the time required to establish accurate synchronization. But in general, it can achieve satisfactory results even when the initial value is large.

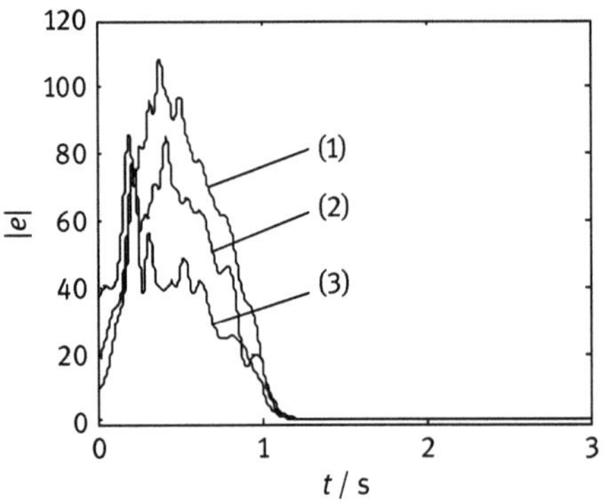

Figure 4.22: Synchronization error curves for different initial values.

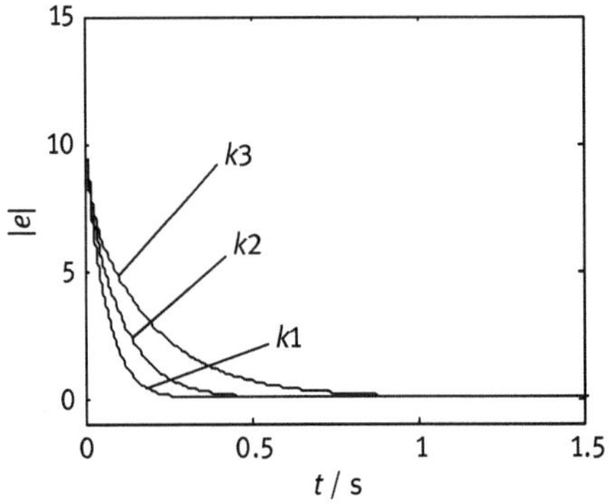

Figure 4.23: Synchronization error curves with different k.

4.6.2.2 Effect of different *k*, *D*, and *h* on synchronization performance

Set $\alpha = 0.8$, $T = 0.1$ s, $D = 0.8$, and $h = 0.0001$, and the initial values are $X(0) = [5, 0, 0]$ and $Y(0) = [-2, 1, 0]$. When the k values are 42, 32, and 27, the synchronization error curves are shown in Figure 4.23. It shows that the smooth degree of the curve is different with different k. For example, the error converges quickly for $k = 27$, and the error converges the fastest for $k = 42$. So the establishment time of the synchronization decreases with the increase of k, which is consistent with the theoretical analysis. In the practical application, considering the cost and the difficulty of the realization, the value of k should not be too large.

To achieve synchronization, the regulation of D is also important. Numerical experiments show that, similar to k, the synchronization setup time decreases with the increase of D. Further experiments prove that the smaller the size of the h is, the smaller the oscillation time is, and the faster the error curve converges.

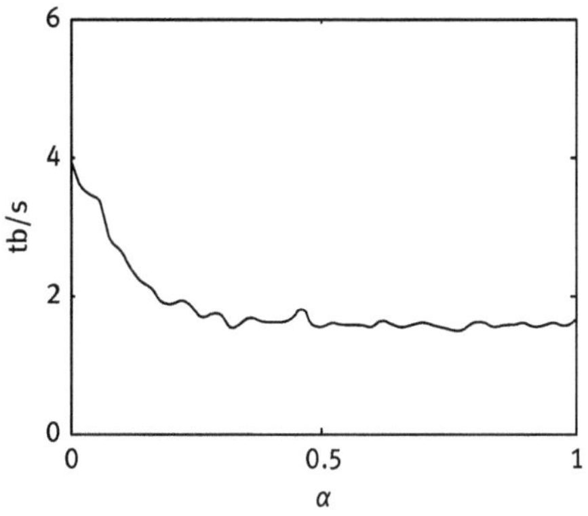

Figure 4.24: Relation curve of synchronization setup time *tb* with parameter α.

4.6.2.3 Effects of different parameter α on synchronization performance

Let $k = 45$, $T = 0.1$ s, $D = 0.6$, and $h = 0.0001$, and the initial values are $X(0) = [15, 0, 0]$, $Y(0) = [-15, -15, -15]$. The relationship of the synchronization setup time *tb* with the parameter α is shown in Figure 4.24. It shows when $\alpha \in [0, 1]$, the synchronization between the unified chaotic system and its deformation system is achieved within about 4 s, and the synchronization setup time decreases with increase of α.

4.7 Synchronization Control Based on State Observer Method

Because this synchronization method does not need to calculate the conditional Lyapunov exponent, it is widely concerned.

4.7.1 Design Principle of State Observer

Consider a class of nonlinear feedback chaotic systems

$$\begin{cases} \dot{x} = Ax + Bu + Ew \\ y = C^{\mathrm{T}}x \\ u = f(x) \end{cases}, \tag{4.80}$$

where $x \in R^n$, $u \in R^k$, and $w \in R^m$ are the state variables, nonlinear feedback input, and random noise of the system, respectively. y is the output in the sender. A, B, C, and E are known constant matrices.

In the receiver, when the nonlinear feedback input u is measurable, then the state observer of the system can be expressed as [102]

$$\begin{cases} \dot{z} = Fz + Gu + Hy \\ \hat{x} = z - Jy \end{cases}, \tag{4.81}$$

where $z \in R^n$ and $\hat{x} \in R^n$. F, G, H, and J are unknown constant matrices.

Suppose I_n is a unit matrix with n order, the error of the state observer is defined as $e = \hat{x} - x = z - (I_n + JC^{\mathrm{T}})x$. Let $M = I_n + JC^{\mathrm{T}}$, then the dynamic equation of the error system is

$$\dot{e} = Fe + (HC^{\mathrm{T}} - MA + FM)x + (G - MB)u - MEw. \tag{4.82}$$

When error system (4.82) satisfies

$$\begin{cases} HC^{\mathrm{T}} - MA + FM = 0 \\ G - MB = 0 \\ (I_n + JC^{\mathrm{T}})E = 0 \end{cases}, \tag{4.83}$$

then the error system becomes $\dot{e} = Fe$. Obviously, if all eigenvalues of the matrix F are less than zero, then $\hat{x}$ converges to x exponentially.

Substitute $M = I_n + JC^{\mathrm{T}}$ into the third term of $HC^{\mathrm{T}} - MA + FM = 0$, then we have

$$F = MA - (H + FJ)C^{\mathrm{T}}. \tag{4.84}$$

Let

$$N = H + FJ \tag{4.85}$$

and substitute eq. (4.84) into eq. (4.85), then we obtain

$$H = N + NC^{\mathrm{T}}J - MAJ. \tag{4.86}$$

At this point, the state observer equation is changed to

$$\dot{z} = (MA - NC^{\mathrm{T}})z + Gu + Hy. \tag{4.87}$$

So, the design problem of the state observer with random disturbance is turned into the problem of how to find the appropriate J, which satisfies $(I_n + JC^{\mathrm{T}})E = 0$, and how to choose the appropriate N, which satisfies the eigenvalues of F are less than zero. As $F = MA{-}NC^{\mathrm{T}}$, the condition that the eigenvalues of F are less than zero must be satisfied by choosing appropriate N. That is to say, in this case, we can always find a suitable N, so that when $t \to \infty$, $\hat{x}$ converges to x exponentially.

From eq. (4.83), we obtain $JC^{\mathrm{T}}E = -E$, and according to matrix knowledge, we have

$$J = -E(C^{\mathrm{T}}E)^{+} + V(I_p - (C^{\mathrm{T}}E)(C^{\mathrm{T}}E)^{+}), \tag{4.88}$$

where $(C^{\mathrm{T}}E)^{+}$ is the generalized inverse matrix of $(C^{\mathrm{T}}E)$. V is the matrix with appropriate dimensions. For the convenience of application, we often let $V = 0$. Then we have $J = -E(C^{\mathrm{T}}E)^{+}$. At the same time, we can obtain $M = I_p + JC^T$.

4.7.2 Design of State Observer for Unified Chaotic System

By applying the state observer method to the unified chaotic system, the sender matrix is described as

$$A = \begin{bmatrix} -25\alpha - 10 & 25\alpha + 10 & 0 \\ 28 - 35\alpha & 29\alpha - 1 & 0 \\ 0 & 0 & -(8+\alpha)/3 \end{bmatrix}, B = \begin{bmatrix} 1 & 0 & 0 \\ 0 & -1 & 0 \\ 0 & 0 & 1 \end{bmatrix}, u = \begin{bmatrix} 0 \\ x_1x_3 \\ x_1x_2 \end{bmatrix}, E = \begin{bmatrix} 0 \\ 1 \\ 0 \end{bmatrix},$$

$$C = [0 \quad 1 \quad 0]^{\mathrm{T}}.$$

The receiver equation is

$$\begin{cases} \dot{z} = Fz + \begin{bmatrix} 1 & 0 & 0 \\ 0 & 0 & 0 \\ 0 & 0 & 1 \end{bmatrix} u + \begin{bmatrix} 25\alpha + 10 \\ 0 \\ 0 \end{bmatrix} y \\ \hat{x} = z - \begin{bmatrix} 0 \\ -1 \\ 0 \end{bmatrix} y \end{cases}, \tag{4.89}$$

where

$$F = \begin{bmatrix} -25\alpha - 10 & 25\alpha + 10 - n_1 & 0 \\ 0 & -n_2 & 0 \\ 0 & -n_3 & -(8+\alpha)/3 \end{bmatrix}. \tag{4.90}$$

Here, n_1, n_2, and n_3 are the three values of the one-dimensional vector N, which can ensure that the eigenvalues of matrix F are less than zero, and these lead the synchronization error for $\dot{e} = Fe$ to zero finally. For simple calculation, we set $n_1 = 25\alpha + 10$, $n_2 > 0$, $n_3 = 0$. When $t \to \infty$, $\hat{x}$ converges to x exponentially.

4.7.3 Simulation and Discussion

Based on Matlab simulation platform, the synchronization of the unified chaotic system is investigated in this section. We set the initial values $x_1(0) = 15$, $x_2(0) = 15$, $x_3(0) = 15$, $\hat{x}_1(0) = 10^{-10}$, $\hat{x}_2(0) = -10$, $\hat{x}_3(0) = 10^{-10}$; $n_1 = 25\alpha + 10$, $n_3 = 0$, and let the synchronization error

$$|e| = \left[\sum_{i=1}^{3} (\hat{x}_i - x_i)^2 \right]^{\frac{1}{2}} . \tag{4.91}$$

Next, we study the synchronization performance of the synchronization system.

4.7.3.1 Synchronization Performance with Different Parameter α

When $n_2 = 3$ and $w(t) = 100\sin(3t)$, we obtain the synchronization error curves for $\alpha = 0.5$, $\alpha = 0.8$, $\alpha = 0.9$, respectively, as shown in Figure 4.25. It can be seen, although the difference between the two initial values is very far, that the system can achieve synchronization within $t = 2$ s for different system parameters. It shows that the system has strong robustness to parameter variation.

4.7.3.2 Effect of Noise Interference on the Performance of System Synchronization

When $n_2 = 3$ and $\alpha = 0.8$, we obtain the synchronization error curves for $w(t) = 5\sin(3t)$, $w(t) = 100\sin(3t)$, and white noise, respectively, as shown in Figure 4.26. It can be seen that the type and size of the noise have little effect on the synchronization error signal. The system can achieve the precise synchronization within $t = 2$ s. It shows the method has strong antinoise performance.

4.7.3.3 Effect of Initial Values on the Performance of System Synchronization

When $n_2 = 3$, $\alpha = 0.8$, and $w(t) = 100\sin(3t)$, we obtain the synchronization error curves for following initial values as shown in Figure 4.27.

(1) $X(0) = [15, 15, 15]$ $\hat{X}(0) = [10^{-10}, -10, 10^{-10}]$,
(2) $X(0) = [50, 50, 50]$ $\hat{X}(0) = [10^{-10}, -10, 10^{-10}]$,
(3) $X(0) = [15, 15, 15]$ $\hat{X}(0) = [10, 5, 10]$.

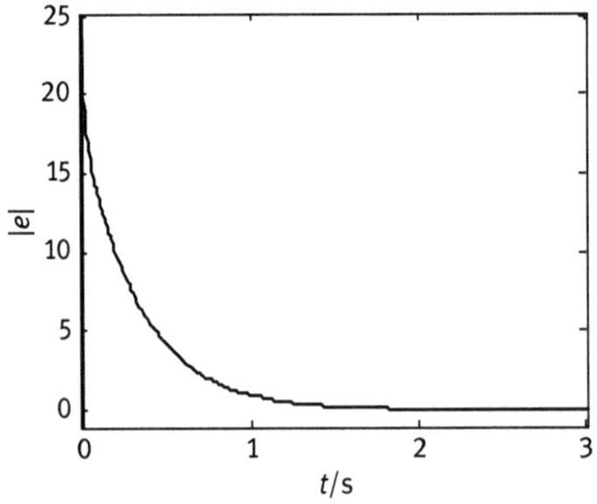

Figure 4.25: Synchronization error curves for different α.

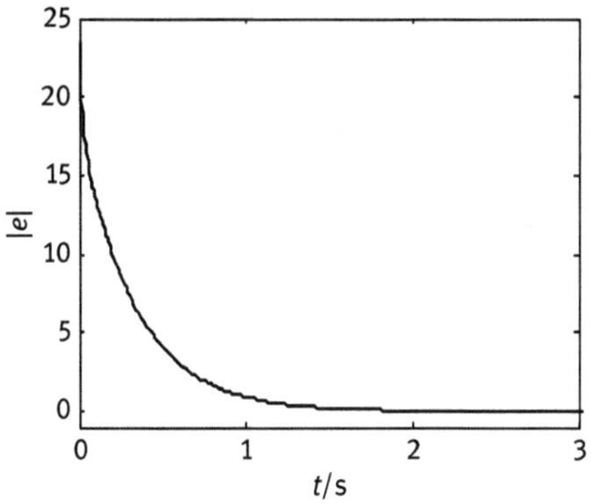

Figure 4.26: Synchronization error curve for different noise interferences.

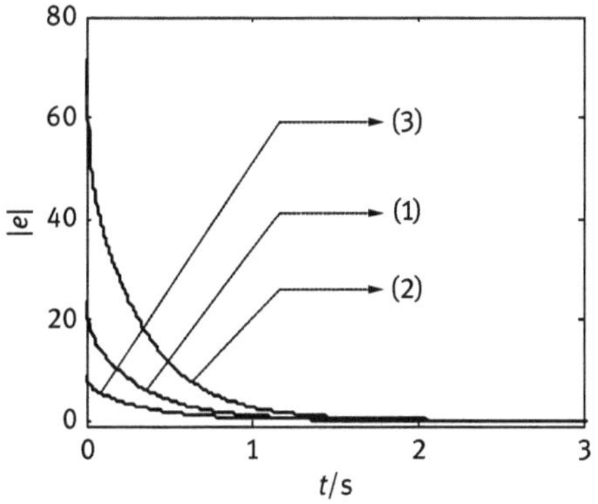

Figure 4.27: Synchronization error curves for different initial conditions.

From Figure 4.27, the setup time for the system synchronization is the longest for (2), and the setup time for the system synchronization is the shortest for (3). Because the synchronization error signal $e(t)$ decays exponentially, the time required to determine the exact synchronization is related to the initial conditions of the transmission end and the receiving end. It is obvious that the difference of the two initial values for (2) between the sender and the receiver is the biggest, so it is the longest time to achieve the exact synchronization.

4.7.3.4 Effect of Different n_2 on the Performance of Synchronization System

When $w(t) = 100\sin(3t)$, $\alpha = 0.8$, we obtain the synchronization error curves for $n_2 = 1$, $n_2 = 5$, and $n_2 = 50$, respectively, as shown in Figure 4.28. Obviously, when $n_2 > 0$, then the larger the n_2 is, the shorter the synchronization time of the system is. But when n_2 exceeds a certain value, then the value of n_2 is increased, and the time needed for synchronization is not significantly reduced.

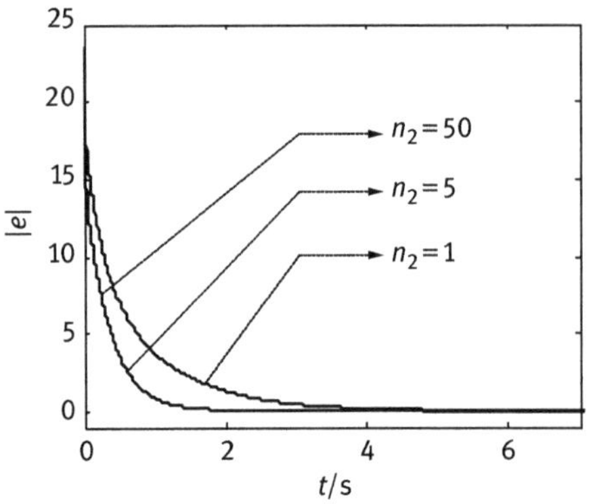

Figure 4.28: Synchronization error curves for different n_2.

4.8 Chaos Synchronization Based on Chaos Observer

4.8.1 Design Principle of Chaos Observer

For the following chaotic system

$$\begin{cases} \dot{x} = F(x) \\ s = h(x) \end{cases}, \tag{4.92}$$

where $x \in R^n$, F: $R^n \to R^n$; s is the scalar output of the system. Set $k(x)$ is a scalar differentiable function, and $V(x)$ is vector field, where V: $R^n \to R^n$. Lee derivative of the vector field $V(x)$ along the scalar differentiable function $k(x)$ is defined as $L_V k$, then $L_V k = <V(x) \bullet \text{grad}\, k(x)>$, where $< \bullet >$ is the Euclidean inner product. The ith Lee derivative of the vector field $V(x)$ along the scalar differentiable function $k(x)$ is defined as $L_V^i k = <V(x) \bullet \text{grad}\, L_V^{i-1} k(x)>$.

Lemma 4.1 [103]: If $\phi = [h(x), L_f h(x), \ldots, L_f^{n-1} h(x)]^{\mathrm{T}}$ is diffeomorphism, then all the state variables of system (4.92) are observable by using scalar output s and it is the ith ($i \leq n-1$) Lee derivative, and $x = \phi^{-1}[s(t), s^{(1)}(t), \ldots, s^{(n-1)}(t)]^{\mathrm{T}}$.

If a state variable x_i is selected as the output of system (4.92), then the remaining state variables of system (4.92) can be expressed as $x_j = g_i(x_i, \dot{x}_i, \ldots, x_i^{(n-1)})$, where $j \neq i$. So system (4.92) is written as

$$\dot{x} = Ax + g(x_i, \dot{x}_i, \ldots, x_i^{(n-1)}), \tag{4.93}$$

where A is a constant matrix, and $g(x_i, \dot{x}_i, \ldots, x_i^{(n-1)})$ is a matrix with $n \times 1$ order, which includes all the nonlinear terms of system (4.92).

Lemma 4.2 [104]: Set the output of system (4.92) as

$$s(x) = g(x_i, \dot{x}_i, \ldots, x_i^{(n-1)}) + B \times [g_1(x_i, \dot{x}_i, \ldots, x_i^{(n-1)}), \ldots, x_i, \ldots, g_n(x_i, \dot{x}_i, \ldots, x_i^{(n-1)})]^{\mathrm{T}}, \tag{4.94}$$

where $B = \operatorname{diag}[b_1, b_2, \ldots, b_n]$. If we select an appropriate diagonal matrix B so that the eigenvalues of the matrix $A–B$ have a negative real part, then system $\dot{y} = F(y) + s(x) - s(y)$ is the chaos observer of chaotic system (4.92), where $s(y) = g(y_i, \dot{y}_i, \ldots, y_i^{(n-1)}) + B \times [g_1(y_i, \dot{y}_i, \ldots, y_i^{(n-1)}), \ldots, y_i, \ldots, g_n(y_i, \dot{y}_i, \ldots, y_i^{(n-1)})]^{\mathrm{T}}$.

4.8.2 Chaos Observer of the Unified Chaotic System

The state observer synchronization scheme is applied to the unified chaotic system, and we have

$$A = \begin{bmatrix} -25\alpha - 10 & 25\alpha + 10 & 0 \\ 28 - 35\alpha & 29\alpha - 1 & 0 \\ 0 & 0 & -(8+\alpha)/3 \end{bmatrix},$$

$$\begin{cases} x_1 = g_1(x_1, \dot{x}_1, \ddot{x}_1) = x_1 \\ x_2 = g_2(x_1, \dot{x}_1, \ddot{x}_1) = \frac{\dot{x}_1}{25\alpha+10} + x_1 \\ x_3 = g_3(x_1, \dot{x}_1, \ddot{x}_1) = \left[\frac{-\ddot{x}_1}{25\alpha+10} + \frac{4\alpha-11}{25\alpha+10}\dot{x}_1 + (27 - 6\alpha)x_1\right] \Big/ x_1 \end{cases}.$$

Here, the drive system is driven by

$$s(x) = g(x_1, \dot{x}_1, \ddot{x}_1) + B \times [x_1, g_2(x_1, \dot{x}_1, \ddot{x}), g_3(x_1, \dot{x}_1, \ddot{x})]$$
$$= \begin{bmatrix} 0 \\ \frac{-\ddot{x}_1}{25\alpha+10} + \frac{4\alpha-11}{25\alpha+10}\dot{x}_1 + (27 - 6\alpha)x_1 \\ \frac{x_1\dot{x}_1}{25\alpha+10} + x_1^2 \end{bmatrix} + B \times \begin{bmatrix} x_1 \\ \frac{x\dot{x}_1}{25\alpha+10} + x_1 \\ \frac{\frac{-\ddot{x}_1}{25\alpha+10} + \frac{4\alpha-11}{25\alpha+10}\dot{x}_1 + (27-6\alpha)x_1}{x_1} \end{bmatrix},$$

where

$$B = \begin{bmatrix} b_{11} & b_{12} & b_{13} \\ b_{21} & b_{22} & b_{23} \\ b_{31} & b_{32} & b_{33} \end{bmatrix}.$$

The state observer system in the receiver can be expressed as

$$\dot{y} = f(y) + s(x) - s(y), \tag{4.95}$$

where $f(y) = Ay + g(y_1, \dot{y}_1, \ddot{y}_1, \ldots, y_1^{(n-1)})$. $s(y)$ is similar with $s(x)$.

By eq. (4.95), when the appropriate B is taken, the characteristic value of the $A-B$ is negative, and then the accurate synchronization between the chaotic system in the sender and the state observer in the receiver can be achieved.

4.8.3 Synchronization Simulation and Performance Analysis

The synchronization simulation of the unified chaotic system is carried out on Matlab, and we set the system's initial conditions $x_1(0) = 1$, $x_2(0) = 1$, $x_3(0) = 1$, $\hat{x}_1(0) = 15$, $\hat{x}_2(0) = 15$, $\hat{x}_3(0) = 15$,

$$B = \begin{bmatrix} 0 & 25\alpha + 10 & 0 \\ 28 - 35\alpha & b_{22} & 0 \\ 0 & 0 & 0 \end{bmatrix},$$

where $b_{22} > 19\alpha - 1$, and let $e_i = (\hat{x}_i - x_i)$.

Next, we will study the synchronization performance of the synchronization system.

4.8.3.1 Effect of Different Parameter α on the Performance of System Synchronization

For simulation, we set $\alpha = 0.5$, $\alpha = 0.8$, and $\alpha = 0.9$ (it means the system is the generalized Lorenz system, the generalized Lü system, and the generalized Chen system respectively). The synchronization error curves are obtained as shown in Figure 4.29. Obviously, for different parameter α, the unified chaotic system and the state observer will achieve accurate synchronization, but the synchronization performance is not exactly the same. This is mainly due to the use of a single parameter transmission, and the function between x_2 and $\hat{x}_2$ is complex. So this calculation will have a greater impact on the synchronization performance between x_2 and $\hat{x}_2$ with different α values. Overall, the synchronization performance is good for different α values.

4.8.3.2 Effect of Initial Values on the Performance of System Synchronization

Setting $b_{22} = 29$ and $\alpha = 0.8$, the changes of the synchronous error signals with the following three different initial conditions are analyzed, respectively. The synchronization error curves of the system are shown in Figure 4.30:

Initial condition (1): $X(0) = [15, 15, 15]$ $\hat{X}(0) = [10^{-10}, -10, 10^{-10}]$.

Initial condition (2): $X(0) = [50, 50, 50]$ $\hat{X}(0) = [10^{-10}, -10, 10^{-10}]$.

Initial condition (3): $X(0) = [15, 15, 15]$ $\hat{X}(0) = [10, 5, 10]$.

From Figure 4.30, when the initial values between the unified chaotic system and the state observer are further away from each other, the oscillation is more obvious, and the time required for synchronization is longer, which is consistent with

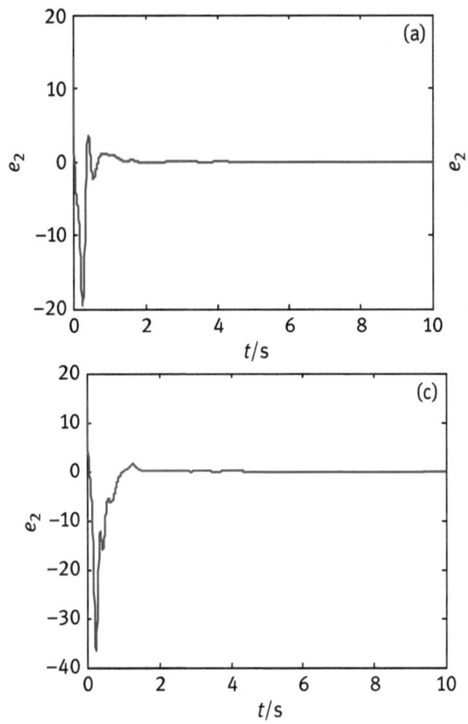

Figure 4.29: Synchronization error curves for different α: (a) $\alpha = 0.5$; (b) $\alpha = 0.8$; and (c) $\alpha = 0.9$.

the conclusion obtained by the previous state observer method. However, even if the initial value of the two is far, the system can still achieve accurate synchronization.

4.8.3.3 Effect of Different b_{22} on the Performance of Synchronization System

When $\alpha = 0.8$, system initial values $x_1(0) = 1$, $x_2(0) = 1$, $x_3(0) = 1$, $\hat{x}_1(0) = 15$, $\hat{x}_2(0) = 15$, $\hat{x}_3(0) = 15$, and $b_{22} = 24$, $b_{22} = 30$, $b_{22} = 50$, respectively, we obtained the synchronization error curves as shown in Figure 4.31. It shows that with the increase of b_{22}, the oscillation amplitude is smaller and smaller before the system reaches synchronization, and the time required to achieve synchronization is also less and less, but the corresponding hardware requirements will be higher. So in practical applications, the appropriate increase of b_{22} is good for improving the performance of system synchronization. In short, the system can achieve accurate synchronization in a certain time, as long as the condition $b_{22} > 29\alpha - 1$ is satisfied.

The chaos synchronization scheme, which is based on the state space reconstruction technique, is relatively simple in design and calculation. In the process of synchronization, the single parameter is transmitted, and the other parameters are derived at the receiving end. In this way, the channel utilization rate is greatly improved.

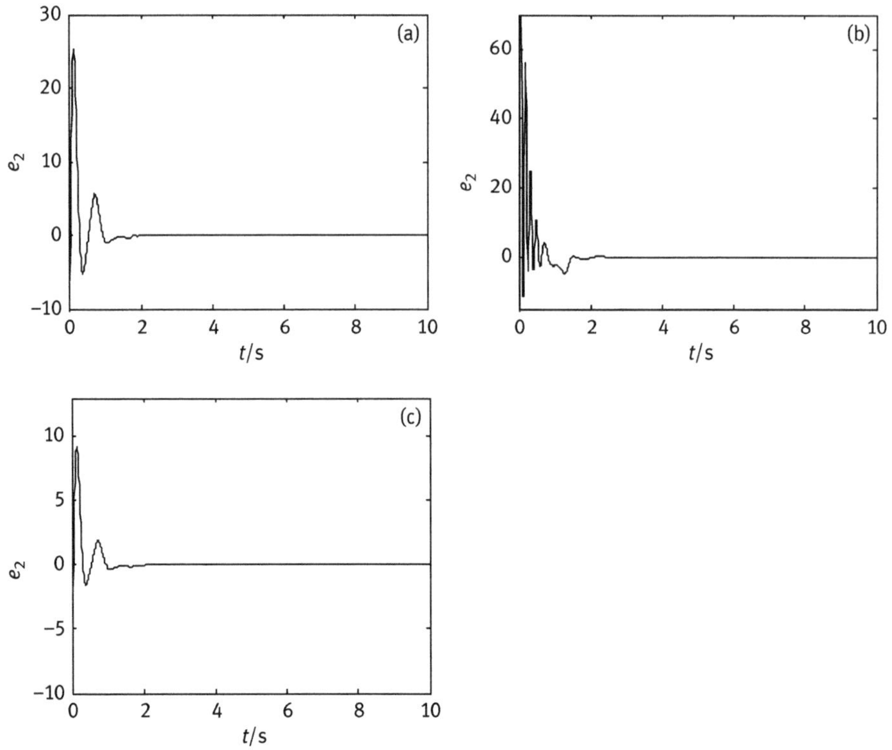

Figure 4.30: Synchronization error curves for different initial conditions: (a) initial value (1); (b) initial value (2); and (c) initial value (3).

As far as the synchronization performance, the proposed scheme has the characteristics of short synchronization time and good synchronization performance. However, the synchronous curve is not monotonous, and the oscillation is obvious; so, in practical application, to make the system synchronization performance optimization, the selection of the system parameters is the key.

4.9 Projective Synchronization

In 1999, Mainieri and Rehacek [28] observed a new synchronization phenomenon in the study of partially linear chaotic systems namely projective synchronization. For some partially linear chaotic systems, when the proper controller is selected, the output phase of the drive–response system state can be locked, and the amplitude of each corresponding state is also evolving according to a certain ratio. The new synchronization phenomenon has been greatly concerned by researchers, and a variety of projection synchronization schemes are proposed. Recently, a function projection synchronization method was proposed [29]. When compared with the general

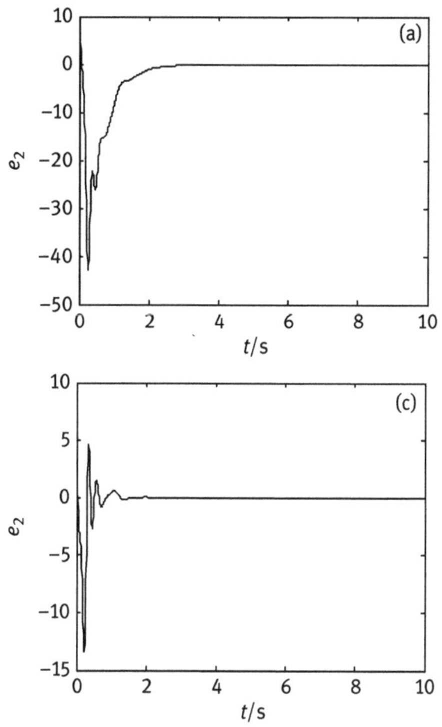

Figure 4.31: Synchronization error curves for different b_{22}: (a) $b_{22} = 24$; (b) $b_{22} = 30$; and (c) $b_{22} = 50$.

projective synchronization, the function projection synchronization means that the drive system and the response system can be synchronized according to a certain function, which is significant for the chaotic secure communication. In engineering practice, the parameters of the system may be unknown. To realize the parameter identification of chaotic systems with unknown parameters, the adaptive synchronization method is applied to the synchronization of chaotic systems with unknown parameters [29–30]. So it is important to study the projective synchronization control of chaotic systems, especially to study the synchronization of chaotic systems by combining adaptive control and function projective synchronization.

4.9.1 Principle and Simulation of Proportional Projection Synchronization

Consider the following continuous chaotic system:

$$\dot{\boldsymbol{X}} = F(\boldsymbol{X}), \tag{4.96}$$

where the state variable $\mathbf{X} = (x_1, x_2, \ldots, x_n)^{\mathrm{T}} \in \mathbf{R}^n$, and the continuous vector function $F\colon \boldsymbol{R}^n \to \boldsymbol{R}^n$. The function $F(\mathbf{X})$ is decomposed into

$$F(\boldsymbol{X}) = \tilde{\boldsymbol{A}}\boldsymbol{X} + \tilde{H}(\boldsymbol{X}), \tag{4.97}$$

where $\tilde{\boldsymbol{A}}\boldsymbol{X}$ is the linear part of $F(\boldsymbol{X})$, and $\tilde{H}(\boldsymbol{X})$ is the nonlinear part of $F(\boldsymbol{X})$. Then, $\tilde{\boldsymbol{A}}\boldsymbol{X}$ is decomposed into

$$\tilde{\boldsymbol{A}}\boldsymbol{X} = \boldsymbol{A}\boldsymbol{X} + \bar{\boldsymbol{A}}\boldsymbol{X}, \tag{4.98}$$

where $\boldsymbol{A}$ is a full rank constant matrix, and the real part of the eigenvalue is negative. Let

$$H(\boldsymbol{X}) = \bar{\boldsymbol{A}}\boldsymbol{X} + \tilde{H}(\boldsymbol{X}), \tag{4.99}$$

then eq. (4.96) can be rewritten as

$$\dot{\boldsymbol{X}} = \boldsymbol{A}\boldsymbol{X} + H(\boldsymbol{X}). \tag{4.100}$$

Now, we take system (4.100) as the drive system and employ the projective synchronization control method. Then the response system is

$$\dot{\mathbf{Y}} = \mathbf{A}\mathbf{Y} + H(\mathbf{X})/\alpha, \tag{4.101}$$

where $\boldsymbol{Y} \in \boldsymbol{R}^n$ is the n-dimensional state vector of the response system, and α is the proportional coefficient. If the synchronization error between drive system (4.100) and response system (4.101) is $\boldsymbol{E} = \boldsymbol{X} - \alpha\boldsymbol{Y}$, then the synchronous error equation is

$$\dot{\boldsymbol{E}} = \dot{\boldsymbol{X}} - \dot{\boldsymbol{Y}} = \boldsymbol{A}(\boldsymbol{X} - \alpha\boldsymbol{Y}) = \boldsymbol{A}\boldsymbol{E}. \tag{4.102}$$

The real part of all eigenvalues of matrix $\boldsymbol{A}$ is negative, so synchronous error system (4.102) is asymptotically stable at the origin according to the stability criterion of linear system. That is, $\lim\limits_{t\to\infty} |\boldsymbol{E}(t)| = 0$. So the synchronization is achieved between response system (4.101) and drive system (4.100).

Applying the projective synchronization method to the simplified Lorenz chaotic system, the drive system becomes

$$\begin{cases} \dot{x}_1 = 10(x_2 - x_1) \\ \dot{x}_2 = (24 - 4c)x_1 + cx_2 - x_1x_3 \\ \dot{x}_3 = x_1x_2 - 8x_3/3 \end{cases}, \tag{4.103}$$

and the response system becomes

$$\begin{cases} \dot{y}_1 = 10(y_2 - y_1) \\ \dot{y}_2 = (24 - 4c)y_1 + by_2 + \frac{1}{\alpha}((c - b)x_2 - x_1x_3) \\ \dot{y}_3 = -8y_3/3 + \frac{1}{\alpha}x_1x_2 \end{cases}, \tag{4.104}$$

where α is the proportional coefficient ($\alpha \neq 0$), and b is the synchronization control parameter. For the synchronization system comprising systems (4.103) and (4.104), the following synchronization theorem is proposed.

Theorem 4.9: For the projection synchronization system composed by drive system (4.103) and response system (4.104), if the synchronization control parameter $b < \min[10, 4c{-}24]$, then the synchronization error system is asymptotically stable at the origin point. That is, response system (4.104) will be synchronized with drive system (4.103) according to the proportional α.

Proof. According to the principle of projective synchronization, for the projection synchronization system comprising drive system (4.103) and response system (4.104), the full rank matrix $\boldsymbol{A}$ and $H(\boldsymbol{X})$ are, respectively,

$$\boldsymbol{A} = \begin{bmatrix} -10 & 10 & 0 \\ 24-4c & b & 0 \\ 0 & 0 & -8/3 \end{bmatrix}, H(\boldsymbol{X}) = \begin{bmatrix} 0 \\ (c-b)x_2 - x_1x_3 \\ x_1x_2 \end{bmatrix}.$$

If λ is the eigenvalue of the matrix $\boldsymbol{A}$, then we have

$$|\lambda \boldsymbol{I} - \boldsymbol{A}| = \begin{vmatrix} \lambda+10 & -10 & 0 \\ 4c-24 & \lambda-b & 0 \\ 0 & 0 & \lambda+8/3 \end{vmatrix} = 0. \tag{4.105}$$

Obviously, one of the eigenvalues $\lambda_1 = -8/3$ and the other two characteristic roots satisfy the equation

$$\lambda^2 + (10-b)\lambda - 10b + 40c - 240) = 0. \tag{4.106}$$

According to the Routh–Hurwitz criterion, if the synchronization control parameter satisfies $b < \min[10, 4c{-}24]$, then the real part of the root of eq. (4.106) is negative. The parameter of the system $c \in [-1.59, 7.75]$, so we get $b < -17.64$. That is, when the control parameter of the synchronization system is $b < -17.64$, the real part of all eigenvalues of the matrix $\boldsymbol{A}$ of the synchronization error system is negative, and the synchronization error is $\lim_{t\to\infty} |\boldsymbol{E}(t)| = 0$. It means response system (4.104) will be synchronized with drive system (4.103) according to the ratio α. ■

On Simulink dynamic simulation platform, we set $c = 5$. According to Theorem 4.9, when $b < -4$, the synchronization will be achieved. Here we set $b = -5$ and $\alpha = -2$, and system initial values $x_1(0) = 4.4356$, $x_2(0) = 6.1771$, $x_3(0) = 7.2330$; $y_1(0) = 2.0883$, $y_2(0) = 0.4535$, $y_3(0) = -5.8578$. The error convergence curve, the evolution of the attractor, and variable of the chaotic synchronization system are shown in Figure 4.32, where the synchronization error is defined by $|\mathbf{E}| = \left(\sum_{i=1}^{3} (x_i - \alpha y_i)^2\right)^{1/2}$. Figure 4.32(a)

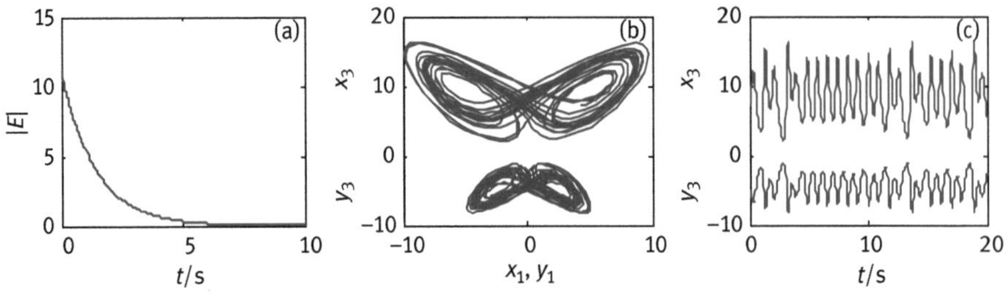

Figure 4.32: Projective synchronization between systems (4.103) and (4.104): (a) synchronization error curve; (b) attractors; and (c) waves of variables x_3 and y_3.

shows that the two chaotic systems with different initial values can achieve the projective synchronization at about 7.0 s, and the dynamic characteristics of the synchronization is good. Figure 4.32(b) shows that the attractor of the response system and drive system is inverse, and the size ratio is 2:1. According to Figure 4.32(c), we can see that the response system variable y_3 is synchronized with the drive system variable x_3 according to the projection ratio.

Now, we analyze the effects of different b values on the synchronization performance of the synchronization system. Set $c = 5$, $\alpha = -2$, and the initial values are same with that of Figure 4.32. Simulations are carried out for $b = -5$, $b = -6$, $b = -15$, respectively. The synchronization error curves are plotted in Figure 4.33. When $b < -4$, the smaller the b, the shorter the time required to achieve synchronization, but when b exceeds a certain value, then b reduces further, and the time required for synchronization cannot be significantly reduced. By formula (4.106), the b value is smaller, and the real part of the characteristic value of the synchronization error system is more negative. The pole is farther from the virtual axis; the faster the convergence of the synchronization system, the shorter the synchronization setup time.

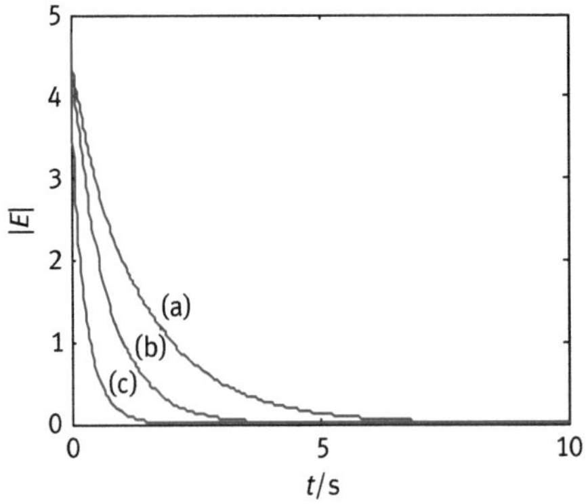

Figure 4.33: Synchronization error curves for different control parameter b: (a) $b = -5$; (b) $b = -6$; and (c) $b = -15$.

4.9.2 Principle and Simulation of Function Projective Synchronization

Consider the following continuous-time chaotic system as the drive system

$$\dot{\boldsymbol{X}} = \boldsymbol{AX} + F(\boldsymbol{X}), \tag{4.107}$$

where $\mathbf{X} = (x_1, x_2, \ldots, x_n)^{\mathrm{T}} \in \mathbf{R}^n$ is the state variable. $\boldsymbol{AX}$ is the linear part, and $F(\mathbf{X})$ is the nonlinear part. Copy eq. (4.107), and add the controller $\boldsymbol{U}$: $\boldsymbol{R}^{2n} \to \boldsymbol{R}^n$ into the system, then the response system is

$$\dot{\boldsymbol{Y}} = \boldsymbol{BY} + G(\boldsymbol{Y}) + \boldsymbol{U}(\boldsymbol{X},\boldsymbol{Y}), \tag{4.108}$$

where $\mathbf{Y} = (y_1, y_2, \ldots, y_n)^{\mathrm{T}} \in \mathbf{R}^n$ is the state vector. Set as the function matrix, and the error function of the synchronization system is $\boldsymbol{E} = \boldsymbol{X} - \Phi(t)Y$. By designing the controller $\boldsymbol{U}(\boldsymbol{X}, \boldsymbol{Y})$, we can get $\lim\limits_{t\to\infty} \|\boldsymbol{E}\| = \lim\limits_{t\to\infty} \|\boldsymbol{X} - \Phi(t)\boldsymbol{Y}\| = 0$, and then the generalized function projective synchronization is achieved between systems (4.107) and system (4.108).

Applying the generalized function projective synchronization method to the simplified Lorenz system, the drive system and response system are, respectively,

$$\dot{\boldsymbol{X}} = \boldsymbol{AX} + F(\boldsymbol{X}) = \begin{bmatrix} -10 & 10 & 0 \\ 24-4c & c & 0 \\ 0 & 0 & -8/3 \end{bmatrix} \begin{bmatrix} x_1 \\ x_2 \\ x_3 \end{bmatrix} + \begin{bmatrix} 0 \\ -x_1x_3 \\ x_1x_2 \end{bmatrix}, \tag{4.109}$$

$$\dot{\boldsymbol{Y}} = \boldsymbol{BY} + G(\boldsymbol{Y}) + \boldsymbol{U}(\boldsymbol{X}, \boldsymbol{Y}) = \begin{bmatrix} -10 & 10 & 0 \\ 24-4c & c & 0 \\ 0 & 0 & -8/3 \end{bmatrix} \begin{bmatrix} y_1 \\ y_2 \\ y_3 \end{bmatrix} + \begin{bmatrix} 0 \\ -y_1y_3 \\ y_1y_2 \end{bmatrix} + \boldsymbol{U}(\boldsymbol{X}, \boldsymbol{Y}), \tag{4.110}$$

where c is the system parameter, and $\boldsymbol{U} = [u_1, u_2, u_3]^{\mathrm{T}}$ is the nonlinear controller. We have the following synchronization theorem for the synchronization system comprising systems (4.109) and (4.110).

Theorem 4.10: For systems (4.109) and (4.110), the control law is $\boldsymbol{U}(\boldsymbol{X}, \boldsymbol{Y}) = \Phi^{-1}(t)[-\dot{\Phi}(t)\boldsymbol{Y} + F(\boldsymbol{X}) - \Phi(t)G(\boldsymbol{Y}) + (\boldsymbol{A} - \boldsymbol{B})\boldsymbol{X} + \boldsymbol{DE}]$, where $\Phi(t)$ is the projective function, and $\boldsymbol{D} = \mathrm{diag}[d_1, d_2, d_3]$ is the control parameter matrix. When the controller parameter satisfies $d_1 + d_2 > c-10$, $d_1d_2 - d_1c + 10d_2 > 240-30c$, and $d_3 > -8/3$, systems (4.109) and (4.110) can realize the function projective synchronization, that is, $\lim\limits_{t\to\infty} \|\boldsymbol{E}\| = 0$ for any initial value.

Proof. For the projection synchronization system comprising drive system (4.109) and response system (4.110), the error function of the synchronization system is

$$\mathbf{E} = \mathbf{X} - \Phi(t)\mathbf{Y}, \tag{4.111}$$

where $\Phi(t)$ is the diagonal function matrix. Then the error system equation is

$$\dot{\mathbf{E}} = \dot{\mathbf{X}} - \dot{\boldsymbol{\Phi}}(t)\mathbf{Y} - \boldsymbol{\Phi}(t)\dot{\mathbf{Y}}. \tag{4.112}$$

Substituting eqs (4.109) and (4.110) into the above formula, then we have

$$\dot{\boldsymbol{E}} = \boldsymbol{AX} + F(\boldsymbol{X}) - \boldsymbol{BY} - G(\boldsymbol{Y}) - \boldsymbol{U}(\boldsymbol{X}, \boldsymbol{Y}). \tag{4.113}$$

If the controller is $\boldsymbol{U}(\boldsymbol{X}, \boldsymbol{Y}) = \Phi^{-1}(t)[-\dot{\Phi}(t)\boldsymbol{Y} + F(\boldsymbol{X}) - \Phi(t)G(\boldsymbol{Y}) + (\boldsymbol{A} - \boldsymbol{B})\boldsymbol{X} + \boldsymbol{DE}]$, then we obtain

$$\dot{\boldsymbol{E}} = (\boldsymbol{B} - \boldsymbol{D})\boldsymbol{E}. \tag{4.114}$$

Construct Lyapunov function $\boldsymbol{V}(t)$ as

$$\boldsymbol{V} = \frac{1}{2}\boldsymbol{E}^{\mathrm{T}}\boldsymbol{E}. \tag{4.115}$$

Then its differential is

$$\dot{\boldsymbol{V}} = \boldsymbol{E}^{T}\dot{\boldsymbol{E}}. \tag{4.116}$$

Substituting eq. (4.114) into the above formula, we have

$$\dot{\boldsymbol{V}} = -\boldsymbol{E}^{\mathrm{T}}(\boldsymbol{D} - \boldsymbol{B})\boldsymbol{E}. \tag{4.117}$$

Set the control matrix $\boldsymbol{D} = \mathrm{diag}[d_1, d_2, d_3]$, when $d_1 + d_2 > c-10$, $d_1d_2 - d_1c + 10d_2 > 240-30c$, and $d_3 > -8/3$, $\boldsymbol{D}-\boldsymbol{B}$ is positive, and then $\dot{\boldsymbol{V}} \le 0$ if and only if $\boldsymbol{E} = 0$ for equality. By the Lyapunov stability theorem, the error system of the drive–response system is asymptotically stable at the origin. That is, $\lim_{t\to\infty} \|\boldsymbol{E}\| = 0$. The function projection synchronization is implemented. ■

On Simulink dynamic simulation platform, we set $c = 5$. According to Theorem 4.10, when the control parameters are $d_1 = 40$, $d_2 = 20$, $d_3 = -2$, the synchronization will be achieved. Set the projective function $\Phi(t) = -3 + 2\sin t$, and the initial values $x_1(0) = 4.4356$, $x_2(0) = 6.1771$, $x_3(0) = 7.2330$; $y_1(0) = -1.8028$, $y_2(0) = -2.3247$, $y_3(0) = -2.2974$. The synchronization error convergence curve, the evolution of the attractor, and variable of the chaotic synchronization system are shown in Figure 4.34. Figure 4.34(a) shows that the two chaotic systems with different initial values can achieve the projective synchronization within 4.8 s, and the dynamic characteristics

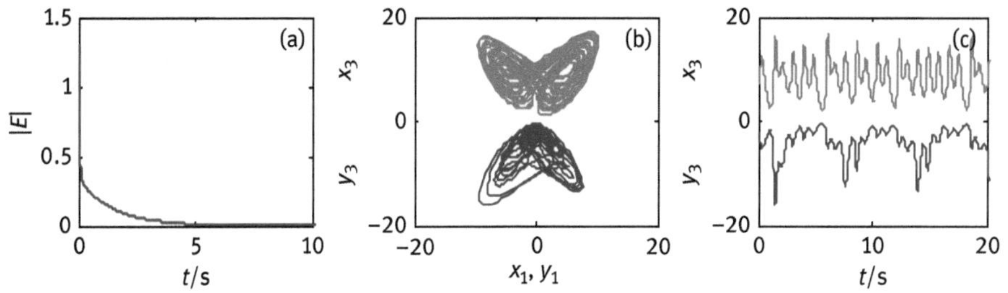

Figure 4.34: Projective synchronization between systems (4.109) and (4.110): (a) Synchronization error curve; (b) attractors; and (c) waves of variables x_3 and y_3.

of the synchronization are good. Figure 4.34(b) shows that the attractor of the response system and drive system is inverse, and the size ratio satisfies the projective function. According to Figure 4.34(c), we can see that the response system variable y_3 is synchronized with the drive system variable x_3 according to the projection function.

Now, we analyze the effects of different control parameter d_1, d_2, d_3 on the synchronization performance of the synchronization system. Set the system parameter $c = 5$, the projective function $\Phi(t) = -3 + 2\sin t$, and the initial values $x_1(0) = 4.4356$, $x_2(0) = 6.1771$, $x_3(0) = 7.2330$; $y_1(0) = -1.8028$, $y_2(0) = -2.3247$, $y_3(0) = -2.2974$.

1. When $d_2 = 20$, $d_3 = -2$, then $d_1 > -7.3$ according to Theorem 4.10. Synchronization simulations are carried out for $d_1 = -7$, $d_1 = -6.7$, $d_1 = -5$, respectively. The synchronization error curves are plotted in Figure 4.35(a). Obviously, when $d_1 > -7.3$, the bigger the d_1, the shorter the time required to achieve synchronization, but when the d_1 exceeds a certain value, and then continue to increase d_1, the time required for synchronization cannot be significantly reduced.
2. When $d_1 = 40$ and $d_3 = -2$, then $d_2 > 5.8$ according to Theorem 4.10. Synchronization simulations are carried out for $d_2 = 6.15$, $d_2 = 6.5$, $d_2 = 7$, respectively. The synchronization error curves are plotted in Figure 4.35(b). Obviously, when $d_2 > 5.8$, the bigger the d_2, the shorter the time required to achieve synchronization,

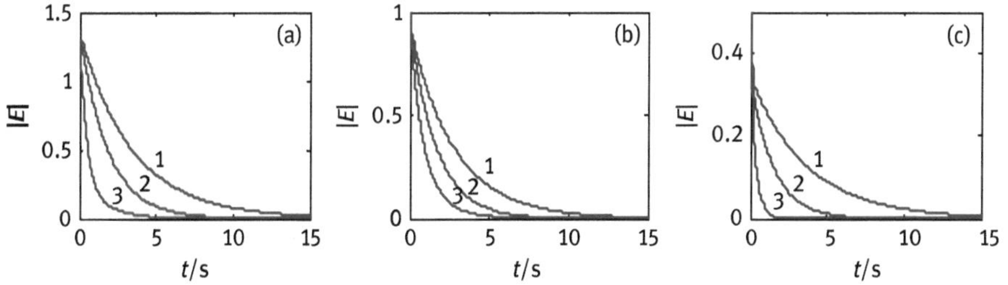

Figure 4.35: Synchronization error curves for different control parameter d. (a) Curve 1: $d_1 = -7$; curve 2: $d_1 = -6.7$; curve 3: $d_1 = -5$. (b) Curve 1: $d_2 = 6.15$; curve 2: $d_2 = 6.5$; curve 3: $d_2 = 7$. (c) Curve 1: $d_3 = -2.3$; curve 2: $d_3 = -2$; curve 3: $d_3 = 0$.

but when the d_2 exceeds a certain value, and continue to increase d_2, the time required for synchronization cannot be significantly reduced.

3. When $d_1 = 40$ and $d_2 = 20$, then $d_3 > -8/3$ according to Theorem 4.10. Synchronization simulations are carried out for $d_3 = -2.3$, $d_3 = -2$, $d_3 = 0$, respectively. The synchronization error curves are plotted in Figure 4.35(c). Obviously, when $d_3 > -8/3$, the bigger the d_3, the shorter the time required to achieve synchronization, but when the d_2 exceeds a certain value, if we continue to increase d_3, the time required for synchronization cannot be significantly reduced.

4.9.3 Principle and Simulation of the Adaptive Function Projective Synchronization

Consider the following drive system and response system:

$$\dot{\boldsymbol{X}} = \boldsymbol{AX} + \boldsymbol{f}(\boldsymbol{X}) + \boldsymbol{F}(\boldsymbol{X})\boldsymbol{\theta}, \tag{4.118}$$

$$\dot{\boldsymbol{Y}} = \boldsymbol{BY} + \boldsymbol{g}(\boldsymbol{Y}) + \boldsymbol{G}(\boldsymbol{Y})\boldsymbol{\delta} + \boldsymbol{U}(\boldsymbol{X}, \boldsymbol{Y}), \tag{4.119}$$

where $\boldsymbol{\theta}$ and $\boldsymbol{\delta}$ are unknown parameter vectors in the drive system and the response system, respectively. $\boldsymbol{f}(\boldsymbol{X})$ and $\boldsymbol{g}(\boldsymbol{Y})$ are vector functions of state variables $\boldsymbol{X}$ and $\boldsymbol{Y}$. $\boldsymbol{F}(\boldsymbol{X})$ and $\boldsymbol{G}(\boldsymbol{Y})$ are function matrices. $\boldsymbol{U}(\boldsymbol{X}, \boldsymbol{Y})$ is the control law.

Supposing the projective function is $\Phi(t)$, and the error function of the synchronization system is $\boldsymbol{E} = \boldsymbol{X} - \Phi(t)\boldsymbol{Y}$, then the synchronization error system equation is

$$\dot{\mathbf{E}} = \dot{\mathbf{X}} - \dot{\Phi}(t)\mathbf{Y} - \Phi(t)\dot{\mathbf{Y}}. \tag{4.120}$$

Due to eqs (4.118) and (4.119), we have

$$\dot{\mathbf{E}} = \dot{\mathbf{X}} - \dot{\Phi}(t)\mathbf{Y} - \Phi(t)\dot{\mathbf{Y}} = \mathbf{M}(\mathbf{X}, \mathbf{Y}) + \mathbf{F}(\mathbf{X})\boldsymbol{\theta} - \dot{\Phi}(t)\mathbf{Y} + \mathbf{BE} - \Phi(t)\mathbf{G}(\mathbf{Y})\boldsymbol{\delta} - \Phi(t)\mathbf{U}(\mathbf{X}, \mathbf{Y}), \tag{4.121}$$

where $\mathbf{M}(\mathbf{X}, \mathbf{Y}) = \mathbf{f}(\mathbf{X}) - \Phi(t)\mathbf{g}(\mathbf{Y}) + (\mathbf{A} - \mathbf{B})\mathbf{X}$. Obviously, the adaptive function projective synchronization problem of drive system (4.118) and response system (4.119) is transformed into the stability problem of eq. (4.121). By designing the controller $\mathbf{U}(\mathbf{X}, \mathbf{Y})$ and the adaptive law, to make $\lim_{t\to\infty} \|\mathbf{E}\| \lim_{t\to\infty} \|\mathbf{X} - \Phi(t)\mathbf{Y}\| = 0$, the adaptive function projective synchronization between systems (4.118) and (4.119) can be realized and the unknown parameters can be identified.

Theorem 4.11: For the chaotic synchronization system with unknown parameter, which is composed of drive system (4.118) and response system (4.119), if the control function of the system is $\mathbf{U}(\mathbf{X}, \mathbf{Y}) = \Phi^{-1}(t)[\mathbf{M}(\mathbf{X}, \mathbf{Y}) + \mathbf{F}(\mathbf{X})\hat{\boldsymbol{\theta}} - \dot{\Phi}(t)\mathbf{Y} - \Phi(t)\mathbf{G}(\mathbf{Y})\hat{\boldsymbol{\delta}} + \mathbf{DE}]$, where $\mathbf{M}(\mathbf{X}, \mathbf{Y}) = \mathbf{f}(\mathbf{X}) - \Phi(t)\mathbf{g}(\mathbf{Y}) + (\mathbf{A} - \mathbf{B})\mathbf{X}$, and $\mathbf{D} = \mathrm{diag}[d_1, d_2, d_3]$. If the appropriate control parameters and the adaptive law of the system parameters are selected as $\dot{\hat{\boldsymbol{\theta}}} = \boldsymbol{F}^{\mathrm{T}}(\boldsymbol{X})\boldsymbol{E} + \Delta\boldsymbol{\theta}$, $\dot{\hat{\boldsymbol{\delta}}} = -\Phi(t)\mathbf{G}^{\mathrm{T}}(\mathbf{Y})\mathbf{E} + \Delta\boldsymbol{\delta}$, where $\bar{\boldsymbol{\theta}} = \boldsymbol{\theta} - \hat{\boldsymbol{\theta}}$, $\bar{\boldsymbol{\delta}} = \boldsymbol{\delta} - \hat{\boldsymbol{\delta}}$. The adaptive function projective synchronization is realized between systems (4.118) and (4.119), and the unknown parameters are identified correctly. That is, for any initial values, we have $\lim_{t\to\infty} \|\mathbf{E}\| = 0$, $\hat{\boldsymbol{\theta}} = \boldsymbol{\theta}$, and $\hat{\boldsymbol{\delta}} = \boldsymbol{\delta}$.

Proof. Choose $\mathbf{U}(\mathbf{X},\mathbf{Y}) = \Phi^{-1}(t)[\mathbf{M}(\mathbf{X},\mathbf{Y}) + \mathbf{F}(\mathbf{X})\hat{\boldsymbol{\theta}} - \dot{\Phi}(t)\mathbf{Y} - \Phi(t)\mathbf{G}(\mathbf{Y})\hat{\boldsymbol{\delta}} + \mathbf{DE}]$ as the controller, then the error system eq. (4.120) becomes

$$\dot{\mathbf{E}} = \mathbf{F}(\mathbf{X})\bar{\boldsymbol{\theta}} - \boldsymbol{\Phi}(t)\mathbf{G}(\mathbf{Y})\bar{\boldsymbol{\delta}} - (\mathbf{D} - \mathbf{B})\mathbf{E}. \tag{4.122}$$

Constructing Lyapunov function

$$\mathbf{V} = \frac{1}{2}\mathbf{E}^{\mathrm{T}}\mathbf{E} + \frac{1}{2}\bar{\boldsymbol{\theta}}^{\mathrm{T}}\bar{\boldsymbol{\theta}} + \frac{1}{2}\bar{\boldsymbol{\delta}}^{\mathrm{T}}\bar{\boldsymbol{\delta}}, \tag{4.123}$$

where $\bar{\boldsymbol{\theta}} = \boldsymbol{\theta} - \hat{\boldsymbol{\theta}}$, $\bar{\boldsymbol{\delta}} = \boldsymbol{\delta} - \hat{\boldsymbol{\delta}}$, then its differential is

$$\dot{V} = \boldsymbol{E}^{\mathrm{T}}\dot{\boldsymbol{E}} + \bar{\boldsymbol{\theta}}^{\mathrm{T}}\dot{\bar{\boldsymbol{\theta}}} + \bar{\boldsymbol{\delta}}^{\mathrm{T}}\dot{\bar{\boldsymbol{\delta}}} = \boldsymbol{E}^{\mathrm{T}}\dot{\boldsymbol{E}} - \bar{\boldsymbol{\theta}}^{\mathrm{T}}\dot{\hat{\boldsymbol{\theta}}} - \bar{\boldsymbol{\delta}}^{\mathrm{T}}\dot{\hat{\boldsymbol{\delta}}}. \tag{4.124}$$

Set the adaptive law as $\dot{\hat{\boldsymbol{\theta}}} = \boldsymbol{F}^{\mathrm{T}}(\boldsymbol{X})\boldsymbol{E} + \bar{\boldsymbol{\theta}}$ and $\dot{\hat{\boldsymbol{\delta}}} = -\Phi(t)\mathbf{G}^{\mathrm{T}}(\mathbf{Y})\mathbf{E} + \bar{\boldsymbol{\delta}}$, and substitute eq. (4.121) into eq. (4.124), then we have

$$\dot{V} = -(\boldsymbol{D} - \boldsymbol{B})\boldsymbol{E}^{\mathrm{T}}\mathbf{E} - \bar{\boldsymbol{\theta}}^{\mathrm{T}}\bar{\boldsymbol{\theta}} - \bar{\boldsymbol{\delta}}^{\mathrm{T}}\bar{\boldsymbol{\delta}}. \tag{4.125}$$

Obviously, if we choose control parameters d_1, d_2, and d_3, to make $(\boldsymbol{D}-\boldsymbol{B})$ positive definite, then we have $\dot{\boldsymbol{V}} \le 0$. If and only if $\boldsymbol{E} = 0$, $\bar{\boldsymbol{\theta}} = \mathbf{0}$, $\bar{\boldsymbol{\delta}} = \mathbf{0}$ for equality. By the Lyapunov stability theorem, the error system of the drive–response system is asymptotically stable at the origin. That is, $\lim_{t\to\infty} \|\mathbf{E}\| = 0$, and the adaptive function projection synchronization is implemented. At the same time, when $t \to \infty$, we have $\hat{\boldsymbol{\theta}} \to \boldsymbol{\theta}$ and $\hat{\boldsymbol{\delta}} \to \boldsymbol{\delta}$. That is, the unknown parameters are identified correctly. ■

Remark 1: If $\boldsymbol{A} = \boldsymbol{B}$, $\boldsymbol{f}(\cdot) = \boldsymbol{g}(\cdot)$, and $\boldsymbol{F}(\cdot) = \boldsymbol{G}(\cdot)$, then the two chaotic systems are homogeneous systems, and vice versa they are heterogeneous systems. Therefore, the adaptive function projective synchronization theorem can be applied to both homogeneous and heterogeneous systems with unknown parameters.

Remark 2: If the parameters of the drive system or response system are known, using the above control method, the parameter adaptation law is removed, and the corresponding generalized function projective synchronization can be realized.

4.9.3.1 Adaptive Function Projective Synchronization and Simulation for Driving System with Unknown Parameters

Applying the adaptive function projective method to the simplified Lorenz system, the drive system and response system are, respectively,

$$\dot{X} = AX + f(X) + F(X)\theta = \begin{bmatrix} -10 & 10 & 0 \\ 0 & 0 & 0 \\ 0 & 0 & -8/3 \end{bmatrix} \begin{bmatrix} x_1 \\ x_2 \\ x_3 \end{bmatrix} + \begin{bmatrix} 0 \\ -x_1x_3 \\ x_1x_2 \end{bmatrix} + \begin{bmatrix} 0 & 0 & 0 \\ x_1 & x_2 & 0 \\ 0 & 0 & 0 \end{bmatrix} \begin{bmatrix} \theta_1 \\ \theta_2 \\ \theta_3 \end{bmatrix}, \tag{4.126}$$

$$\dot{Y} = BY + g(Y) + U(X, Y) = \begin{bmatrix} -10 & 10 & 0 \\ 24-4c & c & 0 \\ 0 & 0 & -8/3 \end{bmatrix} \begin{bmatrix} y_1 \\ y_2 \\ y_3 \end{bmatrix} + \begin{bmatrix} 0 \\ -y_1y_3 \\ y_1y_2 \end{bmatrix} + U(X, Y), \tag{4.127}$$

where c is the system parameter. $\Theta = [\theta_1, \theta_2, \theta_3]^T$ is the unknown parameter of the drive system. $U = [u_1, u_2, u_3]^T$ is the nonlinear controller. If we set the projective function as $\Phi(t) = -3 + 2\sin(t)$, according to Theorem 4.11, the control law and the parameter adaptive law of the synchronization system consisting of systems (4.126) and (4.127) are

$$U = \frac{1}{\Phi(t)} \begin{bmatrix} -\dot{\Phi}(t)y_1 + d_1e_1 \\ (\hat{\theta}_1 - 24 + 4c)x_1 + (\hat{\theta}_2 - c)x_2 - \dot{\Phi}(t)y_2 - x_1x_3 + \Phi(t)y_1y_3 + d_2e_2 \\ -\dot{\Phi}(t)y_3 + x_1x_2 - \Phi(t)y_1y_2 + d_3e_3 \end{bmatrix}, \tag{4.128}$$

$$\dot{\hat{\theta}} = \begin{bmatrix} x_1e_2 + \theta_1 - \hat{\theta}_1 \\ x_2e_2 + \theta_2 - \hat{\theta}_2 \\ 0 \end{bmatrix}. \tag{4.129}$$

There is a following inference for the synchronization system consisting of systems (4.126) and (4.127).

Inference 4.2 For the adaptive function projective synchronization system comprising drive system (4.126) and response system (4.127), when the controller parameters satisfy $d_1 + d_2 > c-10$, $d_1d_2-d_1c + 10d_2 > 240-30c$, and $d_3 > -8/3$, the adaptive function projective synchronization is achieved between systems (4.126) and (4.127). That is, $\lim_{t\to\infty} \|E\| = 0$ for any initial values. (Proof omitted.)

On Simulink platform, we set $c=5$. According to Inference 4.2, when the control parameters are $d_1=40$, $d_2=20$, and $d_3= -2$, the synchronization will be achieved. Set the projective function $\Phi(t)=-3 + 2\sin(t)$, and the initial values $X(0)=[4.4356, 6.1771, 7.2330]^T$, $Y(0)=[0.5431, 0.0995, 2.8096]^T$, $\hat{\theta}_1(0)=-1$, $\hat{\theta}_2(0)=-1$. The synchronization error convergence curve, the evolution of the attractor, and variable waves of the chaotic synchronization system are shown in Figure 4.36. Figure 4.36(a) shows that the two chaotic systems with different initial values can achieve the projective synchronization in 4 s, and the dynamic characteristics of the synchronization are good. Figure 4.36(b) shows that the attractor of the response system and drive system is inverse phase, and the size ratio satisfies the projective function. Figure 4.36(c) shows that in a relatively short period of time (about 2 s),

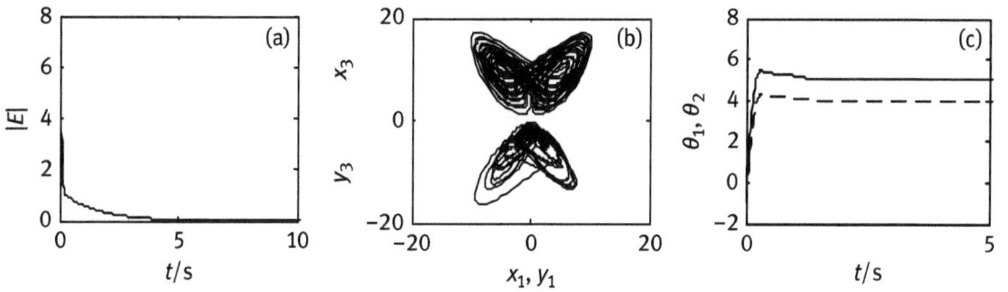

Figure 4.36: Projective synchronization between systems (4.126) and (4.127):
(a) Synchronization error curve; (b) attractors; and (c) waves of the estimated parameters $\hat{\theta}_1$ and $\hat{\theta}_2$.

the estimated parameters $\hat{\theta}_1$ and $\hat{\theta}_2$ converge to the target value and realize the identification of unknown parameters.

Now, we analyze the effects of different control parameters d_1, d_2, and d_3 on the synchronization performance of the synchronization system. When $d_2 = 20$ and $d_3 = -2$, then $d_1 > -7.3$ according to Inference 4.2. Synchronization simulations are carried out for $d_1 = -7$, $d_1 = -6.7$, $d_1 = -5$, respectively. The synchronization error curves are plotted in Figure 4.37. Obviously, when $d_1 > -7.3$, the bigger the d_1, the shorter the time required to achieve synchronization, but when the d_1 exceeds a certain value and continue to increase d_1, the time required for synchronization cannot be significantly reduced. Similarly, when d_1 and d_3 are fixed, d_2 changes, or when d_1 and d_2 are fixed, d_3 changes, and the law is the same.

4.9.3.2 Synchronization Control and Simulation of Homogeneous Chaotic Systems with Unknown Parameters

Applying the adaptive function projective synchronization method to two simplified Lorenz system with unknown parameters, the drive and response systems are, respectively,

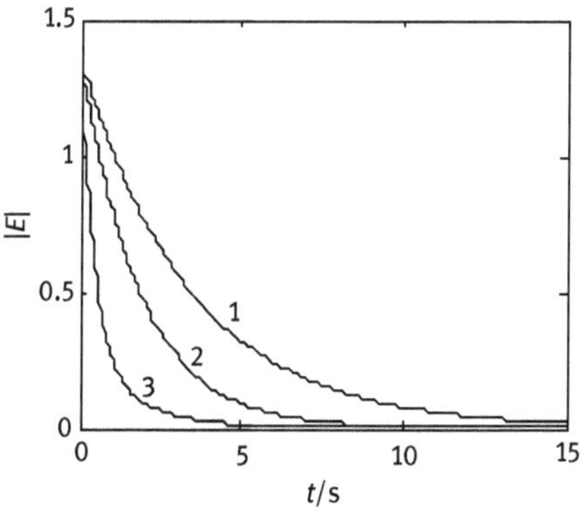

Figure 4.37: Synchronization error curves for different control parameter d_1.

$$\dot{\boldsymbol{X}} = \boldsymbol{AX} + \boldsymbol{f}(\boldsymbol{X}) + \boldsymbol{F}(\boldsymbol{X})\boldsymbol{\theta} = \begin{bmatrix} -10 & 10 & 0 \\ 0 & 0 & 0 \\ 0 & 0 & -8/3 \end{bmatrix}\begin{bmatrix} x_1 \\ x_2 \\ x_3 \end{bmatrix} + \begin{bmatrix} 0 \\ -x_1x_3 \\ x_1x_2 \end{bmatrix} + \begin{bmatrix} 0 & 0 & 0 \\ x_1 & x_2 & 0 \\ 0 & 0 & 0 \end{bmatrix}\begin{bmatrix} \theta_1 \\ \theta_2 \\ \theta_3 \end{bmatrix}, \tag{4.130}$$

$$\begin{aligned}\dot{\boldsymbol{Y}} = \boldsymbol{BY} + \boldsymbol{g}(\boldsymbol{Y}) + \boldsymbol{G}(\boldsymbol{Y})\boldsymbol{\delta} + \boldsymbol{U}(\boldsymbol{X},\boldsymbol{Y}) &= \begin{bmatrix} -10 & 10 & 0 \\ 0 & 0 & 0 \\ 0 & 0 & -8/3 \end{bmatrix}\begin{bmatrix} y_1 \\ y_2 \\ y_3 \end{bmatrix} + \begin{bmatrix} 0 \\ -y_1y_3 \\ y_1y_2 \end{bmatrix} \\ &+ \begin{bmatrix} 0 & 0 & 0 \\ y_1 & y_2 & 0 \\ 0 & 0 & 0 \end{bmatrix}\begin{bmatrix} \delta_1 \\ \delta_2 \\ \delta_3 \end{bmatrix} + \boldsymbol{U}(\boldsymbol{X},\boldsymbol{Y}), \end{aligned} \tag{4.131}$$

where $\boldsymbol{\theta} = [\theta_1, \theta_2, \theta_3]^{\mathrm{T}}$ and $\boldsymbol{\delta} = [\delta_1, \delta_2, \delta_3]^{\mathrm{T}}$ are unknown parameters of the chaotic systems. $\boldsymbol{U} = [u_1, u_2, u_3]^{\mathrm{T}}$ is the nonlinear controller. If the projective function is $\Phi(t) = -3 + 2\sin(t)$, then according to Theorem 4.11, the control law and parameter adaptive law of the synchronization are, respectively,

$$\mathbf{U} = \frac{1}{\Phi(t)}\begin{bmatrix} -\dot{\Phi}(t)y_1 + d_1e_1 \\ -x_1x_3 + \Phi(t)y_1y_3 + x_1\hat{\theta}_1 + x_2\hat{\theta}_2 - \dot{\Phi}(t)y_2 - \Phi(t)(y_1\hat{\delta}_1 + y_2\hat{\delta}_2) + d_2e_2 \\ x_1x_2 - \Phi(t)y_1y_2 - \dot{\Phi}(t)y_3 + d_3e_3 \end{bmatrix}, \tag{4.132}$$

$$\dot{\hat{\boldsymbol{\theta}}} = \begin{bmatrix} x_1e_2 + \theta_1 - \hat{\theta}_1 \\ x_2e_2 + \theta_2 - \hat{\theta}_2 \\ 0 \end{bmatrix}, \tag{4.133}$$

$$\dot{\hat{\boldsymbol{\delta}}} = \begin{bmatrix} -\Phi(t)y_1e_2 + \delta_1 - \hat{\delta}_1 \\ -\Phi(t)y_2e_2 + \delta_2 - \hat{\delta}_2 \\ 0 \end{bmatrix}, \tag{4.134}$$

On Simulink platform, we set the unknown parameter $c = 5$. It means the true value of the estimated parameter is $\boldsymbol{\theta} = \boldsymbol{\delta} = [0, 4, 5]^{\mathrm{T}}$. If the initial conditions of the synchronization system are $\boldsymbol{X}(0) = [4.4356, 6.1771, 7.2330]^{\mathrm{T}}$, $\boldsymbol{Y}(0) = [-0.5431, 0.0995, -2.8096]^{\mathrm{T}}$, $\hat{\boldsymbol{\theta}}(0) = [-1, -1, 0]^{\mathrm{T}}$, $\hat{\boldsymbol{\delta}}(0) = [-2, -2, 0]^{\mathrm{T}}$, respectively, and $d_1 = d_2 = d_3 = 5$, we obtain the synchronization error curves, attractors, and error curves of the estimated parameters as shown in Figure 4.38. Figure 4.38(a) shows that the two chaotic systems with different initial values and unknown parameters can achieve the function projective synchronization in about 2.5 s, and it has oscillation attenuation characteristic. Figure 4.38(b) shows that the attractor of the response system and drive system is inverse, and the size ratio satisfies the projective function. From Figure 4.38(c) and (d), it can be seen that in a short time, the error of the estimated parameters $\hat{\theta}_1$, $\hat{\theta}_2$ and $\hat{\delta}_1$, $\hat{\delta}_2$ converges to zero, which indicates that the unknown parameters are identified. The results also show that the larger the control parameter d_i is, the better

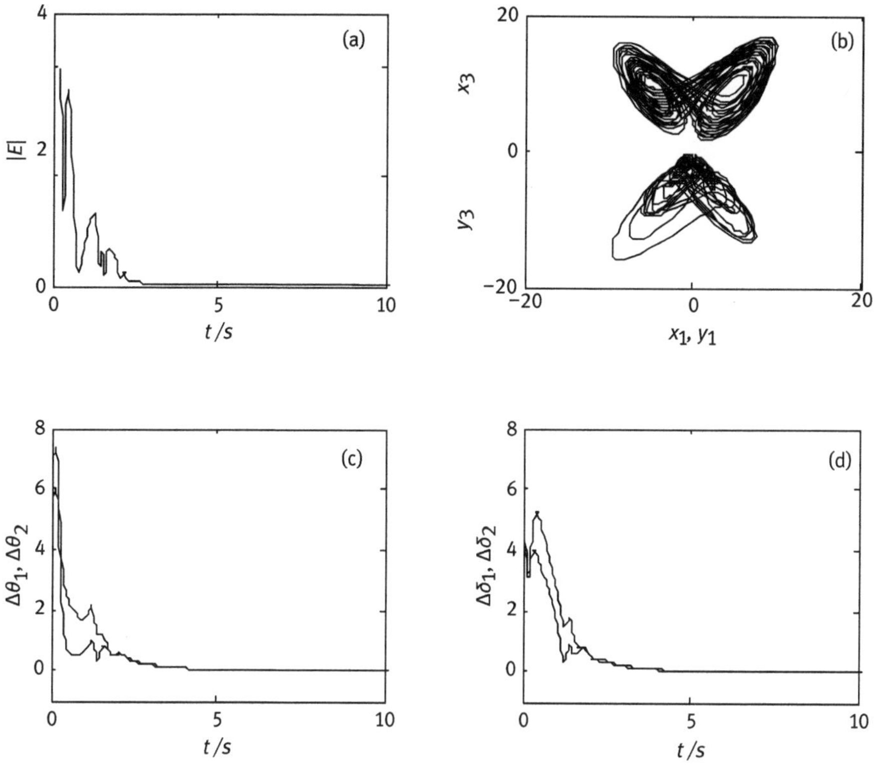

Figure 4.38: Projective synchronization between systems (4.130) and (4.131); (a) synchronization error curve; (b) attractors; (c) error curves for estimated parameters $\hat{\theta}_1$ and $\hat{\theta}_2$; and (d) error curves for estimated parameters $\hat{\delta}_1$ and $\hat{\delta}_2$.

the synchronization performance is, which is important for secure communication application.

4.9.3.3 Synchronization Control and Simulation of Heterogeneous Chaotic Systems with Unknown Parameters

Applying the adaptive function projective synchronization method to the simplified Lorenz system and Chen system with unknown parameters, the drive system and the response system are, respectively,

$$\dot{\boldsymbol{X}} = \boldsymbol{AX} + \boldsymbol{f}(\boldsymbol{X}) + \boldsymbol{F}(\boldsymbol{X})\boldsymbol{\theta} = \begin{bmatrix} -10 & 10 & 0 \\ 0 & 0 & 0 \\ 0 & 0 & -8/3 \end{bmatrix} \begin{bmatrix} x_1 \\ x_2 \\ x_3 \end{bmatrix} + \begin{bmatrix} 0 \\ -x_1x_3 \\ x_1x_2 \end{bmatrix} + \begin{bmatrix} 0 & 0 & 0 \\ x_1 & x_2 & 0 \\ 0 & 0 & 0 \end{bmatrix} \begin{bmatrix} \theta_1 \\ \theta_2 \\ \theta_3 \end{bmatrix}, \tag{4.135}$$

$$\dot{Y} = BY + g(Y) + G(Y)\delta + U(X,Y) = \begin{bmatrix} -10 & 10 & 0 \\ -10 & 0 & 0 \\ 0 & 0 & -8/3 \end{bmatrix} \begin{bmatrix} y_1 \\ y_2 \\ y_3 \end{bmatrix} + \begin{bmatrix} 0 \\ -y_1y_3 \\ y_1y_2 \end{bmatrix} + \begin{bmatrix} y_2 - y_1 & 0 & 0 \\ -y_1 & y_1 + y_2 & 0 \\ 0 & 0 & -y_3 \end{bmatrix} \begin{bmatrix} \delta_1 \\ \delta_2 \\ \delta_3 \end{bmatrix} + U(X,Y), \tag{4.136}$$

where $\theta = [\theta_1, \theta_2, \theta_3]^T$ and $\delta = [\delta_1, \delta_2, \delta_3]^T$ are the unknown parameters of the two systems, and $U = [u_1, u_2, u_3]^T$ is the nonlinear controller. If we set the projective function as $\Phi(t) = -3 + 2\sin(t)$, according to Theorem 4.11, the control law and parameter adaptive law of the synchronization system are

$$U = \frac{1}{\Phi(t)} \begin{bmatrix} -\dot{\Phi}(t)y_1 - \Phi(t)(y_2 - y_1)\hat{\delta} + d_1e_1 \\ 10x_1 - x_1x_3 + \Phi(t)y_1y_3 + x_1\hat{\theta}_1 + x_2\hat{\theta}_2 - \dot{\Phi}(t)y_2 - \Phi(t)[-y_1\hat{\delta}_1 + (y_1 + y_2)\hat{\delta}_2] + d_2e_2 \\ x_1x_2 - \Phi(t)y_1y_2 - \dot{\Phi}(t)y_3 + \Phi(t)y_3\hat{\delta}_3 + d_3e_3 \end{bmatrix}, \tag{4.137}$$

$$\dot{\hat{\theta}} = \begin{bmatrix} x_1e_2 + \theta_1 - \hat{\theta}_1 \\ x_2e_2 + \theta_2 - \hat{\theta}_2 \\ 0 \end{bmatrix}, \tag{4.138}$$

$$\dot{\hat{\delta}} = \begin{bmatrix} -\Phi(t)[(y_2 - y_1)e_1 - y_1e_2] + \delta_1 - \hat{\delta}_1 \\ -\Phi(t)(y_1 + y_2)e_2 + \delta_2 - \hat{\delta}_2 \\ \Phi(t)y_3e_3 + \delta_3 - \hat{\delta}_3 \end{bmatrix}. \tag{4.139}$$

For simulation, we set the unknown parameter of the simplified Lorenz system $c = 5$, which means that the true value of the estimated parameter is $\theta = [0, 4, 5]^T$. At the same time, we choose the true value of the unknown parameter of the Chen system $\delta = [25, 28, 1/3]^T$. If the initial conditions of the system and unknown parameter are $X(0) = [4.4356, 6.1771, 7.2330]^T$, $Y(0) = [-0.62, -0.4261, -4.0838]^T$, $\hat{\theta}(0) = [-1, -1, 0]^T$, $\hat{\delta}(0) = [2, 2, 2]^T$, and $d_1 = d_2 = d_3 = 5$, then we obtain the synchronization error curves, attractors, and error curves of the estimated parameters as shown in Figure 4.39. Figure 4.39(a) shows that the two chaotic systems with different initial values and unknown parameters can achieve the function projective synchronization in about 2 s, and it has oscillation attenuation characteristic. Figure 4.39(b) shows that the attractor of the response system and drive system is inverse phase, and the size ratio satisfies the projective function. From Figure 4.39(c) and (d), we know that in a short time, the error of the estimated parameters $\hat{\theta}_1, \hat{\theta}_2$ and $\hat{\delta}_1, \hat{\delta}_2, \hat{\delta}_3$ converges to zero, which indicates the unknown parameters are identified. The results also show that the larger the control parameter d_i is, the better the synchronization performance is. That is, the setup time of the synchronization is shorter.

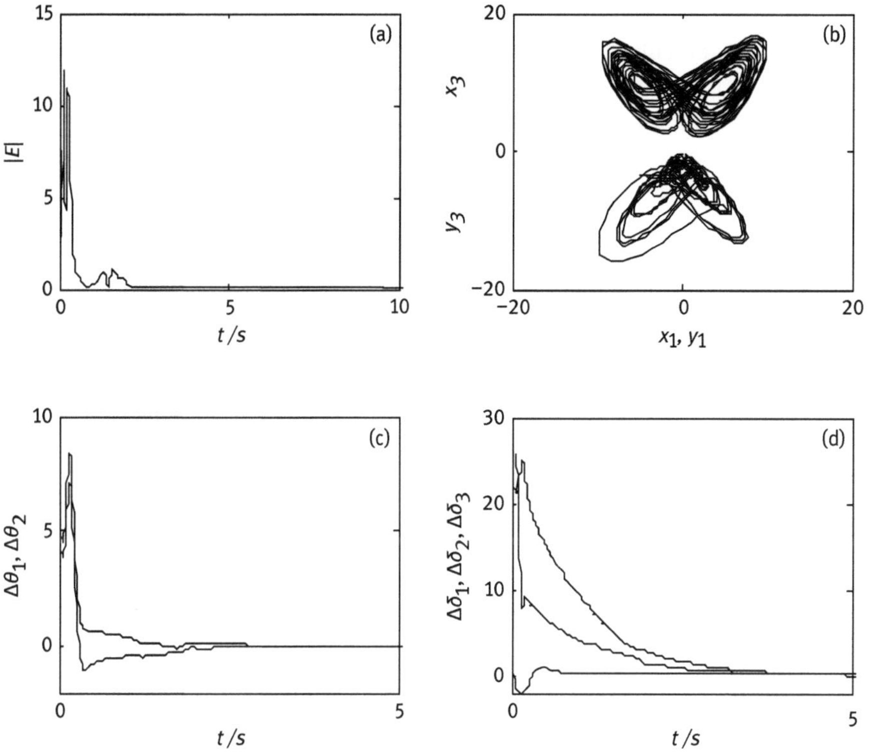

Figure 4.39: Projective synchronization between systems (4.135) and (4.136): (a) synchronization error curve; (b) attractors; (c) error curves for estimated parameters $\hat{\theta}_1$ and $\hat{\theta}_2$; and (d) error curves for estimated parameters $\hat{\delta}_1$, $\hat{\delta}_2$, and $\hat{\delta}_3$.

4.10 Problems and Development of Chaos Synchronization

Combining with the new control strategies or dynamic methods, how to improve the performance of chaotic synchronization is key for the application research of chaos. Under the condition of noise interference, channel distortion, and parameter mismatch, the synchronization performance improvement of chaotic system is an important part of the research of chaos synchronization. In terms of the current research of chaos synchronization, we should combine the robust method of control and the stability of dynamics to improve the robustness and the synchronization performance of the synchronization system.

We should pay more attention on the problem of hyperchaos synchronization, because the secure communication based on hyperchaotic system has better security performance. For example, hyperchaos system is a high-dimensional strange attractor in the appropriate regression map. The modulated signal and the original chaotic signal are mixed together, and it cannot be resolved, so it is difficult to decryption. At present, the research on the hyperchaotic system and its synchronization is still in the early stage. The difficulties and hot topics in the research of chaos synchronization theory include the criteria, the physical mechanism, and the applications of the hyperchaotic synchronization.

It is very important to study the effects of noise interference and channel distortion on synchronization. In the real transmission process, the chaotic signal is inevitably subject to noise interference. In addition, any transmission channel is not ideal. So the transmitted signal will have amplitude attenuation, phase, and nonlinear distortion. This will inevitably have a direct or indirect effect on the synchronization of chaotic systems. It is very important to develop such kind of research for the application of chaotic synchronization in secure communication.

Although chaos synchronization control method has been studied for more than 30 years, and a lot of physics and information scientists have done a lot of work in both theory and experiment, it is still in the exploratory stage. We believe many problems such as synchronization mechanism, anti-interference ability, and so on will be solved in the future.

Questions

1. What are the performance index of chaotic synchronization system and its significance?
2. What are the criteria or guidelines for synchronization of chaotic systems?
3. What are the key issues and challenges facing the chaos synchronization?
4. Try to construct a chaotic synchronization system and analyze its synchronization performances by simulation.

5 Secure Communication Technology Based on Chaos Synchronization

Chaotic system is sensitively dependent on the initial conditions, long-term unpredictability, and anti-interception capability. At the same time, chaotic system itself is deterministic, and it is completely determined by the equations, parameters, and initial conditions of the nonlinear system. Hence, it is easy to be produced and copied. The high randomness, unpredictability, high complexity, wide frequency band character, the deterministic parameters and initial conditions of the chaotic signal, and its easy implementation make it a hot topic in the research of chaos application.

When compared with modern communication, chaos communication has many advantages and differences. The similarities and differences between them are shown in Table 5.1 [18]. When compared with the traditional modern communication, the chaos communication has obvious advantages for its potential applications and development prospects. However, the chaotic secure communication technology, which is still in the early stages of development, needs to solve a series of key technical issues.

Chaotic secure communication is divided into chaos analogy communication and chaos digital communication. Chaos analog communication is usually achieved by nonlinear circuit systems, but it requires very high standards for electrical design and manufacture precision. In addition, it is difficult for synchronization implementation. Among the existing chaotic communication schemes, chaotic masking is used to transmit analog signals, and its disadvantage is that the fidelity and security of the scheme are still not satisfied. Chaotic digital communication has become a hot research topic because of its less dependence on the circuit element, easy implementation in hardware, convenience in computer processing, transmission with less loss of information, strong versatility, and wide application [112]. Chaotic digital communication mainly includes chaos shift keying (CSK), chaos spread spectrum, and chaos modulation.

5.1 Chaotic Masking Communication

Chaotic masking is the simplest form of chaotic communication, where the information signal is superimposed on the chaotic signal directly and it is masked by the chaotic signal [113]. The principle block diagram is shown in Figure 5.1. Synchronization is achieved between the drive subsystem and the response subsystem. At the sender, message signal and chaotic signal with much larger amplitude are superposed to form a noise signal. The chaotic signal completely masks the useful information and realizes the secure communication. At the receiver, the response system is capable of reproducing the chaotic signal. If the received signal is subtracted by the

Table 5.1: Comparison between modern communication and chaotic communication.

Content	Modern communication	Chaotic communication
History	Matured technology for over 100 years	Emerging information technology for over 20 years
Industrial base	Has become the core of the field of information technology	Research hot spot
Emission bandwidth	Information bandwidth and broadband emission	Broadband spectral emission
Relationship between theory and technology	Category obviously, system integrity (from information theory to communication technology, from communication software to circuit hardware, etc.)	Selection of new methods and new schemes to seek a breakthrough. Some approved communication schemes, such as circuit, laser, and computer network, have been proposed
Security technology foundation	Mathematical methods, involving complex mathematical knowledge	Physical methods, involving chaos characteristics
Key security technology	DES, IDEA, 3DES, RSA	Chaos masking, chaos switching, chaos modulation
Confidentiality	Good performance, but there are loophole due to the period of the larger random number. It can be decoded, but it is very time-consuming, and high cost.	aperiodic, unpredictable, fast fading correlation function, and broadband characteristics. Stronger confidentiality
Robustness	Achieved certain robustness	Application of spatiotemporal chaos and hyperchaos can enhance the robustness. Difficult to be attacked
Capacity	Bulk can be achieved, but there are still restrictions	Dynamic storage capability with high capacity
Power	High	Low power and observational ability
Equipment cost	High	Relatively low
The main safety issues of security system	The fatal vulnerabilities of software, inefficient attack protection, and poor random number generator	It is studying and is expected to address key issues, such as noise, channel distortion, and multichannel and multiuser problems

chaotic signal generated by the response system, the message signal can be recovered correctly.

Chaotic masking is realized by utilizing the chaos randomness and broadband power spectrum characteristics. Taking the low-dimensional chaotic synchronization signal as a modulation signal, the transmitted message is superposed on the chaotic signal. From the observation of power spectrum or chaotic time series, the transmitted message is not easy to be detected directly. To accurately extract the message signal,

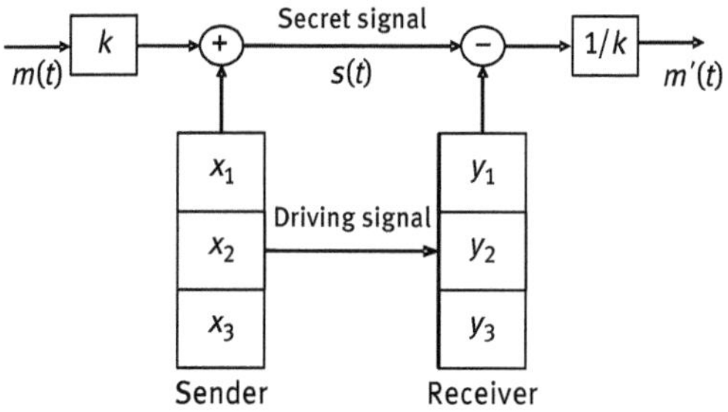

Figure 5.1: Double-channel chaotic secure communication scheme based on drive–response synchronization.

this method requires the sender and receiver chaotic circuit parameters to match exactly. That is, it requires the same initial state and the parameters. Synchronization will be failed for tiny difference. Therefore, it is very difficult for the attacker to analyze the key parameters. On the other hand, the parameters that match exactly put forward very high requirements to the design of the circuit.

At the sender, taking Lorenz chaotic system as an example, the mathematical model of the modulator is described as

$$\begin{cases} \dot{x}_1 = \sigma(x_2 - x_1) \\ \dot{x}_2 = -x_1x_3 + rx_1 - x_2 \\ \dot{x}_3 = x_1x_2 - bx_3 \\ s(t) = km(t) + x_1 \end{cases}, \tag{5.1}$$

where k is a compression ratio constant, which ensures that the useful signal $m(t)$ is much less than the chaotic signal. x_1 is the masking signal, and x_2 is the synchronous driving signal.

At the receiver, the mathematical model of the demodulator is defined as

$$\begin{cases} \dot{y}_1 = \sigma(x_2 - y_1) \\ \dot{y}_2 = -y_1y_3 + ry_1 - y_2 \\ \dot{y}_3 = y_1x_2 - by_3 \\ m'(t) = (s(t) - y_1)/k \end{cases}. \tag{5.2}$$

Obviously, as long as the synchronization between the response system and drive system is achieved completely, the useful information can be demodulated correctly. In order to save the channel resources, this scheme is improved, and the principle is shown in Figure 5.2. In this scheme, the driving signal consists of useful signal and chaotic signal, that is, the drive signal is the sum of the useful signal and chaotic signal, so only one channel is required to realize the secure communication. Here, taking

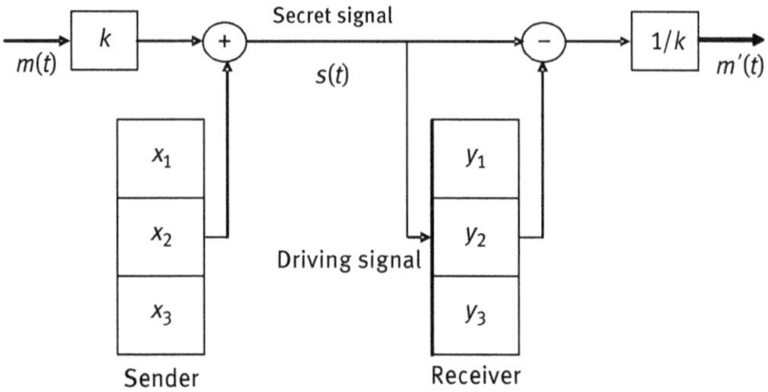

Figure 5.2: Single-channel chaotic secure communication scheme based on drive–response synchronization.

Lorenz chaotic system as an example, the chaotic system equations for the sender and the receiver are given by, respectively,

$$\begin{cases} \dot{x}_1 = \sigma(x_2 - x_1) \\ \dot{x}_2 = -x_1x_3 + rx_1 - x_2 \\ \dot{x}_3 = x_1x_2 - bx_3 \\ s(t) = km(t) + x_1 \end{cases}, \tag{5.3}$$

$$\begin{cases} \dot{y}_1 = \sigma(s(t) - y_1) \\ \dot{y}_2 = -y_1y_3 + ry_1 - y_2 \\ \dot{y}_3 = y_1s(t) - by_3 \\ m'(t) = (s(t) - y_1)/k \end{cases}. \tag{5.4}$$

In this scheme, the channel resource is saved. However, one of the obvious characteristics is that the drive signal is disturbed by the message signal, and synchronization cannot be accurately realized, which leads to the fact that the information signal cannot be fully recovered. Another notable feature is that the information signal doesn't have any impact on the dynamic characteristics of the driving system, so the security of the secure communication system is not high enough, and it cannot effectively resist the third-party attack. To improve the security further, an improved scheme was proposed in Ref. [114] as shown in Figure 5.3. In this scheme, the sum of the information signal and the chaotic signal feedbacks to the drive system, so that the drive system and the response system have the same dynamic characteristics, so it can realize the complete synchronization, and the receiving signal can be demodulated completely. As the information signal is integrated into the chaotic system, it is more difficult for the attacker to obtain useful information from the channel. Here, taking Lorenz chaotic system as an example, the chaotic system equations for the sender and the receiver is written as, respectively,

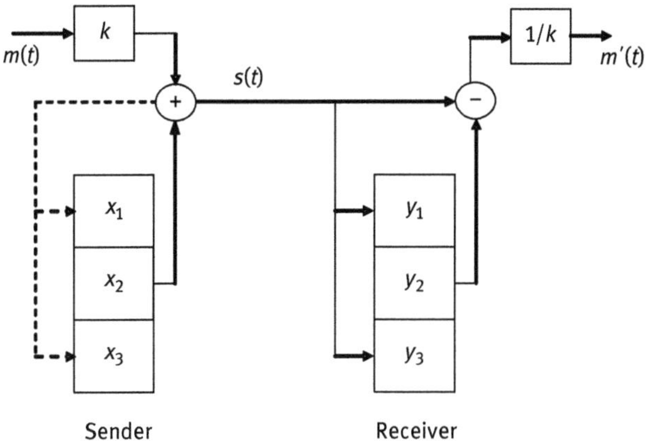

Figure 5.3: Improved single-channel chaotic secure communication scheme based on drive–response synchronization.

$$\begin{cases} \dot{x}_1 = \sigma(s(t) - x_1) \\ \dot{x}_2 = -x_1 x_3 + r x_1 - x_2 \\ \dot{x}_3 = x_1 s(t) - b x_3 \\ s(t) = m(t) + x_2 \end{cases}, \tag{5.5}$$

$$\begin{cases} \dot{y}_1 = \sigma(s(t) - y_1) \\ \dot{y}_2 = -y_1 y_3 + r y_1 - s(t) \\ \dot{y}_3 = y_1 s(t) - b y_3 \\ m'(t) = s(t) - y_2 \end{cases}. \tag{5.6}$$

To improve the secure performance of chaotic masking communication further, Carroll et al. [115] spread this improved chaotic masking scheme to the high-order-linked synchronization system, and its synchronization principle and secure communication scheme are shown in Figure 5.4. In this scheme, the receiver and the sender are composed of two cascaded chaotic systems. Taking Lorenz chaotic system as an example, the equations for the sender and the receiver are presented as, respectively,

$$\begin{cases} \dot{x}_1 = \sigma(s(t) - x_1) \\ \dot{x}_2 = -x_1 x_3 + r x_1 - x_2 \\ \dot{x}_3 = x_1 x_2 - b x_3 \\ \dot{x}'_1 = \sigma(x'_2 - x'_1) \\ \dot{x}'_2 = -x_1 x'_3 + r x_1 - x'_2 \\ \dot{x}'_3 = x_1 x'_2 - b x'_3 \\ s(t) = k m(t) + x'_2 \end{cases}, \tag{5.7}$$

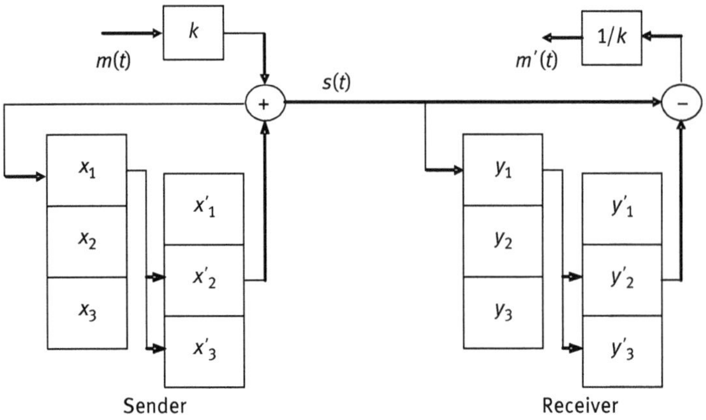

Figure 5.4: Improved single-channel cascade chaotic secure communication based on drive–response synchronization.

$$\begin{cases} \dot{y}_1 = \sigma(s(t) - y_1) \\ \dot{y}_2 = -y_1 y_3 + r y_1 - y_2 \\ \dot{y}_3 = y_1 y_2 - b y_3 \\ \dot{y}_1' = \sigma(y_2' - y_1') \\ \dot{y}_2' = -y_1 y_3' + r y_1 - y_2' \\ \dot{y}_3' = y_1 y_2' - b y_3' \\ m'(t) = (s(t) - y_2')/k \end{cases} . \tag{5.8}$$

5.2 Chaos Shift Keying Communication

First, in 1993, Parlitz et al. [116] and Dedieu et al. [41] proposed the digital modulation method of CSK. Soon, Kolumban et al. [117, 118] successively put forward the improved CSK digital communication system, which mainly includes chaotic on–off keying (COOK), differential chaos shift keying (DCSK), frequency modulation differential chaos shift keying (FM-DCSK), quadrature chaos shift keying (QCSK) [119], and so on. Among these schemes, DCSK and FM-DCSK have better antinoise performance [120]. FM-DCSK was listed as a long-term research program by European Commission. In 2000, *IEEE Transactions on CAS-I* published the 12th special issue about chaos to discuss the performance of a class of digital modulation schemes. Compared with DCSK, QCSK has the same bandwidth and bit error rate (BER) performance, but the data transmission rate is higher. There are two demodulation methods, such as coherent demodulation and noncoherent demodulation.

CSK is a kind of chaotic modulation technology for transmitting digital information. The basic idea is to map the different information code of binary (or multinary) digital signal to the piecewise chaotic waveform. The modulated signal is detected in the receiver based on the correlation reception theory. It needs a chaotic basis function or multiple chaotic basis functions for CSK modulation. A basis function of CSK has two types at least, such as COOK and CSK. The antinoise performance of bipolar coherent demodulation of CSK is better than that of COOK. If the bit energy remains constant and the chaotic basis function is recovered accurately at the receiver, the antinoise performance of the CSK is equivalent to that of Binary Phase Shift Keying (BPSK). The noncoherent detection cannot detect the bipolar CSK, and the antinoise performance of the noncoherent COOK is worse than that of the coherent COOK, but the COOK can achieve the best antinoise performance when the bit energy keeps constant. The decision threshold of the demodulation of reverse-phase CSK and COOK depends on the signal-to-noise ratio (SNR). When two basis functions are adopted, this problem can be solved. When symbols "0" and "1" are mapped to two different chaotic waveforms, the CSK of the two basis functions is formed, and the antinoise performance of coherent demodulation is better than that of the noncoherent demodulation. Especially, through choice of orthonormal basis functions, the decision threshold level becomes zero, and the performance has nothing to do with the SNR, and its performance is similar with that of coherent Frequency Shift Keying (FSK).

5.2.1 COOK Communication

COOK is the most simple CSK communication, and the structure of basic principle is shown in Figure 5.5. $\{b_i\}$ is a binary information signal with "1" and "−1." When $b_i = 1$, the switch is turned on. When $b_i = -1$, the switch is turned off. At the receiver, the information signal is recovered by employing the correlation algorithm. If the input signal is of equal probability distribution, and the average bit energy for a single symbol to E_b, then the code energies of "1" and "−1" are $2E_b$ and 0, respectively. Obviously, the larger the bit energy gap, the better the antinoise performance of the signal. The structure of this scheme is simple. For example, it only needs a chaotic oscillator, but the security of the system is not high.

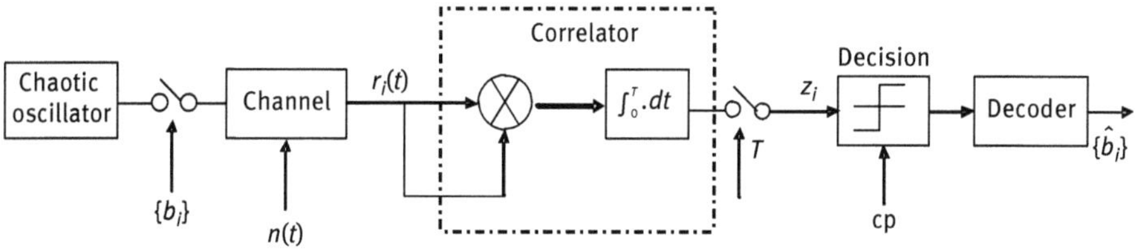

Figure 5.5: System principle block diagram of noncoherent COOK modulation/demodulation.

5.2.2 CSK Communication

Next, we introduce the principle of CSK communication based on two chaotic signal generators. The structure of the transmitter is shown in Figure 5.6. $\{b_i\}$ is a binary sequence of information signal, which only includes "1" and "−1". After modulated by the modulator, the signal transmitted in the channel is a noise-like chaotic signal, whose mathematical description is

$$s_i(t) = \begin{cases} g_1(t), & b_i = 1 \\ g_2(t), & b_i = -1 \end{cases}. \tag{5.9}$$

In the receiver, coherent or noncoherent mode is used to recover the information signal, and the principle of coherent demodulation is shown in Figure 5.7. The output of the correlator is

$$z_{i1} = \int_{T_S}^{T} r_i(t)\hat{g}_1(t)dt = \int_{T_S}^{T} [g_i(t) + n(t)]\hat{g}_1(t)dt = \int_{T_S}^{T} g_i(t)\hat{g}_1(t)dt + \int_{T_S}^{T} n(t)\hat{g}_1(t)dt, \tag{5.10}$$

$$z_{i2} = \int_{T_S}^{T} r_i(t)\hat{g}_2(t)dt = \int_{T_S}^{T} [g_i(t) + n(t)]\hat{g}_2(t)dt = \int_{T_S}^{T} g_i(t)\hat{g}_2(t)dt + \int_{T_S}^{T} n(t)\hat{g}_2(t)dt. \tag{5.11}$$

Obviously, if $b_i = 1$, then $s_i(t) = g_1(t)$, $r_i(t) = g_1(t) + n(t)$. We obtain $z_{i1} > z_{i2}$, and the recovered signal is $\hat{b}_i = 1$. If $b_i = -1$, then $s_i(t) = g_2(t)$, $r_i(t) = g_2(t) + n(t)$. We have $z_{i1} < z_{i2}$, and the recovered signal is $\hat{b}_i = -1$.

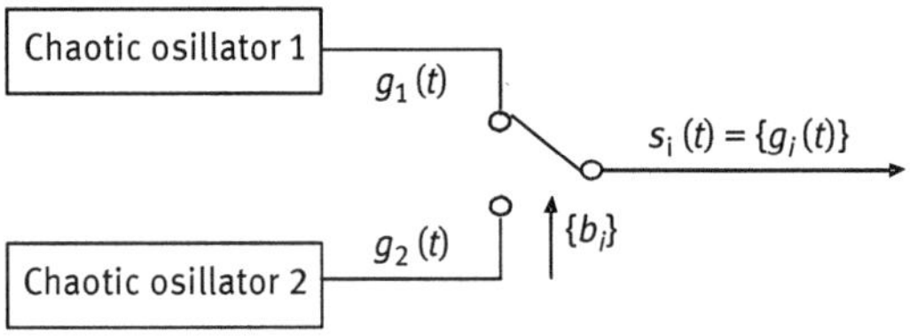

Figure 5.6: Block diagram of CSK modulator.

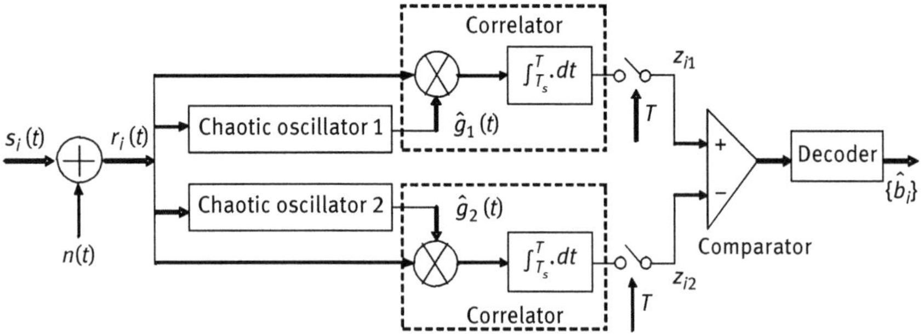

Figure 5.7: Principle block diagram of CSK coherent receiver.

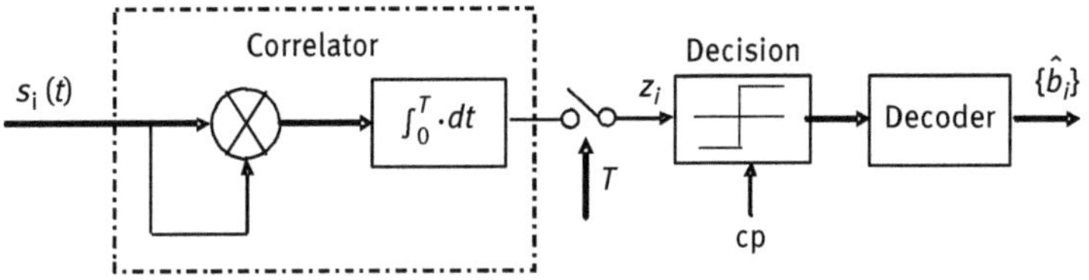

Figure 5.8: Principle block diagram of CSK noncoherent receiver.

By using the noncoherent demodulation, the demodulation principle is shown in Figure 5.8, and the output of the correlator is

$$z_i = \int_0^T s_i^2(t)dt = \int_0^T [g_i(t) + n(t)]^2 dt = \int_0^T g_i^2(t)dt + 2\int_0^T g_i(t)n(t)dt + \int_0^T n^2(t)dt. \quad (5.12)$$

For the CSK noncoherent demodulation, the receiver need not recover the chaotic carrier, and the detector is based on the bit energy of the transmission signal. If the decision threshold is set to the median of the two bit energies, then the received signal can be demodulated correctly. Obviously, the selection of the optimal threshold is closely related to the SNR of the system. So the threshold shift is the main defect of the noncoherent demodulation.

Next, we take CSK digital secure communication system based on the unified chaotic system as an example to illustrate, and the system structure is shown in Figure 5.9. The chaotic systems of the transmitter and receiver are unified chaotic system. The chaotic system 11 in the transmitting end and the chaotic system 21 in the receiving end have the same parameter α_1, and the chaotic signal generated by them are $x_{11}(t)$ and $x_{21}(t)$, respectively. The chaotic system 12 in the transmitting end and the chaotic system 22 in the receiving end have the same parameter α_2, and the chaotic signals generated by them are $x_{12}(t)$ and $x_{22}(t)$, respectively. By employing the synchronization strategy, the unified chaotic systems 11 and 12 at the sending end are synchronized with the unified chaotic systems 21 and 22 at the receiving end, respectively, then we have $x_{11}(t) = x_{21}(t)$, $x_{12}(t) = x_{22}(t)$. The binary digital signal $a(t) = \{a_n\}$ is used to control the chaotic signal. When $a(t)$ is of high voltage, $x_{11}(t)$ is sent. When $a(t)$ is of low voltage, $x_{12}(t)$ is sent. The transmitted signal in the channel is expressed as

$$s(t) = x_{11}(t) \cdot a(t) + (1 - a(t)) \cdot x_{12}(t) = \begin{cases} x_{11}(t), & a(t) = 1 \\ x_{12}(t), & a(t) = 0 \end{cases}. \quad (5.13)$$

Obviously, the signal transmitted in the channel is chaotic. If the noise in the channel is Gaussian white noise $n(t)$, then the signal at the receiving end is expressed as

$$r(t) = s(t) + n(t) = x_{11}(t) \cdot a(t) + (1 - a(t)) \cdot x_{12}(t) + n(t) = \begin{cases} x_{11}(t) + n(t), & a(t) = 1 \\ x_{12}(t) + n(t), & a(t) = 0 \end{cases} \quad (5.14)$$

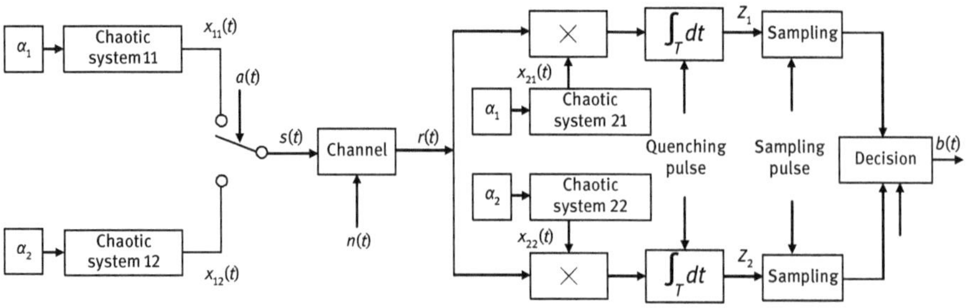

Figure 5.9: Principle block diagram of CSK digital secure communication system based on unified chaotic system.

For bipolar CSK, due to the correlation integral, two signal energies cannot be distinguished, so we can only use coherent demodulation method to demodulate the signal.

The chaotic signals are broadband nonperiodic signals, so the chaotic signals obtained by the same chaotic system with different initial conditions or generated by the same system with different system parameters are almost orthogonal. The correlation properties of chaotic signals are similar to those of Gauss white noise. Generally, the cross-correlation of the chaotic signal approximates to zero, and the autocorrelation approximates to the impulse function. The correlation properties of chaotic signals generated by the unified chaotic system are shown in Figure 5.10.

The receiver is composed of a chaotic synchronization system, a correlator (including multiplier and integrator), and a sampling decision device. If the outputs of correlator 1 and correlator 2 are z_1 and z_2, respectively, and their outputs are

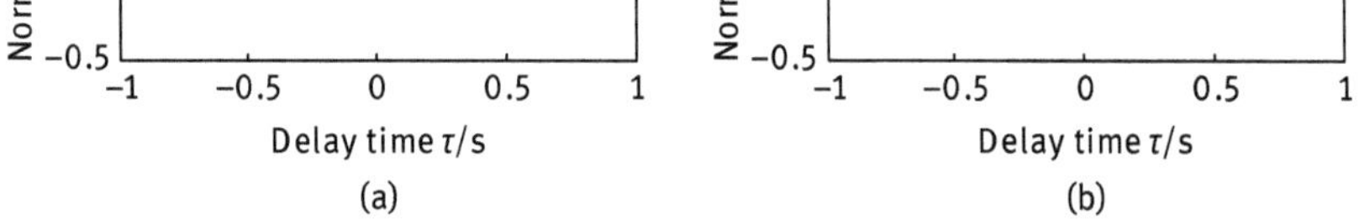

Figure 5.10: Correlation characteristic of chaotic signals generated by the unified chaotic system: (a) autocorrelation and (b) cross-correlation.

$$z_1 = \int_T r(t)x_{21}(t)dt = \begin{cases} \int_T (x_1(t) + n(t))x_{21}(t)dt = E_b + \int_T n(t)x_{21}(t)dt, & a(t) = 1 \\ \int_T (x_2(t) + n(t))x_{21}(t)dt = \rho E_b + \int_T n(t)x_{21}(t)dt, & a(t) = 0 \end{cases}, \quad (5.15)$$

$$z_2 = \int_T r(t)x_{22}(t)dt = \begin{cases} \int_T (x_1(t) + n(t))x_{22}(t)dt = \rho E_b + \int_T n(t)x_{22}(t)dt, & a(t) = 1 \\ \int_T (x_2(t) + n(t))x_{22}(t)dt = E_b + \int_T n(t)x_{22}(t)dt, & a(t) = 0 \end{cases}, \quad (5.16)$$

where $\rho = \left(\int_T x_{11}(t)x_{12}(t)dt\right)/E_b$ is the correlation coefficient of chaotic signals. When the two signals are completely orthogonal, $\rho = 0$. $E_b = \int_T x_{11}^2(t)dt = \int_T x_{12}^2(t)dt$ is the bit energy of the chaotic signal. Obviously, if the correlator output signals, z_1 and z_2, are sampled to be decided, then the original signals can be recovered by the decision rules

$$b(t) = \begin{cases} z_1 > z_2 & \text{decision for "1"} \\ z_1 < z_2 & \text{decision for "0"} \end{cases}. \quad (5.17)$$

Obviously, ignoring the effect of noise, the noise tolerance of coherent demodulation in the receiver system is $N_c = (1 - \rho)E_b$. The output signal of the decision device is $b(t)$. It can be seen that the demodulation performance of the system is related to the characteristics of the chaotic signal. Because the autocorrelation of the chaotic signal generated by the unified chaotic system with different parameters is close to $\delta(t)$, and its cross-correlation is very weak, the system has a good correlation demodulation performance and the original signal can be recovered, that is $b(t) = a(t)$.

Next, we discuss the BER performance and security performance of CSK secure communication based on unified chaotic system. Let the two possible deterministic signals arriving at the receiver be $x_1(t)$ and $x_2(t)$, respectively, and their durations are $(0, T)$, and have the same energy E_b. The noise $n(t)$ at the input of the receiver is Gauss white noise with zero mean, and its unilateral power spectral density is n_0. According to the principle of the best reception, in the case of equal priori probabilities, the limit BER performance is

$$P_e = \frac{1}{2}\left[1 - \operatorname{erf}\left[\sqrt{\frac{E_b(1-\rho)}{2n_0}}\right]\right], \quad (5.18)$$

where erf(x) is the error function, and ρ is the cross-correlation coefficient between signals $x_1(t)$ and $x_2(t)$, and its range is $(-1, 1)$. Because $x_1(t)$ and $x_2(t)$ are orthogonal signals, the cross-correlation coefficient is zero. The BER performance of the system is rewritten as

$$P_e = \frac{1}{2}\left[1 - \operatorname{erf}\left[\sqrt{\frac{E_b}{2n_0}}\right]\right] = \frac{1}{2}\operatorname{erfc}\sqrt{\frac{E_b}{2n_0}}. \quad (5.19)$$

The BER simulation result of the CSK system is shown in Figure 5.11. When the SNR is greater than 10 dB, the BER is less than 10^{-6}.

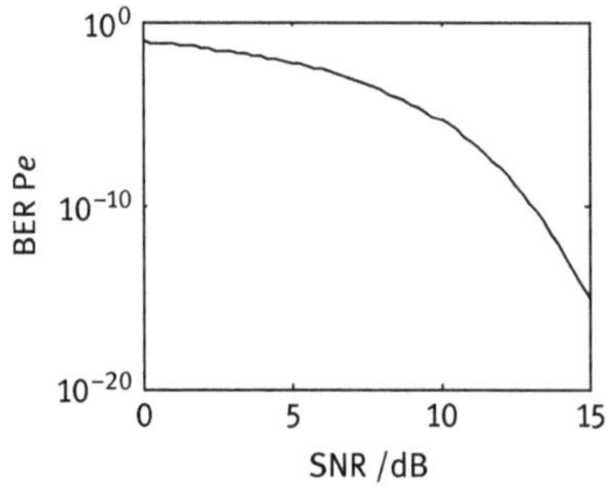

Figure 5.11: BER performance of the CSK system.

The security of CSK communication system is good, and it is mainly manifested in the following aspects. First, the system has strong anticrack ability, because the signal transmitted in channel is chaotic. With the evolution of the chaotic system, different chaotic signals represent the same two value signals. It is equal to "one time pad," so it is unbreakable. From the view of cryptography, the system has a huge key space. The chaotic system is sensitively dependent on initial conditions. The unified chaotic systems with different parameters and different initial conditions can produce different chaotic signals, so chaotic system parameters and initial value can be used as keys. Different chaotic dynamical equations can produce different chaotic signals, and different chaotic variables also can be modulated. In addition, different synchronous mode also affects the demodulated signal. So the key space of the system is composed of parameters, initial values, chaotic systems, chaotic variables, and synchronization modes. The larger the key space is, the higher the security performance is.

It is worth noting that the good anti-interference and security of the system is at the cost of the complexity of the system. In practice, we should consider the trade-off between these two aspects. Because chaotic signal is nonperiodic, the bit energy of output information signal is not entirely consistent. Even in the noise-free case, there is also a deviation for the results related to the estimation, so it will affect the BER performance of the system. To solve this problem, we can use the frequency modulation (FM) to modulate CSK, namely FM-DCSK modulation.

According to the principle shown in Figure 5.9, the simulation model is obtained as shown in Figure 5.12. If the parameter $\alpha = 1$ for the chaotic system 11 and the chaotic system 21, then initial values are $(-5, -5, -5)$. At the same time, if the parameter $\alpha = 0.3$ for the chaotic system 12 and the chaotic system 22, then initial values are (10, 10, 10). The chaotic variable x is modulated by the information signal $a(t)$. When $a(t) = 1$, the modulated signal $s(t)$ transmitted in the channel is the chaotic signal x_{11} generated by the unified system 11. When $a(t) = 0$, the modulated signal $s(t)$ transmitted in the channel is the chaotic signal x_{12} generated by the unified system 12. By using the chaos synchronization control strategy, the synchronization between the unified chaotic systems is achieved as shown in Figure 5.13. Bernoulli random binary sequence is used as

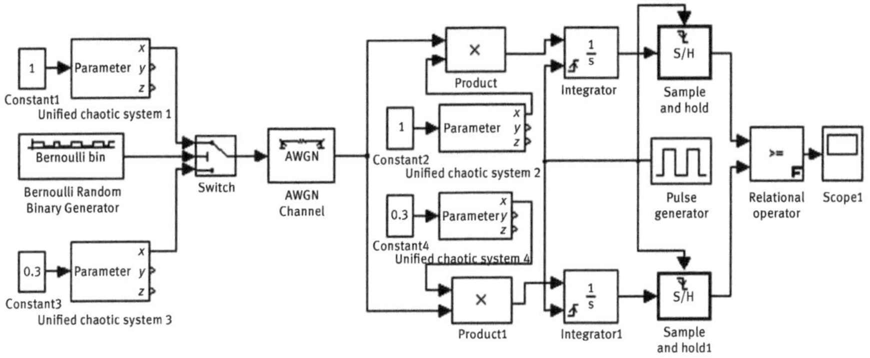

Figure 5.12: Simulation model of CSK digital secure communication based on unified chaotic system.

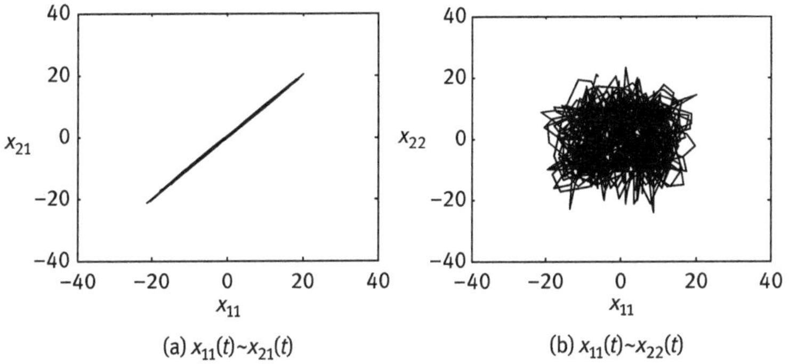

Figure 5.13: Synchronization phase diagram of the variables between the sender and receiver: (a) $x_{11}(t)$–$x_{21}(t)$ and (b) $x_{11}(t)$–$x_{22}(t)$.

the information source. The integrator is triggered by the rising edge of external clock pulse. The sampling–holder is triggered by the falling edge of external clock. Decision is realized by the relevant logic comparator. The simulation results are shown in Figure 5.14, where $T(t)$ is the system clock signal; $a(t)$ is the modulation signal; $s(t)$ is the modulated signal transmitted in channel; and $b(t)$ is the recovered information signal at the receiver. Obviously, $b(t) = a(t)$, that is, chaos signal demodulates correctly. Simulation results show that the original binary digital signal can be demodulated in this CSK secure communication scheme correctly.

5.2.3 DCSK Communication

In 1996, Kolumban et al. [117] proposed the DCSK scheme, and it solved the problem of CSK decision threshold value depending on the SNR. The basic principle is that

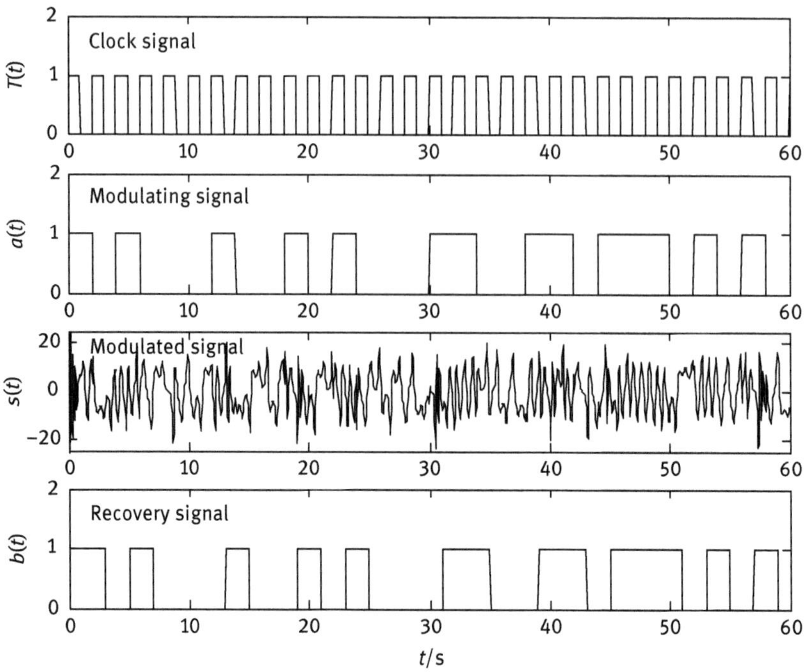

Figure 5.14: Simulation results of the CSK secure communication system.

each digital symbol is represented by two chaotic segments signal. The first segment is called “reference signal,” and the second segment is named “useful signal.” When we send “1,” two same segments are sent (same phase). When we send “0,” the two inverse phase segments are sent (inverse phase). In the receiver, the correlation of these two segments is calculated, and the information signal is recovered according to the correlation value. When the output of the correlator is more than 1, the output signal is “1.” When the output of the correlator is less than 1, the output signal is “0.” The principle of DCSK modulator is shown in Figure 5.15.

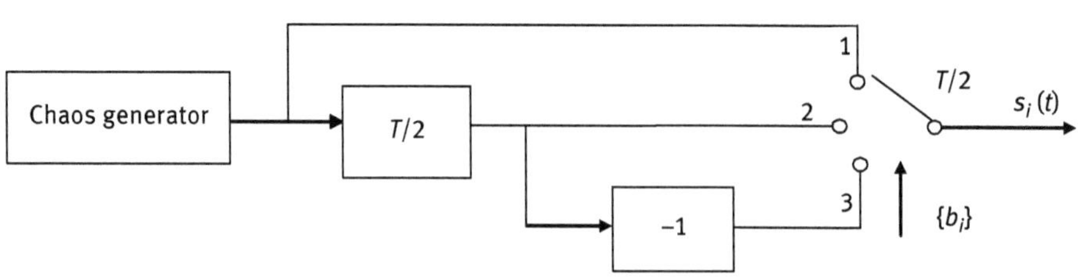

Figure 5.15: Principle block diagram of the DCSK modulator.

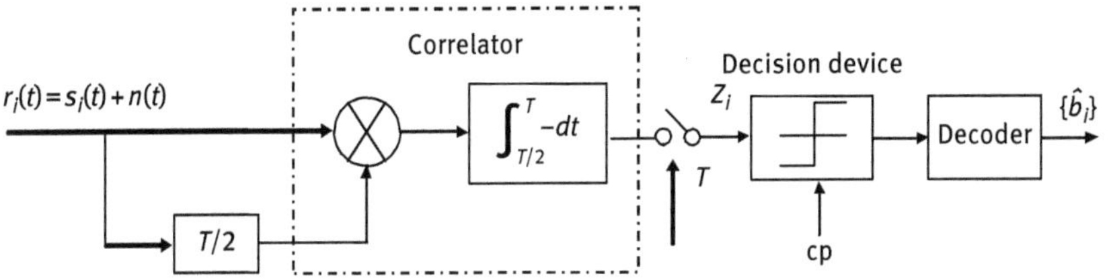

Figure 5.16: Principle block diagram of the DCSK coherent demodulation.

From Figure 5.15, the signal transmitted in channel is presented as

$$s_i(t) = \begin{cases} x(t), & t_k \le t \le t_k + T/2 \\ +x(t - T/2), & t_k + T/2 < t \le t_k + T \end{cases} \quad \text{for } b_i = 1, \tag{5.20}$$

$$s_i(t) = \begin{cases} x(t), & t_k \le t \le t_k + T/2 \\ -x(t - T/2), & t_k + T/2 < t \le t_k + T \end{cases} \quad \text{for } b_i = 0. \tag{5.21}$$

Because each bit is mapped to two continuous segments with the length $T/2$, the information signal at the receiver can be recovered by coherent demodulation. The demodulation principle of the DCSK scheme is shown in Figure 5.16. In this demodulation scheme, the output signal of the correlator is

$$\begin{aligned} z_i &= \int_{T/2}^{T} r_i(t)r_i(t - T/2)dt = \int_{T/2}^{T} [s_i(t) + n(t)][s_i(t - T/2) + n(t - T/2)]dt \\ &= \int_{T/2}^{T} s_i^2(t)dt + \int_{T/2}^{T} s_i(t)(n(t) + n(t - T/2)dt + \int_{T/2}^{T} n(t)n(t - T/2)dt \end{aligned} . \tag{5.22}$$

Compared with CSK and COOK, the advantage of the DCSK scheme is that the decision threshold is zero whether there is noise or not. The reference signal and carrying information signal go through the same channel, so it is not sensitive to channel distortion, and it has a good antinoise performance. The shortage of DCSK is that it can only be used in the chaotic digital communication with low bit rate. To solve this problem, in 1998, Kolumban et al. proposed a DCSK scheme based on FM-DCSK [118].

5.2.4 FM-DCSK Communication

FM-DCSK is constituted by DCSK and a frequency modulator. The output of FM is a chaotic signal with band-limited and uniform power spectrum density. The structure of FM-DCSK modulator is shown in Figure 5.17. It employs the chaotic analog phase-locked loop (APLL) to generate band-pass chaotic signal, and then FM is applied for chaotic signals. The subsequent parts are consistent with DCSK, which transmits

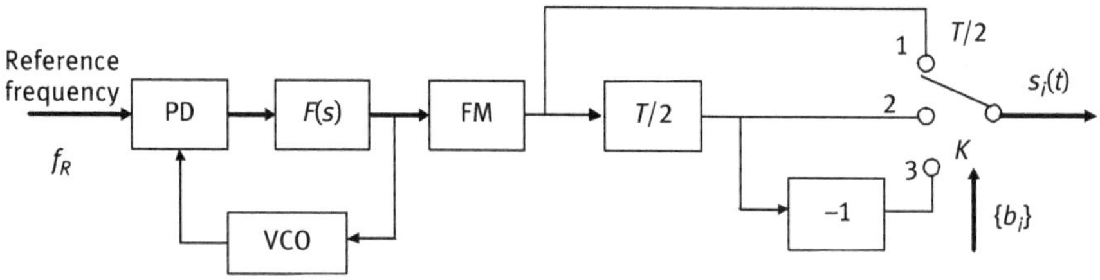

Figure 5.17: Principle block diagram of FM-DCSK modulation.

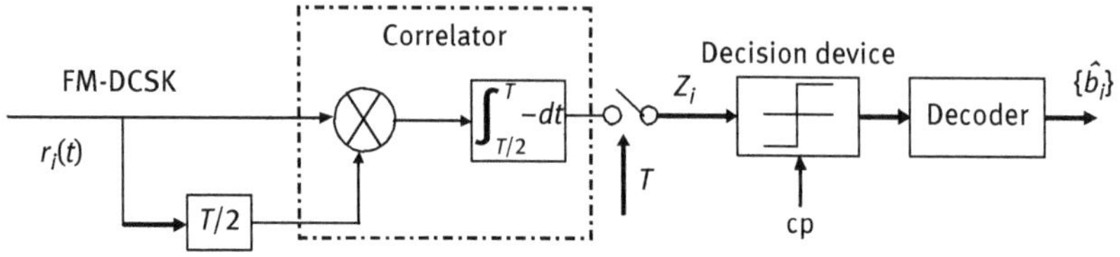

Figure 5.18: Principle block diagram of FM-DCSK demodulation.

reference signal in $[0, T/2]$ and transmits same phase chaotic FM signal or inverse phase chaotic FM signal in $[T/2, T]$.

The demodulation principle of the FM-DCSK is shown in Figure 5.18. Similarly with that of DCSK, it employs the correlation receiving principle to demodulate the received information $\{b_i\}$. For noise channel, the output of the FM-DCSK correlator has zero variance. Its data rate is not limited by the characteristics of the chaotic signal, and its decision threshold is identically zero. Thus, FM-DCSK can improve the symbol rate and reduce errors. When APLL is designed correctly, we can obtain the chaotic band-limited signal with uniform power density, which makes the output (bit energy) of the receiver is constant, regardless of the data rate. Thus, FM-DCSK has higher data rates, and its antinoise performance is the same with that of DCSK.

When compared with other chaotic modulation schemes, FM-DCSK and DCSK have better robustness to multipath interference and nonideal channel. FM-DCSK modulation is suitable for wireless local area networks, indoor wireless communications, mobile communications, and others which are sensitive to multipath interference.

5.2.5 QCSK Communication

QCSK is a kind of chaotic communication scheme, which is similar to the conventional quadrature phase modulation. When compared with DCSK, it has the same bandwidth, approximate BER characteristics, but higher data transmission rate. The basic principle of QCSK is to generate the orthogonal chaotic signal at the specified time interval. In QCSK, the linear combination of two chaotic basis functions is used for coding four symbols. QCSK can be regarded as a combination of two DCSK systems. A double data rate is obtained by increasing complexity of the QCSK.

To explain the principle of QCSK, first, the concept of orthogonal chaotic signal is presented. If two independent chaotic signals $c_1(t)$ and $c_2(t)$ (or even different segments of the same chaotic signal) exhibit very low cross-correlation, they are orthogonal in a time interval $[0, \tau]$, namely

$$\int_0^{\tau} c_1(t)c_2(t)dt = 0. \tag{5.23}$$

If $x(t)$ is a chaotic reference signal defined in the interval $t \in [0, \tau]$, and it has a zero mean. In the interval $t \in [0, \tau]$, $x(t)$ can be expanded as follows by Fourier series ($A_0 = 0$):

$$x(t) = \sum_{k=1}^{\infty} A_k \sin(k\omega t + \varphi_k), \tag{5.24}$$

where $\omega = 2\pi/\tau$. Correspondingly, the average power of $x(t)$ in the interval $t \in [0, \tau]$ is

$$p_{x(t)} = \frac{1}{\tau}\int_0^{\tau} x^2(t)dt = \frac{1}{2}\sum_{k=1}^{\infty} A_k^2. \tag{5.25}$$

The orthogonal signal $y(t)$, in the interval $t \in [0, \tau]$, can be obtained by each Fourier frequency component of $x(t)$ with the phase shift $\pi/2$, namely

$$y(t) = \sum_{k=1}^{\infty} A_k \sin(k\omega t + \varphi_k - \pi/2), \tag{5.26}$$

Then the signals $x(t)$ and $y(t)$ are orthogonal over the interval $t \in [0, \tau]$ and their power is equal, i.e.

$$x \perp y \Leftrightarrow \frac{1}{\tau}\int_0^{\tau} x(t)y(t)dt = 0, \tag{5.27}$$

$$p_{x(t)} = p_{y(t)} \Leftrightarrow \frac{1}{\tau}\int_0^{\tau} x^2(t)dt = \frac{1}{\tau}\int_0^{\tau} y^2(t)dt. \tag{5.28}$$

The above properties are equivalent to

$$\frac{1}{\tau}\int_0^{\tau} A_k \sin(k\omega t + \varphi_k - \alpha)A_m \sin(m\omega t + \varphi_m - \beta)dt = \begin{cases} \frac{1}{2}A_k^2 \cos(\alpha + \beta), & k = m \\ 0, & k \neq m \end{cases}. \tag{5.29}$$

According to definition formula (5.26) of the reference quadrature signal $y(t)$, the $x(t)$ and $y(t)$ are extended into the periodic signal with the periodic τ. It is found that $y(t)$

is the Hilbert transform of $x(t)$, and the Hilbert transform of the given signal can be calculated by the following methods:

1. Frequency domain method: It is basically the same as the method of generating Hilbert transform in Amplitude Modulation - Single Side Band modulation. A fast Fourier transform (FFT) is applied to a chaotic sequence with length k, and then remove the negative frequency (signal energy) corresponding to FFT coefficients. It is obtained by inverse FFT transformation. At present, the algorithm can be realized by a special digital signal processing (DSP) chip.
2. Time domain method: It is achieved by the design of digital Hilbert converter with finite impulse response (FIR) or infinite impulse response (IIR). A simple method for calculating the Hilbert transformation of a given discrete signal is to design an FIR filter which is approximated by Hilbert transform. The FIR filter can be implemented by the ideal impulse response with appropriate width:

$$h(k) = \begin{cases} \frac{2}{\pi}\frac{\sin^2(\pi k/2)}{k}, & k \neq 0 \\ 0, & k = 0 \end{cases}. \tag{5.30}$$

In the interval $I = [0, \tau]$, a chaotic sampling orthogonal base function is defined as

$$c = [c_x(t), c_y(t)] = \left[\frac{1}{\sqrt{E_\tau}}x(t), \frac{1}{\sqrt{E_\tau}}y(t)\right], \tag{5.31}$$

where $E_\tau = P_x\tau$ is the energy of x or y over the interval I. So we have

$$\int_0^\tau c_x^2(t)dt = \int_0^\tau c_y^2(t)dt = 1. \tag{5.32}$$

The orthogonal condition (5.32) indicates that the energy of c_x and c_y is constant at each interval I. The constant is one in special case. The chaotic signal can be represented by a linear combination of c_x and c_y, namely $m_s(t) = a_s c_x(t) + b_s c_y(t)$, where s represents a signal space symbol. Equivalently, the symbol s can be expressed in plural form, namely, $m_s = a_s + jb_s$ and $\int_0^\tau m_s^2(t)dt = 1$.

Being similar to DCSK and QCSK, the modulator structure is shown in Figure 5.19. It includes a chaotic generator, which generates the signal c_x at each interval $[0, T/2]$, and the corresponding quadrature signal c_y is generated by the filter. For simplicity, it is assumed that the system has no time delay. According to the symbol s, the encoder generates a linear combination of c_x and c_y. Information symbol is provided by a bit/symbol converter, and the converter maps each pair of information bits into a symbol. Thus, each symbol period transmits 2 bits information. Then, the QCSK signal is formed by the two-channel analog multiplexer with the switching time $T/2$. Note that although there is no filter in the figure, the QCSK scheme must contain a low-pass band-limiting filter to limit the bandwidth of transmission signal. The chaotic

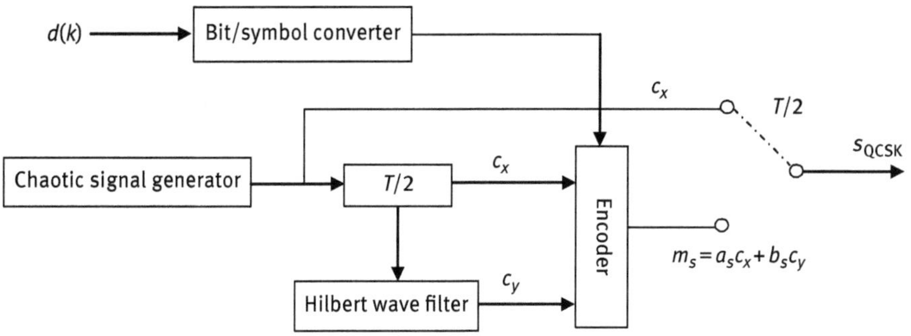

Figure 5.19: Principle block diagram of QCSK modulator.

reference signal segment $r(t) = E_b c_x(t)$ is transmitted in the first half symbol period of the transmission symbol s, and the information segment $i(t) = E_b m_s(t)$ is transmitted in the second half of the symbol period, where s represents the symbol that needs to be transmitted, while E_b represents the energy per bit. The orthogonal condition (5.32) shows that the energy per transmitted bit is a constant. If the symbol period is $T = 2\tau$, then the QCSK signal is

$$S_{\mathrm{QCSK}}(t) = \begin{cases} \sqrt{E_{\mathrm{b}}}c_x(t), & 0 \le t < T/2 \\ \sqrt{E_{\mathrm{b}}}(a_s c_x(t - T/2) + b_s c_y(t - T/2)), & T/2 \le t < T \end{cases}, \tag{5.33}$$

where $s = 0, 1, 2, 3$.

The simplified block diagram of QCSK demodulator is shown in Figure 5.20. QCSK signal may be demodulated by the method of differential coherent detection. During the process of demodulation, the symbol period is assumed to be synchronized. That is, the beginning of each symbol is taken as the starting of the coherent process at the receiver.

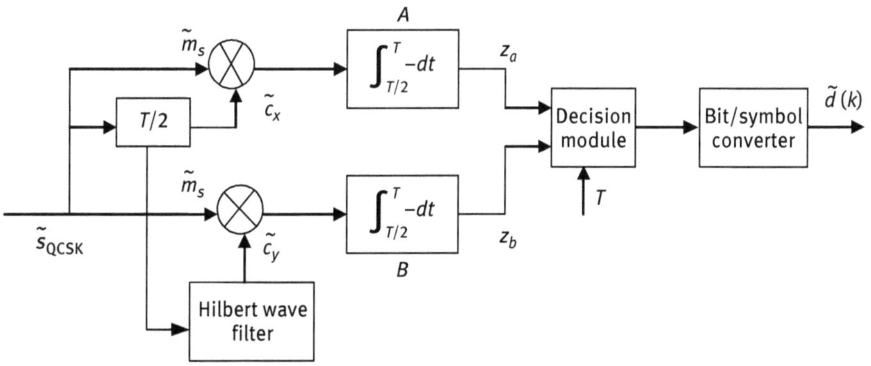

Figure 5.20: Principle block diagram of QCSK demodulation.

In theory, by using the chaos basis signals c_x and c_y, carrier information segment $i(t) = E_b m_s(t)$ is processed in the interval $[T/2, T]$, and we can recover the plural $m_s = a_s + jb_s$. In fact, according to formulas (5.27) and (5.28), we can obtain

$$a_s = \int_0^{T/2} m_s(t) c_x(t) dt, \tag{5.34}$$

$$b_s = \int_0^{T/2} m_s(t) c_y(t) dt. \tag{5.35}$$

At the receiver, we can observe noise, filtered reference signal $r(t)$, and information segment $i(t)$. If the signal is only affected by the additive Gaussian white noise, then the input of the correlator A in Figure 5.20 is

$$\tilde{r}(t - T/2) = \sqrt{E_b}(c_x(t - T/2)) + n(t - T/2), \quad t \in [T/2, T], \tag{5.36}$$

$$\tilde{i}(t - T/2) = \sqrt{E_b}[a_s c_x(t - T/2) + b_s c_y(t - T/2)] + n(t), \quad t \in [T/2, T]. \tag{5.37}$$

The estimated value of the quadrature signal $H[\tilde{r}]$ is generated by the disturbed chaotic reference signal, where H is the Hilbert transform operator:

$$H[\tilde{r}(t - T/2)] = H[\sqrt{E_b} c_x(t - T/2) + n(t - T/2)] = \sqrt{E_b} c_y(t - T/2) + n'(t - T/2), \tag{5.38}$$

where $n'(t) = H[n(t)]$. With the ideal Hilbert transform, the statistical properties of $n'(t)$ are the same as that of $n(t)$.

Considering the orthogonality of c_x and c_y, the output of the correlator is

$$\begin{aligned} z_a = \int_{\frac{T}{2}}^{T} \tilde{r}\left(t - \frac{T}{2}\right) \tilde{i}(t) dt = a_s \sqrt{E_b} + \sqrt{E_b} a_s \int_{\frac{T}{2}}^{T} c_x \left(t - \frac{T}{2}\right) n \left(t - \frac{T}{2}\right) dt \\ + \sqrt{E_b} b_s \int_{\frac{T}{2}}^{T} c_y \left(t - \frac{T}{2}\right) n \left(t - \frac{T}{2}\right) dt + \sqrt{E_b} \int_{\frac{T}{2}}^{T} c_x(t) n(t) dt + \int_{\frac{T}{2}}^{T} n(t) n \left(t - \frac{T}{2}\right) dt \end{aligned}, \tag{5.39}$$

$$\begin{aligned} z_b = \int_{\frac{T}{2}}^{T} H(\tilde{r}\left(t - \frac{T}{2}\right) \tilde{i}(t) dt = b_s \sqrt{E_b} + \sqrt{E_b} a_s \int_{\frac{T}{2}}^{T} c_y(t - \frac{T}{2}) n' \left(t - \frac{T}{2}\right) dt \\ + \sqrt{E_b} b_s \int_{\frac{T}{2}}^{T} c_y \left(t - \frac{T}{2}\right) n' \left(t - \frac{T}{2}\right) dt + \sqrt{E_b} \int_{\frac{T}{2}}^{T} c_y(t) n(t) dt + \int_{\frac{T}{2}}^{T} n(t) n' \left(t - \frac{T}{2}\right) dt \end{aligned} \tag{5.40}$$

CSK belongs to chaotic digital secure communication. When the basis function is strictly recovered, the noise performance of the coherent reception is better than that of the noncoherent reception. If the orthogonal basis function is chosen, the chaotic keying and the traditional digital modulation can achieve similar noise performance.

However, at present, it is difficult to meet the requirements of the strict recovery basis function, and it is more difficult to regenerate the chaotic basis function than to regenerate the periodic basis function. As a result, the performance of the coherent chaotic keying communication is slightly less than that of the same kind of traditional digital communication. But CSK belongs to broadband communication, and the performance of chaos communication shows more superiority in multipath communication environment. Among CSK communication schemes, QCSK and FM-DCSK have better performance and application prospect.

5.3 Chaos Parameter Modulation

Chaos parameter modulation is also known as wide spectrum transmission. The basic idea is that an information signal is injected into the transmitter, and thus it changes the dynamic characteristics of the original chaotic system. So the information signal is modulated. Compared with CSK and chaotic masking, chaos parameter modulation has the following advantages. First, it employs the whole range of the chaotic signal spectrum to hide information. Second, it increases the sensitivity to parameter variation, thereby enhancing confidentiality.

For CSK technology, no matter when the value of binary information signal changed, there is always a synchronization waiting time because of the difference of both initial conditions. While for chaos parameter modulation, receiver tracks transmitter continuously, and two chaotic system states would surely be the same. Compared with chaotic shift keying, it is expected to get a higher bit rate. In addition, with further development of chaos modulation technology, the direct chaotic spectrum expansion technique is also used as a chaos modulation technique.

Research results show that the main drawback of chaos communication scheme is being susceptible to channel noise and distortion. The traditional masking communication requirements of the power information signal are far lower than that of chaotic signal, which makes chaotic masking communication more difficult to be used in practice. A new method of chaotic pulse position modulation (CPPM) is proposed in Ref. [121], which shows good performance in the case of channel distortion. This scheme is based on the chaotic pulse sequence, and the interval between the pulses is determined by the chaotic characteristics of the nonlinear function. The pulse sequence of chaotic pulse interval modulation is used as a carrier. Binary information is modulated onto a carrier by pulse position modulation (PPM). As a result, each pulse is unchanged or delayed some time depending on the transmission information "0" or "1." The pulse duration of "0" or "1" is estimated by chaotic pulse sequence synchronization at the receiver, and thus the transmission information is demodulated in this way.

The block diagram of chaotic pulse position modulator is shown in Figure 5.21. The chaotic pulse signal generated by chaotic pulse generator is expressed as

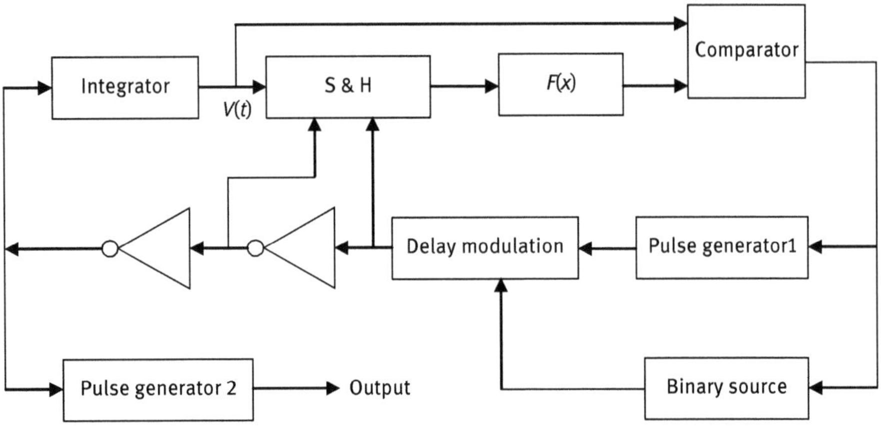

Figure 5.21: Principle block diagram of chaotic pulse position modulator.

$$U(t) = \sum_{j=0}^{\infty} W(t - t_j), \tag{5.41}$$

where $W(t - t_j)$ represents a pulse waveform generated at time t_j, and t_j is expressed as

$$t_j = t_0 + \sum_{n=0}^{j} T_n, \tag{5.42}$$

where T_n is the time interval between the nth pulse and the $(n-1)$th pulse. The time interval sequence T_i represents the iteration of the chaotic process. For the design convenience, we consider the chaos is generated by a one-dimensional map $T_n = F(T_{n-1})$, where $F(\cdot)$ is a nonlinear function.

The information is encoded in the chaotic pulse signal by adding the delay time in the pulse interval T_n. Thus, the resulting pulse sequence can be defined as

$$T_n = F(T_{n-1}) + d + mS_n, \tag{5.43}$$

where S_n is the information signal. Only binary information is considered here, so S_n is equal to 0 or 1. m represents the modulation degree, and the parameter d is the time delay constant for the modulation and demodulation. When we design chaotic pulse generator, the choice of the nonlinear function $F(\cdot)$, and parameters d and m should ensure the chaotic performance of the map.

The transmission signal is a CPPM signal in the channel, which is denoted by

$$U(t) = \sum_{j=0}^{\infty} W\left(t - t_0 - \sum_{n=0}^{j} T_n\right), \tag{5.44}$$

where T_n is generated by formula (5.43), and we assume that the duration of each pulse $W(t)$ of the pulse sequence is much less than the minimum value of the pulse interval T_n. When detecting information at the receiver, the demodulator is triggered by receiving pulse $U(t)$. After measuring successive time intervals T_n and T_{n-1}, the information signal is recovered as

$$S_n = (T_n - F(T_{n-1}) - d)/m. \tag{5.45}$$

If the nonlinear function $F(\cdot)$, parameters d and m at the receiver are same with those of the sender, then the encrypted information S_n can be easily recovered. When the matching accuracy of nonlinear functions is not sufficient, large decoding error occurs.

The demodulator principle is shown in Figure 5.22. When the demodulator is synchronized with the pulse modulator, to detect a single bit of the transmission information, the demodulator must simply determine whether the desired position of the transmitted pulse is delayed. In the figure, the integrator, sample and hold circuit, and a nonlinear threshold function module are reset or triggered by the pulse from the transmitter, not by the feedback loop within the receiver. To be exact, only when the input signal $U(t)$ from the input channel exceeds the threshold, they are triggered. For each received pulse, the result of $V(t_{n+1}) - F(v_n)$ is calculated and used to determine whether the pulse is delayed. Assuming that the difference is less than the reference value $\beta(d + m/2)$, the data bit S_{n+1} is determined to be "1," otherwise, it is determined to be "0."

Another important part of the receiver is the window selection module. Once the receiver has correctly received two consecutive pulses, the receiver can predict the fastest time to receive the next pulse. It means that the input of the demodulator can be

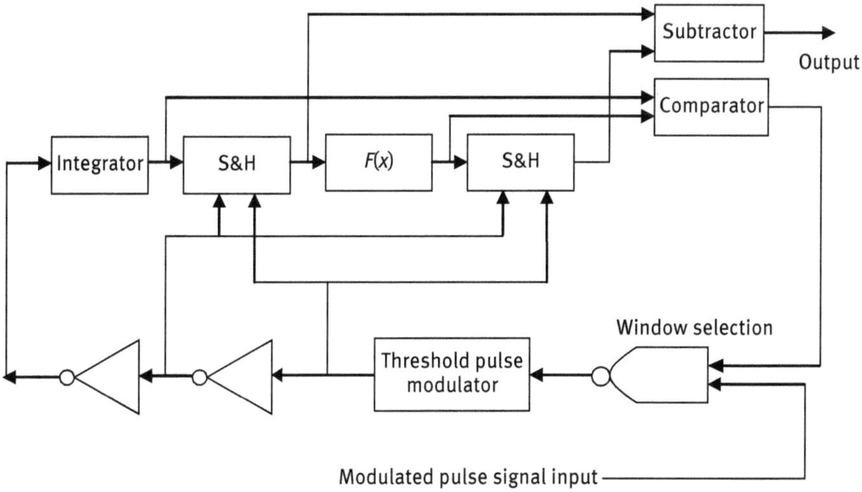

Figure 5.22: Principle block diagram of the chaotic pulse position demodulator.

masked at this time, which is done by the window selection module. In the experiment circuit, the input of the receiver is opened by a window control pulse generated by the comparator at the moment $t'_{n+1} = t_n + \beta F(v_n)$, and the input end is turned on until the demodulator is triggered by the first received pulse. So the window selection module greatly reduces the probability of the receiver triggered by the noise, interference, or the pulse of other users.

Although the performance of CPPM is slightly worse than that of BPSK, noncoherent FSK, and ideal PPM, it has the following characteristics:

1. Low probability intercept and low probability decoding ability.
2. Although the complexity of the circuit is increased slightly, the security of communication system is improved.
3. Compared with the majority of the communication schemes based on chaotic transformation, the system presents very good performance.
4. CPPM can be used for multiuser communication.
5. Compared with other pulse system, it eliminates any traces of periodicity of the transmission signal spectrum due to its independent period clock.

5.4 Chaotic Spread Spectrum Communication

Chaotic spread spectrum communication is divided into direct sequence spread spectrum (DS/SS) and frequency hopping spread spectrum (FH/SS). The principle is to use the chaotic sequence as the pseudorandom sequence. Common pseudorandom sequences include *m* sequence, Gold sequence, and Bent function sequence, but these sequences have a certain periodicity, limit numbers, and poor anti-interception ability. Chaotic sequences are nonperiodic sequences, which have the statistical properties of approaching Gauss white noise. The chaotic sequences generated by different systems or with different phases are different sequences.

5.4.1 Principle of Chaotic DS/SS Communication

A chaotic DS/SS communication system is built instead of pseudorandom codes with the chaotic sequences [122], and its principle diagram is shown in Figure 5.23.

Here, assume each user is synchronized, spread spectrum codes are synchronized, and the signal power of each user is same. If there are M users working at the same time totally, and the message data of each user is $a_i(k)$ $(i = 1, 2, \ldots, M)$, the original data signal of the ith user is expressed as

$$a_i(k) = \sum_{n=KN}^{KN+(N-1)} d_i(n), \tag{5.46}$$

where $d_i(n)$ is the information bit value of the ith user, $d_i(n) \in \{-1, 1\}$. Let $x_i(n)$ be the chaotic spread spectrum sequence of the sender of the ith user, and its expression is

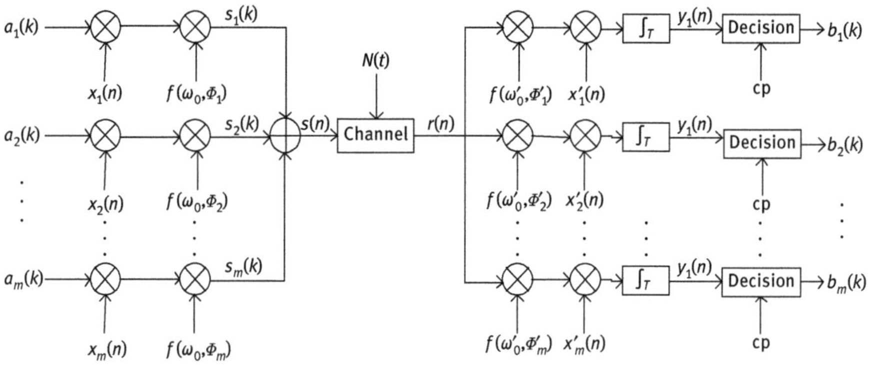

Figure 5.23: Principle block diagram of multiuser chaotic sequence spread spectrum communication system.

$x_i(n) = \sum_{n=KN}^{KN+(N-1)} g_i(n)$, where $g_i(n)$ is the spread spectrum characteristic sequence of the ith user.

The input signal $r(n)$ of the receiver is expressed as

$$r(n) = \sum_{k=1}^{M} A_k(n)d_k(n)g_k(n) + N(n), \tag{5.47}$$

where $A_i(n)$ is the carrier signal amplitude of the ith user. $N(n)$ is the zero mean Gauss white noise. So, the output of the correlator of the receiver of the ith user is

$$y_j(k) = \sum_{n=KN}^{KN+(N-1)} r(n)x_i'(n), \tag{5.48}$$

where $x_i'(n)$ is the chaotic spread spectrum sequence of the receiver of the ith user, and the system synchronizes between the sender and the receiver, that is, $x_i(n) = x_i'(n)$. Let T_b be the time width of each original information bit, then

$$\begin{aligned} y_i(k) &= \frac{1}{T_\text{b}} \int_{T_\text{b}} r(t)g_i(t)dt = A_i d_i + \sum_{\substack{i=1\\ i\neq k}}^{K} \rho_{i,k} A_i d_i + \frac{1}{T_\text{b}} \int_{T_\text{b}} n(t)g_i(t)dt \\ &= A_i d_i + MAI_i + Z_i \end{aligned} \tag{5.49}$$

For formula (5.49), the first item is the desired information component. The second item is the recovery data obtained by the received signal with the kth signal itself, and multiple access interference (MAI) caused by other users. Obviously, the MAI term is related to the user number k, the amplitude of the signal of other user, and the cross-correlation coefficient $\rho_{i,k}$, where $\rho_{i,k} = \frac{1}{T_\text{b}} \int_0^{T_\text{b}} g_i(t)g_k(t)dt$. The third item is noise interference.

After the integral (correlation operation) of the receiver for the ith user, the signal is decided by the decision threshold, which is zero. The data is demodulated by the ith user at the receiver according to

$$b_i(k) = \begin{cases} -1, & y_i(k) < 0 \\ 1, & y_l(k) > 0 \end{cases}. \tag{5.50}$$

To sum up, whether the user information is correctly recovered is greatly related with the characteristics of the spread spectrum code.

5.4.2 Generation of Chaotic Spread Spectrum Code

Taking three typical chaotic maps, including logistic, Chebyshev, and Tent map as examples, the binary chaotic spread spectrum sequence is generated [123]. In the multiuser chaotic sequence spread spectrum communication system, spread spectrum sequence is a binary sequence with value ±1, namely, the real values of chaotic map sequences are quantized into binary to generate spread spectrum sequence.

5.4.2.1 Logistic Chaotic Map

Logistic map equation is

$$x_{n+1} = rx_n(1 - x_n), \quad 0 < x_n < 1, \tag{5.51}$$

where r is the system parameter, and $r \in (0, 4]$. When $3.5699\ldots < r \le 4$, the system is chaotic. The mean value of the sequence is 0.5.

5.4.2.2 Chebyshev Chaotic Map

The ω-order Chebyshev chaotic map is defined as

$$x_{k+1} = \cos(\omega \cos^{-1} x_k), \quad -1 \le x_k \le 1, \tag{5.52}$$

where ω is usually taken as an integer power of 2, and the mean value of the sequence is 0.

5.4.2.3 Tent Chaotic Map

The Tent chaotic map is defined as

$$x_{k+1} = \begin{cases} x_k/a, & 0 < x_k < a \\ (1 - x_k)/(1 - a), & a \le x_k \le 1 \end{cases}, \tag{5.53}$$

where $0 < a < 1$, and the mean value of the sequence is 0.5.

Next, we discuss the balance and correlation of the three kinds of chaotic sequences.

5.4.2.3.1 Balance of the Sequences

The balance of a sequence is determined by the statistical distribution about "1" and "–1" in the sequence. If P and Q denote the number of "1" and "–1" in the chaotic sequence, respectively, then the balance degree of the sequence E is defined as

$$E = (P - Q)/N, \tag{5.54}$$

where N is the total number of symbol sequence. The sequence balance is related with the carrier suppression degree of the DS/SS system. The imbalance of spread spectrum sequence will lead the leakage of carrier information, or be prone to error or lost. Thus, it is important to study the balance of the direct spread spectrum system (DS/SS).

Applying formula (5.54) to analyze the impact of the initial value of three kinds of chaotic maps, the results are shown in Figure 5.24. When $r = 4$ for the logistic map, the initial value x_0 is one of the 9,999 different numbers between 0 and 1. From Figure 5.24(a), we can obtain that the effect of x_0 on the balance is small, but there are three peaks, and the initial value $x_0 = 0.25$, 0.5, and 0.75, respectively. When $\omega = 20$ for the Chebyshev map, the initial value x_0 belongs to one of the 9,999 different numbers between 0 and 1. From Figure 5.24(b), we found that the effect of x_0 on the balance is very small, and there are no peaks. When $a = 0.49$ for the Tent map, the initial value x_0 forms the 9,999 different numbers between 0 and 1. From Figure 5.24(c), we find that the effect of x_0 on the balance is small, and there are three peaks, and the initial value $x_0 = 0.2401$, 0.49, and 0.7501, respectively. It is worth pointing out that the peak point of Tent map is related with the initial value selection. If the accuracy of the initial value is not enough, the peak of this map may be missed.

5.4.2.3.2 Correlation of the Sequence

From formula (5.49), in the case of same SNR, and without consideration of multipath interference, MAI becomes the major factor in the transmission performance of the system. Root mean square (RMS) of cross-correlation of the sequence characterizes the size of the MAI of the spreading sequence, so RMS of a sequence is important for analyzing the transmission performance. As a result, the actual simulation of the chaotic

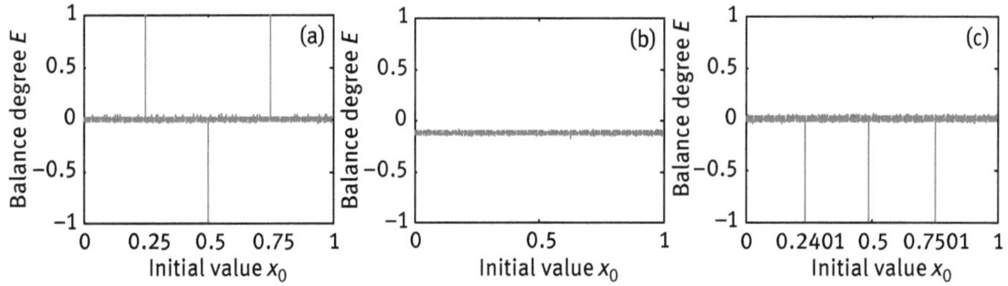

Figure 5.24: Relationship between the balance E and the initial value x_0, for different maps: (a) Logistic map; (b) Chebyshev map; and (c) Tent map.

sequence has a deviation with the theoretical value. Here, we change the length of the chaotic sequence N and maintain the relevant range M, to study the effect of the length of the sequence on the deviation. The RMS σ_{R12} is defined as

$$\sigma_{R12} = \sqrt{\frac{1}{2M+1} \sum_{m=-M}^{M} [R_{12}(m)]^2}. \tag{5.55}$$

Here, we use formula (5.55) to calculate the RMS values of the cross-correlation of the sequence with length $N = 2{,}000$ generated by the above three kinds of chaotic maps. The results are shown in Table 5.2.

From Table 5.2, after the frame processing, RMS value of cross-correlation for each chaotic sequence has been greatly affected, and its change is different. Besides, different frame lengths correspond to different RMS values, which destroy the original relevant characteristics. From Table 5.2, after the frame processing, the cross-correlation of three chaotic sequences becomes smaller, and it also shows that the chaotic sequence with a certain length has a better cross-correlation compared with the original sequence. But there is no fixed relationship between the frame length and the RMS value, that is to say, after the frame processing, chaotic sequence destroys the original conclusion. That is, the longer the sequence, the smaller the deviations of correlation.

5.4.3 Simulation Module Design of Multiuser Chaotic Spread Spectrum System

Multiuser chaotic spread spectrum system involves the design of multiple modules, including the spread spectrum sequence generator module, the correlation receiving module, the multipath channel module, the polarity transformation module, and the frame sampling module. The next sections introduce each of these.

5.4.3.1 Spread Spectrum Sequence Generator Module

The chaotic spread spectrum code is generated by Chebyshev, improved logistic, logistic, Tent, and other iterative chaotic maps. In addition, the real value is transformed into the binary sequence by setting a decision threshold ξ, and the real value is

Table 5.2: RMS values of three chaotic sequences.

Frame length	Chebyshev	Tent	Logistic
64	0.2450	0.2769	0.2783
950	0.2034	0.2292	0.2306
1,000	0.2031	0.2300	0.2307
Processing without frames	0.2488	0.2816	0.2824

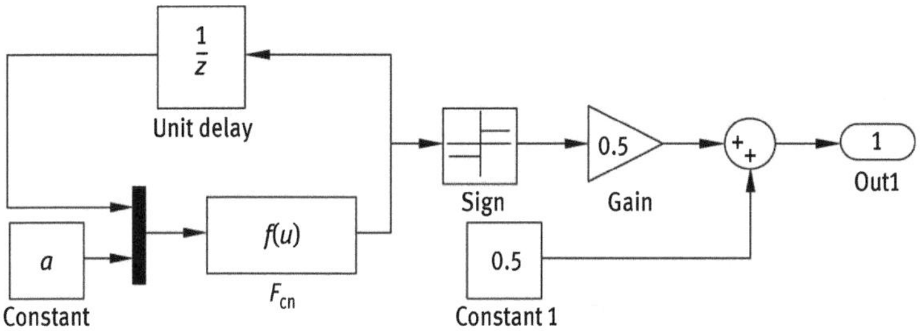

Figure 5.25: Simulation model of chaotic sequence generator based on Tent map.

converted by the decision function $T(x) = [1 + \mathrm{sgn}(x - \xi)]/2$. Here, two design methods of the chaotic sequence generator module are introduced. One is the design of spread spectrum sequence generator based on Tent sequence, and the other is the design of spread spectrum sequence generator module based on series chaotic map.

The simulation model of the spread spectrum sequence generator module based on Tent chaotic sequence is shown in Figure 5.25. In the figure, the initial value and the sampling time of the whole sequence generator are controlled by the initial value and the sampling time of the unit delay module, respectively. The value of the constant module corresponds to the bifurcation parameter of the Tent chaotic map. The parameter of the F_{cn} module indicates the chaotic map, and finally the chaotic sequence is converted to binary sequence by the sign quantization module. It is worth noting that the Tent map from the mathematical model to the computer simulation model is more difficult to achieve, because the expression of the chaotic map is a piecewise function, and the bifurcation parameters are divided into two segments. Therefore, the expression F_{cn} is

$$\begin{aligned} F_{cn} &= u(1) \cdot (u(1) \geq 0) \cdot (u(1) < u(2))/u(2) \\ &\quad + (1 - u(1)) \cdot (u(1) \geq u(2)) \cdot (u(1) \leq 1)/(1 - u(2)) \end{aligned}, \tag{5.56}$$

where $u(1)$ represents x_n, and $u(2)$ represents the bifurcation parameter a.

The simulation model of the spread spectrum sequence generator module based on series chaotic map is shown in Figure 5.26. The first chaotic sequence of series sequence adopts improved logistic map, and the second chaotic sequence uses Chebyshev map. Generally, the sampling time of the first chaotic sequence is $1/n$ times the sampling time of the second chaotic sequence. In this system, we set the sampling time of the first chaotic sequence as 1/192 s, and then the sampling time of the second chaotic sequence is 1/192,000 s. Let the sampling time be 1 s, namely $x_1 = y_1$, $x_2 = y_{1,000}, \ldots, x_{192} = y_{191,000}$, that is, the first chaotic sequence should control the second chaotic sequence to generate 1,000 chaotic sequences per second. For simulation, the initial values of the two chaotic sequences are set first. Because $x_1 = y_1$, the two chaotic sequences must be set to the same initial value in the series simulation,

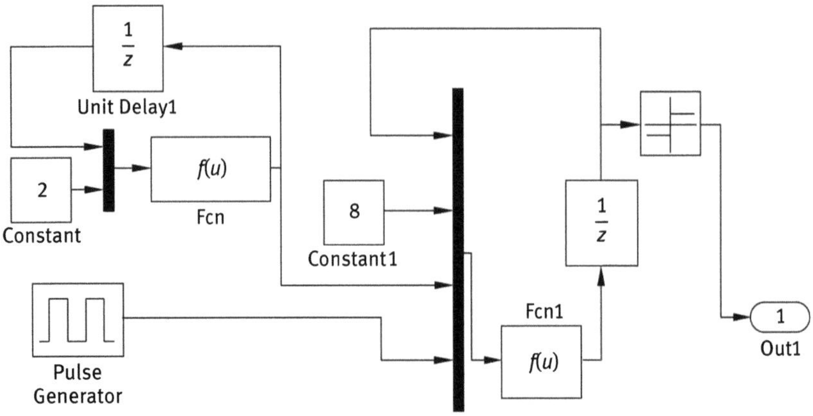

Figure 5.26: Simulation model of spread spectrum sequence generator based on series chaotic map.

namely, the initial values of two unit delay modules are the same. Multiple n of the sampling time of the two-stage sequence is controlled by a periodic sequence generator, and the sampling time of the periodic sequence generator can be the same as the sampling time of a chaotic sequence generator. In the multiuser chaotic spread spectrum communication system, when the frames are sampled, it is required that the values of chaotic sequences are discrete. So when the 192nd value controlled by clock arrives, we obtain the chaotic sequence y_n, rather than y_{n+1}, and at this moment $y_n = y_{192,000}$. Otherwise chaotic sequence generator cannot work due to the value of the continuous chaotic sequence. *Fcn*1 in Figure 5.26 is expressed as

$$Fcn1 = (\cos(u(2) \cdot a\cos(u(1)))) \cdot (abs(u(4) - 1)) + \cos(u(2) \cdot a\cos(u(3))) \cdot u(4), \quad (5.57)$$

where $u(1)$ represents x_n of the Chebyshev map, $u(2)$ is the bifurcation parameter q of the Chebyshev map, $u(3)$ represents x_{n+1} of improved logistic map which belongs to the first chaotic sequence, and $u(4)$ is the sampling time of multiple n.

5.4.3.2 Correlation Receiving Module

The block diagram of traditional correlation detection at receiver is shown in Figure 5.27. The dashed box in the figure corresponds to a match filter.

Integral module is the core part of the spread spectrum system, which consists of unbuffer module, integral sampling module, zero-order hold module, and frame sampling module, as shown in Figure 5.28. It works as follows. Each frame data $(N * 1)$ from *In*1 module is decomposed to $(1 * N)$ sequences by unbuffer module, and then do sampling integral. The zero-order hold module makes the time duration, hold time, and information code sampling time of each binary symbol to be the same after the integration. Finally, they are converted into frame format by the sampling module. The integrator parameter settings are as follows: Output sampling time is the sampling

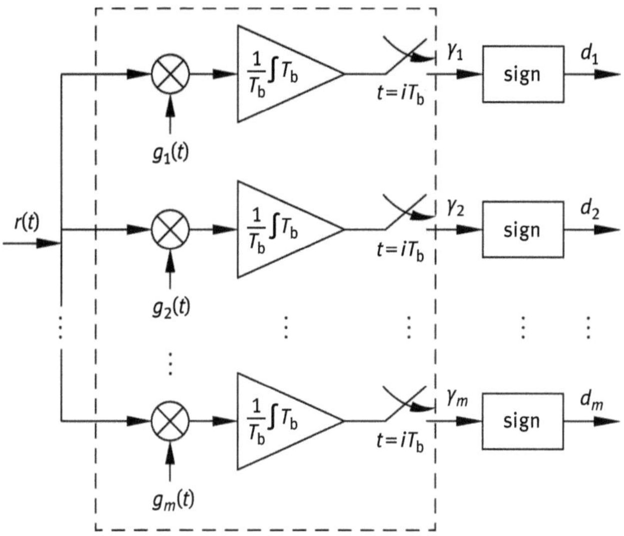

Figure 5.27: Principle block diagram of the detector of the traditional DS-Code Division Multiple Access system.

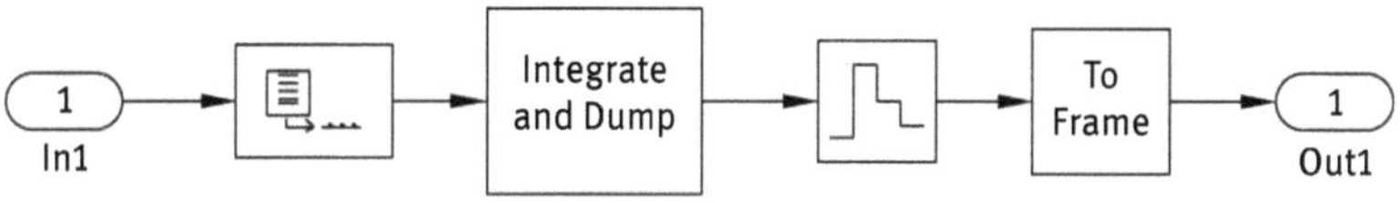

Figure 5.28: Simulation block diagram of correlation receiver.

time of information code, and it is the same with simulation step. The sampling number of the buffer module is N.

5.4.3.3 Rake Receiving Module

In the environment of multipath transmission, the performance of the conventional correlation receiver is not ideal, while the rake receiver is a good choice. Rake receiver technology is a multipath diversity reception technology. It collects all the delay components (multipath components) of the original transmitted signals by the relevant receiving for each multipath signal. Therefore, multipath diversity is also known as code diversity.

The principle of rake receiver is that the relevant receiving units are applied to a correlation receiver for each path, where the relevant receiver is relevant to a delayed version of the signal (that is the multipath components of desired signal) with the same expectations (received). Then it is weighted based on the relative intensity of output (called rake tooth output) of these correlation receivers, and all the weighted outputs from various quarters are synthesized into a single output. The selection principle of weighted coefficient is to make the output SNR maximum. The theoretical basis is that when the multipath time delay exceeds a chip period, multipath signal is taken as uncorrelated.

More specifically, we assume that the L correlators in the CDMA receiver are used to capture the L strongest multipath components of the desired codes, and then a weighted network is used to obtain the combination of the outputs of the L correlators, which is applied as bits detection of the desired codes as shown in Figure 5.29.

From Figure 5.29, the outputs $c_i(t)$ of L correlators are multiplied by weighted coefficients $d_i(t)$ (where $i = 1, 2, \ldots, L$), respectively, and they are summed to obtain the total output signal, which is denoted as

$$e(t) = \sum_{i=1}^{L} d_i(t) c_i(t), \tag{5.58}$$

where the weighted coefficients $d_i(t)$ are defined as

$$d_i(t) = \frac{c_i^2(t)}{\sum_{i=1}^{L} c_i^2(t)}, \quad i = 1, \ldots, L \tag{5.59}$$

Obviously, when the correlator (e.g. correlator i) fades, the energy of output c_i is small, and the corresponding weighted coefficients d_i also decrease automatically, and it leads the component in formula (5.59) to decrease.

The simulation model of the Rake receiver is shown in Figure 5.30. Assuming that the data transmitted in channel appears four multipath transmissions, and the time delay is 0, 2, 6, and 7, respectively. At the receiver, these multipath signals are done in the time period corresponding 0, 2, 6, and 7 in advance, and then sum them. Finally, the equal-gain combining will decrease the BER significantly. The equal-gain combining means that the various branch signals are summed in the same phase, and the weighted factors of each branch are equal.

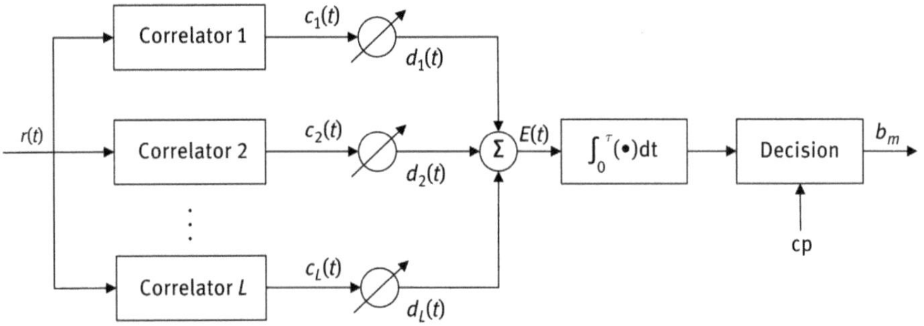

Figure 5.29: Structure block diagram of rake receiver with L channels.

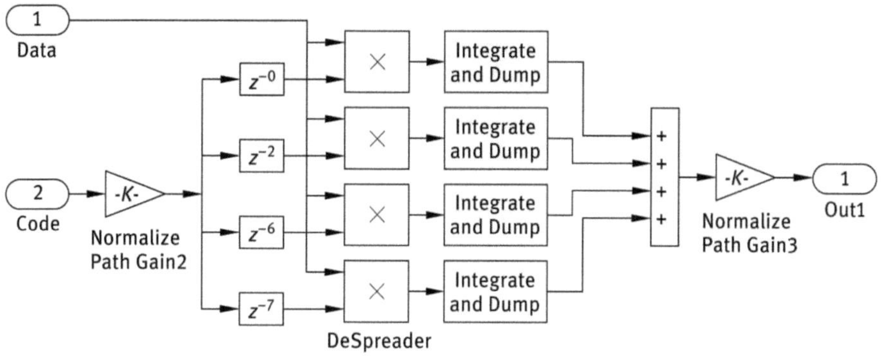

Figure 5.30: Simulation model of Rake receiver.

5.4.3.4 Multipath Channel Module

The tapped delay line model is adopted for the multipath fading channel, and its impulse response is

$$h(t) = \sum_{l=1}^{L} \alpha_l \delta(t - \tau_l) e^{j\phi_l}, \tag{5.60}$$

where L is the number of decomposable paths in the multipath channel. α_l, ϕ_l, τ_l are the gain, the phase, and the time delay of the lth decomposable channel, respectively. $\delta(t)$ is the delta function. Under the condition of Rayleigh channel, α_l obeys the Rayleigh distribution, and ϕ_l obeys the uniform distribution in $[0, 2\pi)$.

The simulation model of multipath channel is shown in Figure 5.31. Through the multipath channel, signal becomes four columns delayed signal, of which the time delay is 0, 2, 6, and 7, respectively, and the frame sampling format is used. Note that the sampling time of multipath channel should be the same as that of information code.

5.4.3.5 Polarity Transformation Module

The operator of spreading spectrum and de-spreading spectrum can be realized by the XOR operator of the unipolar binary code, but it is difficult to process when the result is zero sometimes. When the signal is superimposed noise, it is no longer a binary

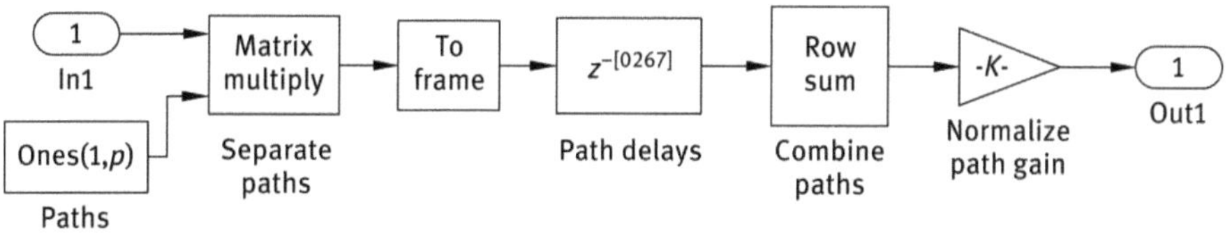

Figure 5.31: Multipath channel simulation model.

code, and we cannot use XOR operator at this time. Employing multiplication of bipolar binary symbols can also realize spreading and de-spreading operation, and can overcome the shortcomings of the above-mentioned method. During simulation, the polarity conversion is realized by bipolar to unipolar converter module and unipolar to bipolar converter module.

5.4.3.6 Frame Sampling Module

Taking into account the inherent requirements of the DSP and data communication expenses, the frame-based processing is needed. Obviously, the signal processing based on frame is much more complex than that based on sampling. The sampled signal processes a sampled point at each time step, while a frame-based signal processes one frame with *N* sample points at each time step. But by using the matrix feature of MATLAB, Simulink greatly improves the processing efficiency. The communication between the blocks is reduced by using the frame-based processing, and the simulation is much faster than that of the signal based on the sampling. In short, the simulation speed is improved by using the frame-based signal. Therefore, we employ the frame-based method.

In Simulink, the conversion between the sampled signal and the frame signal is achieved by the buffer module (buffer). There are two functions in buffer module. The first is to accept the sampling signal input and generate a frame input with certain frame size. The second is to accept the frame input and modify the size of the frame, and the buffer module must be used in this situation. Both cases involve the overlap between frames, and the problem of setting the initial value of the frame. When a frame is generated by sampling, the buffer will use the input scalar to generate a column vector. If you need to generate a sample from a frame signal, you should use the unbuffer module.

5.4.4 Design and Simulation of Multiuser Chaotic Spread Spectrum System Based on Rake Receiver

According to Figure 5.23, multiuser chaotic spread spectrum communication simulation system with five users is built based on the rake receiver as shown in Figure 5.32. In the figure, the subsystem of users 2–5 at the sender is formed by multiplication of user information with spread spectrum code. There are corresponding subsystems of users 2–5 at the receiver. User 1 and receiver 1 within the dotted line are composed of unpackaged modules.

In addition, according to the setting of the parameters for the multiple paths module, the type of spread spectrum communication can be changed flexibly. Setting parameters are shown in Table 5.3.

All the initial value of Chebyshev, improved logistic, logistic, tent, and series chaotic sequence is set to 0.63. Parameters r, q, and a are equal to 4, 8, and 0.49,

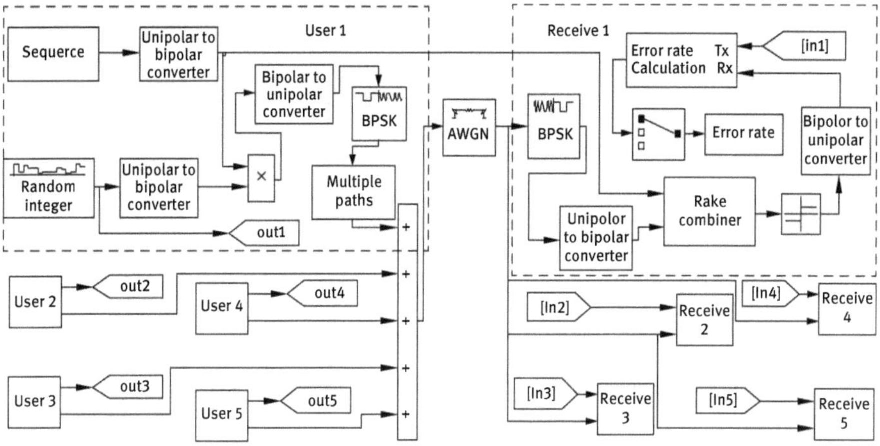

Figure 5.32: Simulation model of multiuser chaotic spread spectrum communication system.

Table 5.3: Setting parameter of the multiple paths module.

Number of paths	Path delays	Types of spread spectrum communication system
0	0	Synchronous chaotic spread spectrum communication system
3	[0, 3, 5]	Asynchronous chaotic spread spectrum communication system

respectively. The output range of the random integer generator is 2, and sampling time is 1/192 s, and the sampling number for each frame is 1. The sampling time of chaotic spread spectrum code is 1/192,000 s, and the sampling number for each frame is 1,000. The multipath number is 3, and the delay time is 0, 3, and 5, respectively, and the symbol period time is 1/192 s. The symbol period of additive white Gaussian noise channel is 1/192 s, and the delay number in the receiver is 1. The number of users is 5, and the simulation time is 10 s. By simulation of the chaotic spread spectrum communication system, the results are shown in Figure 5.33.

In Figure 5.33, the transmitted signal (a) and received signal (b) are almost identical, except a little bit delay for the received signal. It is shown that the simulation system can receive signal correctly.

5.4.5 Performance Analysis of Multiuser Chaotic Spread Spectrum System Based on Rake Receiver

Next, we analyze the performance of the multiuser spread spectrum communication system from the aspects of the initial value, the SNR, the number of users, the length of the spreading sequence, and so on.

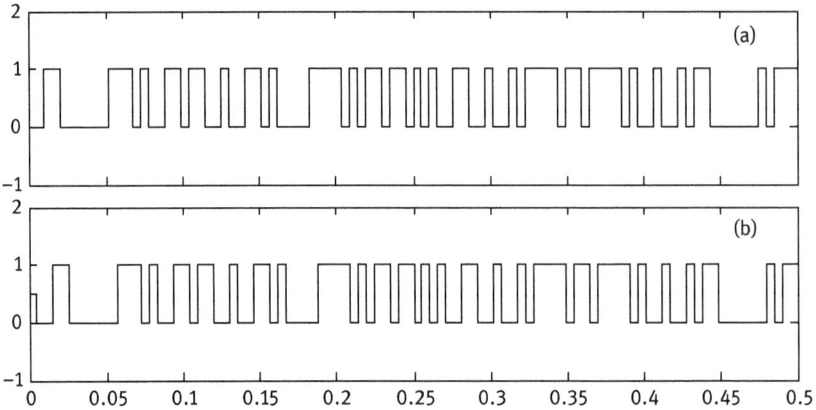

Figure 5.33: Waveforms of (a) transmitted signal and (b) received signal.

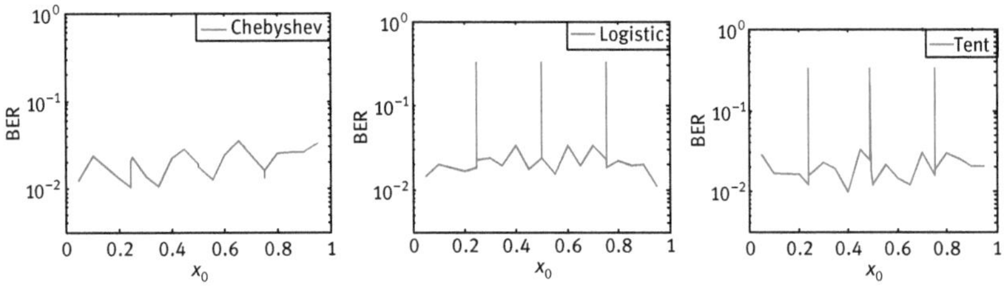

Figure 5.34: Relationship curves between the BER of spread spectrum system and the initial value x_0 of different maps.

5.4.5.1 Influence of Initial Value x_0 of Chaotic Map on System Error Rate

Without loss of generality, we set SNR E_s/E_0 to 1 dB. The initial value changes between 0.0001 and 0.9999 as shown in Figure 5.34. For Chebyshev chaotic map, the error rate is less affected by the initial value, and there is no peak point.

The peak value of the balance degree of the Logistic chaotic map (corresponding to the initial values are 0.25, 0.5, and 0.75, respectively) results in the peak value of BER of the system. When the other initial values are selected, the change of the BER is relatively flat. Thus, when we choose the initial value, the three initial values of 0.25, 0.5, and 0.75 should be avoided.

From Figure 5.34, we find that there are three peak values of the balance degree of the tent chaotic map, and the corresponding initial values are 0.2401, 0.49, and 0.7501, respectively. When $x_0 = 0.49$, the fractal parameter is also 0.49. If the value 0.49 is substituted into formula (5.53), we find all the elements are zeros except the first element 1 in the real value sequence. In this case, the result of formula (5.54) is $E = -0.998$. It means there is a peak point at the initial value 0.49, which has a great impact on the balance degree, as shown in Figure 5.24(c). But we cannot simply say that the element of a sequence should avoid this point when constructing the sequence. From eq. (5.53), it can be seen that the range of sequence element is (0, 1], namely the sequence element of the tent chaotic map cannot be zero. Obviously, the obtained binary sequence is meaningless at $x_0 = 0.49$, so we do not discuss the case of $x_0 = 0.49$. However, when the initial value $x_0 = 0.2401$ and $x_0 = 0.7501$, there are peaks that show that the tent map does have a balance degree peak point and it verifies the conclusion in Figure 5.24(c). So when the initial value of the sequence is selected, the two values 0.2401 and 0.7501 should be avoided.

5.4.5.2 Influence of SNR on BER

Relationship between BER and SNR of spread spectrum system with five kinds of spreading codes is shown in Figure 5.35. Obviously, when the value of SNR E_s/E_0 varies from 0 to 8, the BER performance of multiuser logistic, Chebyshev, and tent chaotic spread spectrum communication system is similar, and is better than that of *m* sequence and *Gold* sequence. It indicates that the capability of the anti-MAI of chaotic spread spectrum signal is better than that of the *m* sequence and *Gold* sequence. When the SNR is smaller, the system exhibits mainly noise interference. At this time, the BER performance of Chebyshev map is the best, and it possesses good antinoise performance. When the SNR is larger, the system exhibits mainly MAI. Then, logistic map has good resistance to MAI.

5.4.5.3 Influence of User Number *i* on BER

Without loss of generality, the user number $i = 1$–10, and we set the SNR E_s/E_0 1 dB. From Figure 5.36, it can be seen that the BER of chaotic spread spectrum system with

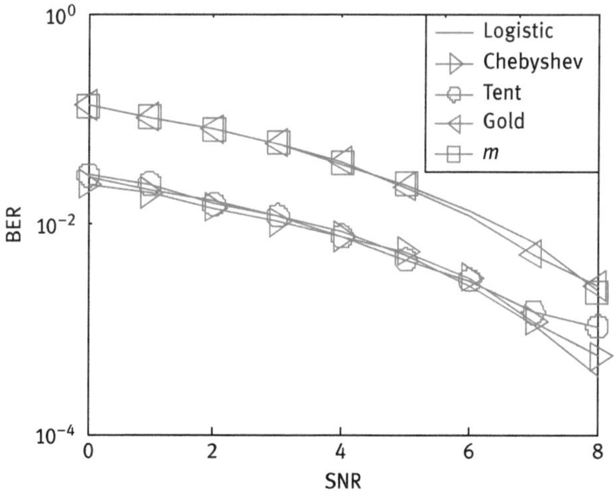

Figure 5.35: BER of spread spectrum system with five kinds of spreading codes.

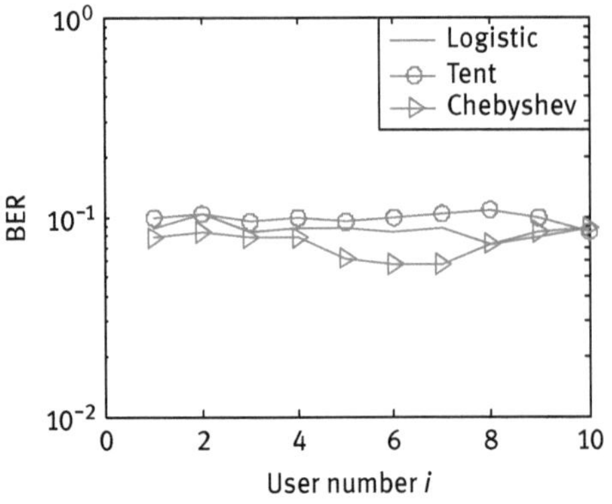

Figure 5.36: BER of the system with three kinds of chaos spreading code for different users.

three kinds of spreading codes has little change with the increase in the number of users. Where the BER of spread spectrum system of Chebyshev map is minimum, the BER of tent map is maximum. It indicates that the system based on Chebyshev map has good antinoise and anti-MAI.

5.4.5.4 Influence of the Sequence Length *N* on BER

Without loss of generality, the SNR is 8 dB and the sequence length N varies from 100 to 3,500. The other conditions remain the same. When N varies from 0 to 1,050, Figure 5.37 shows that the effect of logistic map on the BER of the system is better than that of Chebyshev map or tent map. When N varies from 1,050 to 3,500, the BER of Chebyshev map is the best, and logistic map follows, while tent map is the worst. Also, from Figure 5.37, it is easy to see that the BER of the logistic map changes suddenly at $N = 1,100$. The main reason is that the correlation properties of chaotic maps may

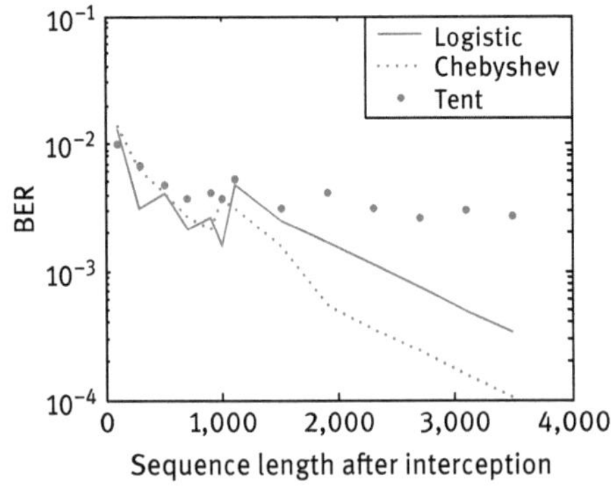

Figure 5.37: BER of the system for different sequence lengths.

be destroyed after the sequence is truncated, which also confirms the conclusion of Table 5.2.

The simulation results show that the multiuser chaotic spread spectrum communication system based on rake receiver has the characteristics of strong anti-interference ability, low BER, high subscriber capacity, good security and easy to implement, and it is flexible for model transformation.

The BER performance of the chaotic spread spectrum communication system in series chaotic map is better than that of a single chaotic sequence and m sequence of IS-95 standard. Further, the generation of series chaotic sequence is simple, and it has rich source codes, high security, strong antijamming performance, and similar random states, which are good to increase the system capacity and reduce the system complexity.

Questions

1. What are the key technologies in chaotic secure communication?
2. What are the factors that affect the performance of chaotic secure communication?
3. Try to compare the principle and characteristics of different chaotic switched communication schemes.
4. Describe the construction and simulation of a multiuser chaotic spread spectrum communication system.

6 Data Encryption Based on Chaotic Sequence

Chaotic signal has great potential in cryptography with special features like unpredictability, high complexity, easy implementation, and so on. Compared to other encryption methods, chaotic encryption is a kind of dynamic encryption. The processing speed is not related with the key length, so it has high efficiency. It can be applied in real-time signal processing or static encryption.

Chaos theory began to attract the attention of cryptographic community in the late 1980s. The idea of chaotic encryption was proposed by British mathematician Matthews [49]. On one hand, due to the loss of information in the iteration, the amount of information is gradually lost to zero, so it is impossible to predict the chaotic sequences correctly. On the other hand, the chaotic synchronization is the key to achieve the correct decryption. Because the computer system can accurately reproduce all the starting states, the synchronization algorithm of chaotic iterative system can be realized by computer process synchronization. This is a feature that a physical chaos model does not have. There is no need to use complex physical chaotic synchronization and control method. Therefore, since the 1990s, the chaotic sequence encryption has become an important research frontier of modern cryptography [112, 124–127].

There are two research directions in chaotic cryptography. One is the chaotic secure communication based on chaotic synchronization. The other is to construct a new stream cipher and block cipher using chaotic system. In this chapter, we first discuss the theory of chaotic cryptography, and then we study the data encryption algorithm based on chaotic sequences and explore the application of chaotic sequences in digital watermarking and image encryption.

6.1 Chaos Cryptography

6.1.1 Traditional Cryptography

According to different methods of the text data encryption, traditional cryptography can be divided into the sequence cipher system and the block cipher system [128].

For sequence cipher, the original plaintext is transferred into plaintext data sequence first, and then the ciphertext sequence is obtained by employing the XOR operation with the key sequences. The generated ciphertext sequence is sent to the receiver. At the receiver, the plaintext sequence is recovered by using the XOR operation again. The sequence cipher encryption model is shown in Figure 6.1. There is no data expansion and error transmission in sequence cipher. Its advantages include good real-time performance and easy implementation for encryption

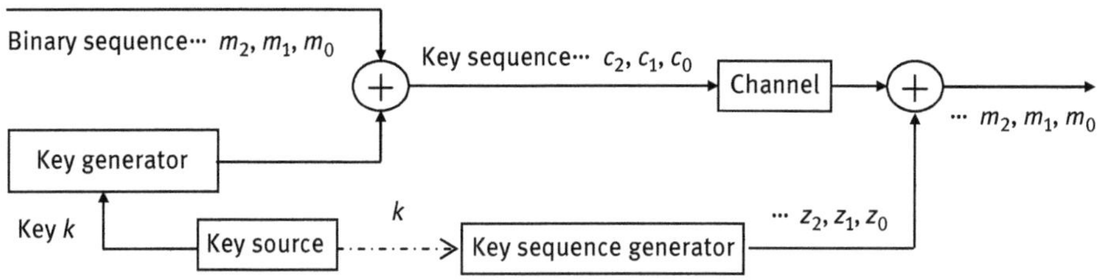

Figure 6.1: Principle diagram of sequence cipher system.

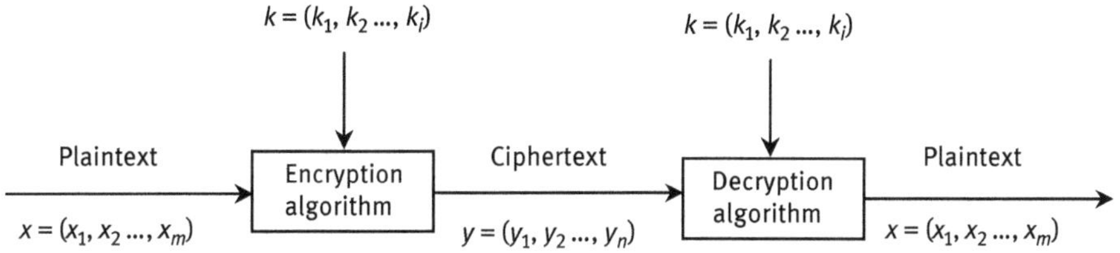

Figure 6.2: Principle diagram of block cipher system.

and decryption. The security of the sequence is mainly dependent on the key sequence. If the key sequence is completely random, it is "one time pad," which is not to be deciphered. However, in practice, one-time pad is very difficult to be realized, so the general encryption and decryption key sequence is pseudo-random sequence. One-time pad requires the key sequence has a sufficiently long period, good random statistical property, and greater linear unpredictability, and it also needs to meet the cipher principle about "diffusion" and "confusion" in the algorithm design.

For block cipher system, the data sequence $x_1, x_2, \ldots,$ is divided into $x = (x_1, x_2, \ldots, x_m)$ groups with the length m, and each group (a vector with the length m) is converted into equal length data sequence $y = (y_1, y_2, \ldots, y_n)$ (a vector with the length n) with the control of key sequence $k = (k_1, k_2, \ldots, k_t)$. The principle diagram of block cipher is shown in Figure 6.2. The difference between block cipher and the sequence cipher is that each bit of the output is not only related to the input bit of the corresponding time, but also related to the plaintext digital with the length m. The block cipher has the following advantages. First, the block cipher is easy to be standardized. Second, synchronization is easy to be achieved for block cipher. A transmission error of a group data will not affect the others. The loss of a plaintext–ciphertext group will not affect the decryption of its subsequent groups. However, the block cipher cannot hide data patterns, and it cannot resist the attack of replay, embedding, and deletion. General block cipher algorithms include IDEA and DES.

6.1.2 Relationship between Chaos Cryptography and Traditional Cryptography

Relation between chaos theory and traditional cryptography is summarized in Figure 6.3. The chaotic system is corresponding to cryptography algorithm. The extreme sensitivity of chaotic system to initial conditions and control parameters is corresponding to the diffusion of cryptography. Random and long periodic orbits of chaotic system can replace the pseudo-random signal in the cipher system. An initial value of chaotic map is spread to the whole phase space through the iteration, while the cryptographic algorithm generates the expected diffusion and confusion. Diffusion and confusion are the basic principles of design encryption algorithm. Diffusion means that the effects of each plaintext numbers are spread to multiple output cipher as much as possible so that the statistics characteristic of the plaintext can be concealed easily. Confusion means the statistics relationship between the plaintext and the ciphertext is complex. The parameters and initial values of the chaotic system are equivalent to the keys in the cipher algorithm. One important difference between chaotic system and cryptographic algorithm is that the traditional cryptographic algorithm is defined on finite set, while chaos cryptograph is defined on the real set, which means the traditional chaos theory is adapted to the continuous system, and this is the essential difference between them. In the chaotic system, Lyapunov exponent is a measure of the initial condition sensitivity, and Roland pointed out the possible link between it and security [129].

Chaotic encryption is different from the traditional encryption algorithm. The main difference is that the random signal used in chaotic encryption is generated by chaotic system. Chaotic signals are classified into continuous chaotic flows and discrete chaotic sequences. The former is used for the analog chaotic secure communication, and the latter is used for encrypted communication. The continuous signal

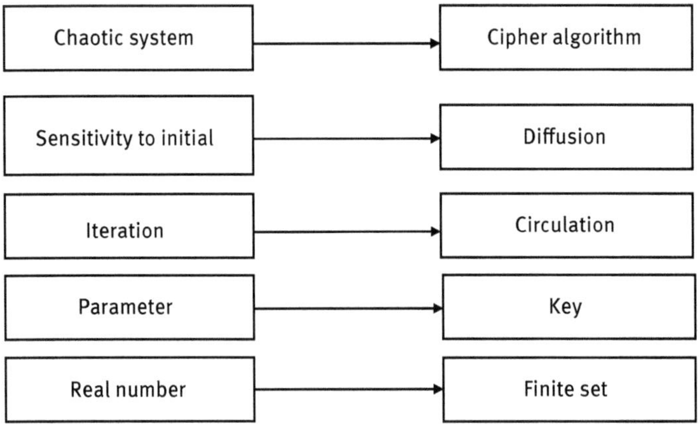

Figure 6.3: Relationship between chaos theory and traditional cryptography.

generated by chaotic system can also be used as a discrete signal after it is made discrete, and the new Field Programmable Gate Array (FPGA) technology is used to design the hardware encryption system, which has become a new research hotspot [130]. In this section, we will discuss the information encryption based on discrete chaotic system.

6.1.3 Principle of Chaos Sequence Encryption

Chaos sequence encryption belongs to the sequence cipher system, and its encryption principle is similar to that of the sequence cipher encryption. The difference is that the general sequence cipher uses the pseudo-random sequence generated by the circuit based on the shift registers, while the chaos sequence encryption uses the chaos sequence generated by chaotic system. In the early stage, the pseudo-random sequence generator is used to generate the key stream. An attacker can use the correlation analysis method to analyze and reconstruct the pseudo-random sequence effectively. Now, chaotic system is used to replace the pseudo-random sequence generator, which destroys the adaptive condition of the correlation analysis, and the security of the encryption algorithm is enhanced. The encryption principle is shown in Figure 6.4.

Encryption process is

$$Y = K \oplus X, \tag{6.1}$$

and decryption process is

$$X = K \oplus Y, \tag{6.2}$$

where K is the key stream generated by chaotic system. X is plaintext, and Y is ciphertext, and $\oplus$ is XOR operation.

It is a good encryption method by taking the chaotic sequence as key stream and using a one-time pad cryptosystem, but due to computer finite precision effect, chaotic sequence will eventually be periodic. How to produce a longer cycle, better random chaotic sequence is the key for an encryption algorithm, therefore, researchers have proposed a variety of schemes [131–133].

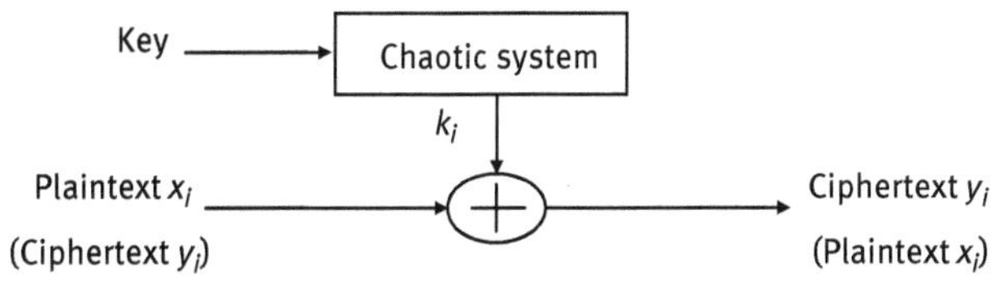

Figure 6.4: Block diagram of chaos sequence encryption/decryption.

The security of chaotic sequence encryption is mainly dependent on the key stream. In the chaotic encryption system, the random sequence $\{x_i\}$ generated by the chaotic system is taken as the key flow $\{k_i\}$, and it does XOR operation with the plaintext data flow $\{m_i\}$; thereby the encrypted data stream $\{c_i\}$ is generated. The plaintext data stream is binary, while there are two ways to produce the key stream $\{k_i\}$. One is to use the original chaotic sequence $\{x_i\}$ directly, and the other is to use the processed data of the sequence $\{x_i\}$. The former is actually the direct simulation of chaotic modulation communication technology, and the latter is the improvement of the former. There are a variety of processing methods for the original chaotic data by employing computer technology. For example, thousands of bytes at the beginning are discarded to overcome the transition process, or the sequence is obtained from the chaotic sequence according to a certain law, or a compound iterative algorithm is used to generate random sequences. So the randomness of the processed sequence will be greatly improved, which improves the anticracking ability of the encryption system.

The security performance of the encryption system is directly related to the size of the key space. For chaotic encryption system, many values can be used as a key, including nonlinear equations, initial values, system parameters, data-processing mode, and data encoding way. Obviously, the key space is huge, and the security is good.

6.2 Data Encryption Algorithm Based on Chaotic Sequence

6.2.1 Basic Requirements of Chaotic Encryption Algorithm

A good encryption algorithm not only can resist the brute-force attacks, but also has the ability of anti-plaintext attack, and superior temporal and spatial performance.

Resisting brute-force attack. The encryption system should have enough key space to resist the brute-force attack. Taking into account the extreme case, the user password is assumed to be a byte of all values, and then the number of possible values of the password is only 256. Assuming that a possible password needs only a second, it needs only 256 seconds to try all possible passwords. Obviously, the greater the password scope is, the more difficult the brute-force attack is.

Anti-chosen plaintext attack. Many cryptanalysts previously obtain the plaintext and ciphertext pairs through a variety of ways. Especially for data transmission on the network, it is easy to speculate a few hundred bytes of encrypted TCP data packets, which contain user name and password. Because a few beginning bytes of the data packets are often fixed, the encryption algorithm requires that it is still difficult to analyze the remaining plaintext and key, even when the cryptanalyst has a certain number of plaintext/ciphertext pairs.

Superior temporal and spatial performance. If an encryption algorithm consumes a lot of time, or consumes a large amount of computer memory, then it is not desirable. In fact, research results show that longer encryption time does not necessarily lead to greater strength of the encryption.

To ensure the security of the algorithm, we think that if a chaotic system is used in the encryption algorithm, it must be secure chaotic system.

Definition 6.1 Secure chaotic system: A chaotic system with discrete, zero correlation, mixed, robust and large parameter sets is defined as a secure chaotic system. It should be pointed out that for a chaotic encryption algorithm, the encryption system based on a secure chaotic system with the above properties is not necessarily safe, but when a chaotic system without the above properties is used for data encryption, the chaotic encryption system is bound to be weak. Next, we discuss it in detail.

1. **Discrete.** It means that the chaotic system should be a discrete map, and the phase trajectory is not continuous. If the phase trajectory is continuous, the evolution trend of the system will be implied by any small neighborhood of the track, which may reduce the security of the algorithm.
2. **Zero correlation.** It means that the cross-correlation of chaotic sequence is zero, and it is useful to resist the cipher correlation analysis. Zero correlation is associated with statistical properties of the system, and it indicates that the probability density of the system is uniform.
3. **Mixed.** If plaintext is considered to be the initial condition, the mixed means that the effect of a single plaintext symbol is spread to many ciphertext symbols. Obviously, this property corresponds to the diffusion property in cryptography. A chaotic system with mixed property has better statistics, and when iteration number increases, the statistics of ciphertext is not dependent on that of plaintext. The statistics of ciphertext cannot be obtained by that of plaintext.
4. **Robustness.** Robustness means that chaotic system can still keep the characteristics with small parameter perturbation. Most chaotic attractors are not stable, so a chaos encryption system based on the chaotic system without robustness has weak key property.
5. **Large parameter sets.** Chaotic system should have large parameters and initial value space, and a global chaotic system is the best choice. In addition, an important index of cryptographic system is the Shannon entropy, which is used to measure the key space. For discrete time chaotic system, it is approximately $\log_2 k$, where k is the number of keys. The larger the parameter space of the dynamical system is, the greater the k of discrete system is.

In summary, when we design a chaotic encryption system, the discrete chaotic system with robustness, zero correlation, mixed property, and large parameter set should be given priority. Although these five conditions cannot contain all security content, it is very high threshold for chaotic encryption. Strictly speaking, among the existing chaotic systems, no one is the safe chaotic system. Sheng et al. [88] proposed the elliptic chaotic system based on tangent delay, which has the above five conditions. However, this chaotic system is too complicated, and its mechanism also needs to be studied further. At present, Logistic map and its improvement have very good security features.

6.2.2 Logistic Map and Its Statistics

One-dimensional discrete time nonlinear system is defined as

$$x_{n+1} = T(x_n), \tag{6.3}$$

where $x_n \in V$ and $n = 0, 1, 2, 3, \ldots, x_n$ is the system state. T: $V \to V$ is a map. If the initial value is x_0, then we can get a sequence $\{x_n, n = 0, 1, 2, 3, \ldots\}$ by iteration operation.

In this section, Logistic map is employed to design an encryption algorithm, and its equation is

$$x_{n+1} = f(x_n) = \mu x_n(1 - x_n), \tag{6.4}$$

where $\mu \in (0, 4]$, $x_n \in (0, 1)$. The Lyapunov exponent of the system with the parameter varying is shown in Figure 3.2. When, $\mu \in (3.5699\ldots, 4]$, it is chaotic.

For an encryption communication, both the initial value sensitivity and randomness of the signal are important. The stronger sensitivity and better randomness can lead to higher security. These characteristics can be described by probability statistics, mean value, autocorrelation, and cross-correlation, as shown in Section 3.1.1.

For Logistic map, when $\mu = 4$, the mean value of the chaotic sequence is constant, and the autocorrelation is delta function and the cross-correlation is zero. Its probability statistics property is similar to that of the white noise with zero mean (as shown in Figure 6.5). So the Logistic map is suitable for information encryption. In addition, Logistic map basically satisfies the definition of the secure chaotic system. Therefore, Logistic map can be applied in digital communication and multimedia security.

6.2.3 Chaotic Encryption Model and Algorithm

To encrypt digital signal, we need to turn the real number sequence $\{x_k\}$ to pseudo-random sequence [134], which is taken as a key sequence. The simplest way is to obtain a new integer by selecting a few digits after the decimal point of the x_k. The encryption/decryption model based on Logistic map is shown in Figure 6.6.

Based on the model above, an encryption algorithm is designed, and the flow chart is shown in Figure 6.7. Taking into account the randomness and encryption speed, we chose the interval of pseudo-random sequence number $N = 5$. y_i is 4th,

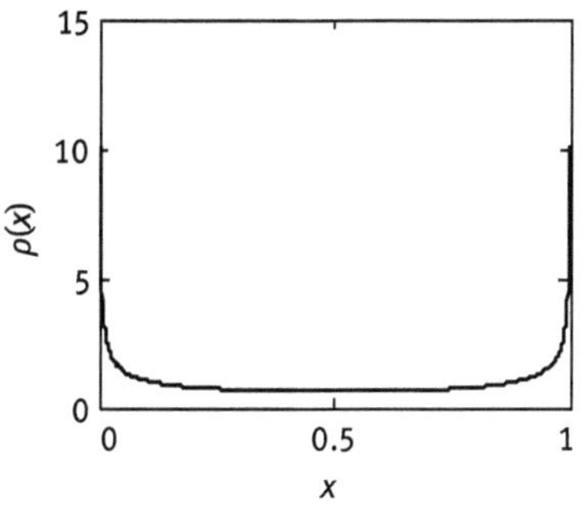

Figure 6.5: Probability density distribution of Logistic map.

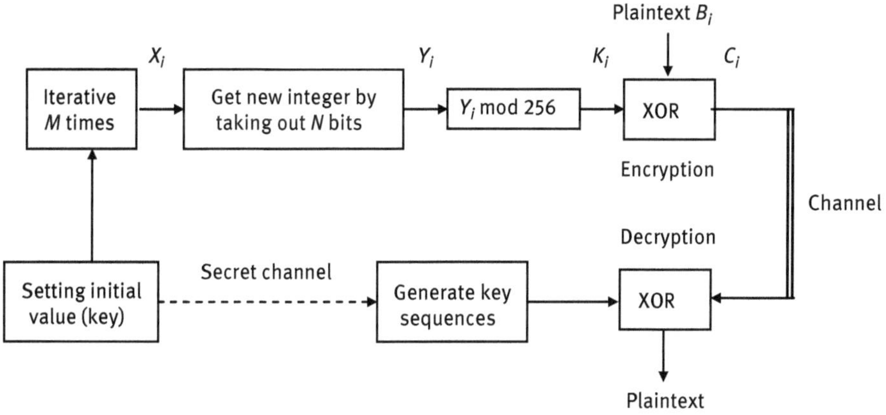

Figure 6.6: Chaotic encryption/decryption model.

5th, 6th, bit following the decimal point of x_i, which can improve the ability to resist plaintext attack. In order to decrease the influence of the transition process, it is appropriate to discard the previous 2,000 iterations. In addition, it is difficult for users to remember the password with two floating-point values, so the first step includes a process of mapping the strings remembered by the user to x_0 and μ. The encryption algorithm is presented as follows:

1. Initialize the Logistic map, and set the key x_0 and μ.
2. Iterate Logistic map for 2,000 times.
3. y_i is gotten by taking the 4th–6th bit of x_i.
4. Key sequence is obtained by y_i mod 256.
5. Do XOR operation with the plaintext or ciphertext.
6. Judge whether encryption/decryption is complete; if it is incomplete, repeat steps 3–5 after 5 iterations of Logistic map. Otherwise, exit encryption system.

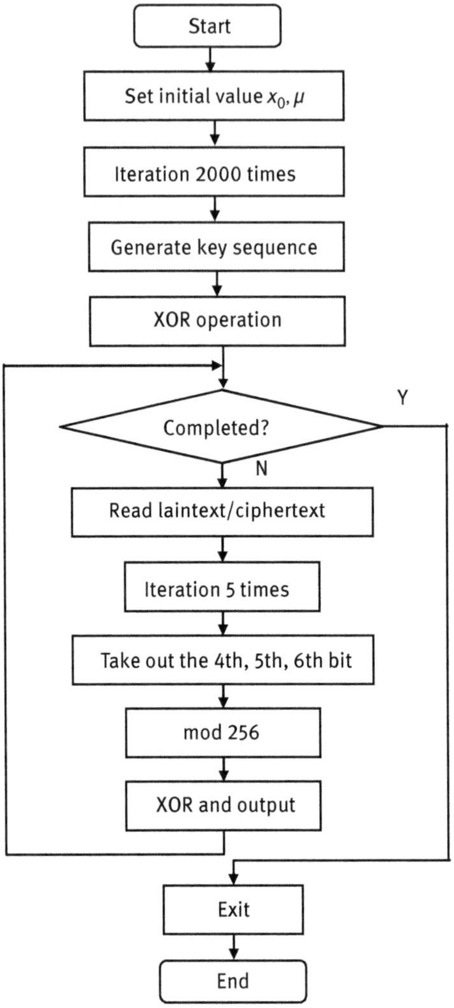

Figure 6.7: Flow chart of chaotic encryption/decryption algorithm.

6.2.4 Implement of Encryption Algorithm Based on Chaotic Sequence

Based on the proposed encryption model and encryption algorithm, we developed the data encryption software by using VC.Net in Visual Studio.NET tool. VS.Net is a complete tools for the generation of Web ASP applications, Web services XML, desktop applications, mobile applications, etc. Programming languages use the functions of the .NET framework, which provides access to key technologies to simplify the development of Web ASP applications and Web services XML. VS.Net employs MFC (Microsoft Foundation Classes) program method. It makes full use of the advantages

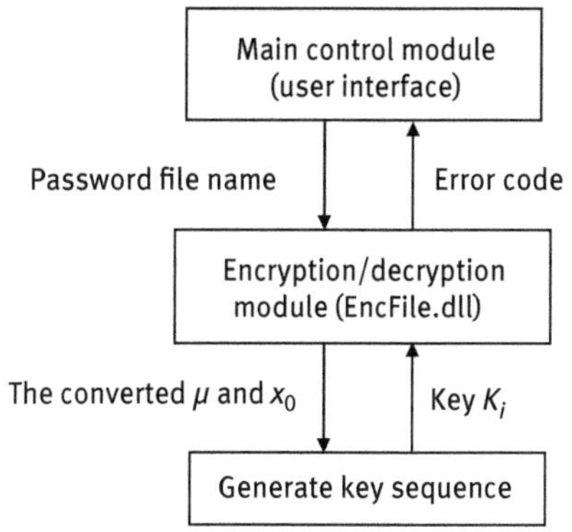

Figure 6.8: Relationship diagram of encryption software modules.

of oriented object technology for which there is little need to be concerned about the implementation details of an object when we are programming. At the same time, the strong function of the various objects in the class library can complete most of the program functions, and it effectively ensures that the program is good for debugging.

We use Microsoft Visual C++ to develop the core algorithm. The software size is controlled within 1.5M for facilitating users to download. This software is composed of the main control module, the chaotic key sequence generation module, and the file encryption/decryption module. The secret relationship among the modules is shown in Figure 6.8.

6.2.4.1 File Encryption/Decryption Module

This module provides three functions: TranslateKey, EncryptFile, and DencryptFile. The three interfaces provide a certain degree of abstraction to the main control module. The main control module does not need to be concerned about the details of the underlying file encryption, and it only needs to get the encrypted file and the user password. The specific encryption/decryption process is completed by the file encryption module.

The test procedure of the parameters is accomplished by the algorithm using these parameters, and several standard error codes are defined by the main control program. When an error occurs, the algorithm can return the standard error code and its specific error information code. The format of the error code is DWORD iError. The high sixteen bits are standard error code, which can be extracted by the HIWORD of the main control program, while the low 16 bits are custom error information codes. In addition, a standard procedure for obtaining the error description information from the error information code is also included in order to report to the user and realize the software's "self description."

After the standard encryption/decryption file method is provided, it is free to change the underlying encryption algorithm and will not affect the main control module, which makes the software version more flexible to upgrade.

6.2.4.2 Chaotic Sequence Generation Module

This module includes two functions: MapKey and GetNextKey. The former can set two parameters of the logistic map and check the parameters to ensure they are within reasonable range. The latter is used to get the next encryption key. This module is transparent to the main control module, and it is also the general module of chaotic encryption/decryption. It contains two global variables to record the parameters of logistic map set by the encryption/decryption module. One key is obtained by 5 iterations of the generating function. Experiments show that the design is very successful.

6.2.4.3 Main Control Module

It is the most challenging part in the whole design because the software does not contain too many user interaction processes, and users need not submit too much information. The software is suitable to design to be anti-traditional with relatively new interface form. At the same time, because this program is user oriented, and it costs a long time when large data files are encrypted, the multithread technology should be employed to separately implement the user interface process and the encryption process. The interface display thread can query the working state and progress of the underlying work thread at scheduled time and report them to the user.

On the Windows platform, the basic unit of interaction with the user is the application window, so we design a corresponding window in each interactive stage. This independent design for each state can make the process of design and implementation relatively intuitive, and maintenance and optimization of the process will become easier.

The interaction between the users and the main control module includes the following several aspects: the users start the application; the users choose the file encryption or the file decryption; the users select the plain text file and the cipher text file; the users enter the password; and the encryption/decryption process is completed.

The window procedure mentioned above can be described by a state transition diagram as shown in Figure 6.9. Each state can be represented by a window, and each state conversion can be achieved with displaying or hiding the window. For the three groups, key input 1 and key input 2, file input 1 and file input 2, as well as the encryption progress 1 and the encryption progress 2, each group can be achieved with the same window, and we only need to record the current state. To display and hide the window, two functions (FadeIn() and FadeOut()) are added to each window class. FadeIn() is used to display the window, while FadeOut() is used to hide the window. The window fade and other special effects can be achieved by modifying the two standard processes.

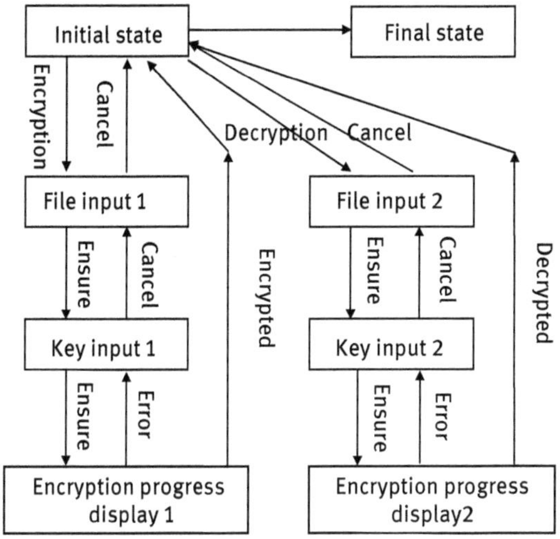

Figure 6.9: State transition diagram of the main control module.

6.2.5 System Performance Analysis

6.2.5.1 Analysis of Anticracking Ability

The key of cryptanalysis is to obtain the initial value x_0 or parameter μ set by the user. The two values will be taken as floating-point number. If the effective value of a computer floating-point number is 16, then 30 bits are uncertain for the two numbers, and the possible combinations are 10^{30}, while key combination of DES encryption algorithm is 2^{56}. Obviously $30 > 56 \log 2$. So the proposed algorithm can effectively resist brute-force attack.

To improve the adaptability of chosen plaintext attack, we use interval sampling method. If the interval number is one, then the relationship between x_n and x_{n+1} is

$$z(i) = (x_1(i) + x_2(i)) \bmod 2^{32}. \tag{6.5}$$

It is a quadratic equation with one unknown. If the interval number is N, then the relationship between x_n and x_{n+1} becomes one-dimensional equation of N degrees.

6.2.5.2 Statistics Distribution of Ciphertext

Whether the statistics distribution of encrypted data is uniform is an important index to evaluate the encryption algorithm.

Theorem 6.1: If the chaotic sequence is completely random, the encrypted data is subject to the uniform distribution.

Proof. The plaintext and chaotic (key) sequence is considered as a byte of data stream, and we take a byte of plaintext m and key k randomly from them. Supposing that the value of m is different, that is, the probability is unequal for one bit to be "0" or "1," then the appearance of "0" or "1" is independent. If the probability of "1" is p, then the probability of "0" is $1–p$. The chaotic (key) sequence satisfies the performance of white noise for the ideal state; that is, "0" or "1" appearing at each bit has the same probability of 0.5, and then the probability of appearing "1" in the ith bit of ciphertext c is

$$P_{(Ci=1)} = P_{(mi=1,ki=0)} + P_{(mi=0,ki=1)} = p \times 0.5 + (1-p) \times 0.5 = 0.5. \quad (6.6)$$

The probability of appearing "0" in the ith bit of ciphertext c is

$$P_{(ci=0)} = 1 - P_{(ci=1)} = 0.5. \quad (6.7)$$

Obviously, the probability distribution of the ciphertext is uniform. ∎

6.2.5.3 Experiment Results

The chaotic encryption software is designed by employing modular design method and multithread technology. If both communication sides have the same key and the encryption software, they can realize the encryption transmission of the information.

The software can encrypt not only the text file, but also the bitmap file. Figure 6.10 is the result of a bitmap encryption. Figure 6.10(a) is the plaintext image. Figure 6.10(b) is the encrypted image. Figure 6.10(c) is the decrypted image. Compared with Figure 6.10(a), Figure 6.10(c) is clear and accurate without distortion. Figure 6.10(d) is the decryption image with small deviation of key, and obviously it cannot be decrypted correctly.

Data encryption standard (DES) is a relatively good symmetric encryption algorithm. Rivest-Shamir-Adleman (RSA) proposed by Rivest, Shamir, and Ademanis is an outstanding representative of the public key algorithm. Now, the chaotic encryption algorithm is compared with the DES and RSA encryption algorithm by encrypting/decrypting Lena image, and the results are shown in Table 6.1. Obviously, chaotic encryption has advantages in encryption speed and security.

It should be pointed out that the software is mainly suitable for static data encryption. In order to realize the real-time data stream encryption, we must solve the problems, such as fault tolerance, synchronization, delay, and so on, which will be researched in the seventh chapter.

6.3 Binary Watermark Image Encryption Algorithm Based on Chaos

Chaotic sequence is a kind of pseudo-random sequence. It has a simple form and is very sensitive to initial conditions and has statistical properties of white noise. These

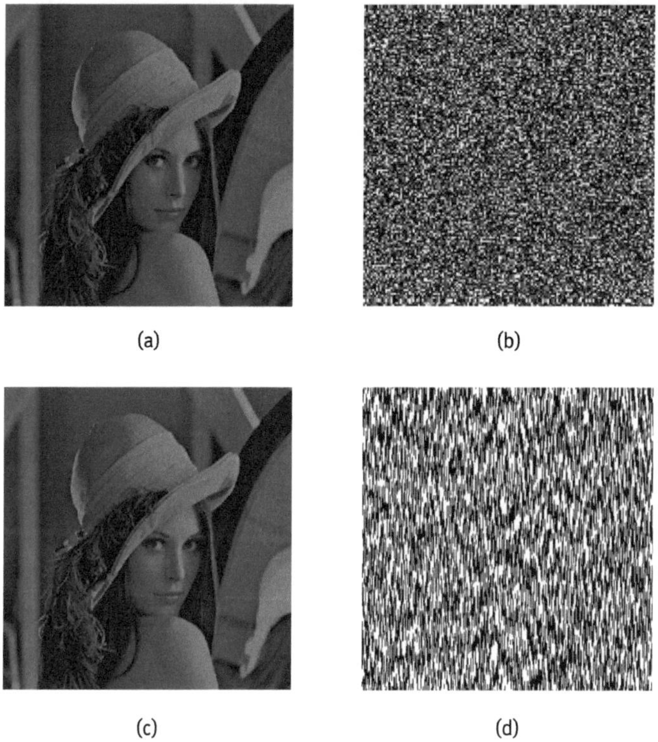

Figure 6.10: Image encryption results: (a) original image; (b) encrypted image; (c) correct decryption image; and (d) decryption image with key change for 10^{-14}.

Table 6.1: Performances of DES, RSA, and chaotic encryption algorithm.

Algorithm	Encryption/ decryption time (s)	Advantages	Defects
DES	0.12/0.15	Implementation by software and hardware; complexity of encryption/decryption is same	Small key space; insecurity
RSA	92/120	High security; large key space	Slow encryption speed; decryption time is longer than encryption time; used in digital signature with small data
Chaos	1.9/2.1	Implementation by software and hardware; complexity of encryption/decryption is same; large key space	Immature algorithm; further study about security

characteristics can meet the requirements of the security of digital watermarking technology. Therefore, chaos-based digital watermarking is becoming a hot topic.

6.3.1 Improvement of Chaotic Pseudo-Random Sequence Generator

Aiming at the defects of various chaotic sequence generators, many researchers have put forward the improvement schemes, which mainly include the following two methods.

6.3.1.1 Improvement of Traditional Chaotic Sequence Generator

This method focuses on improving the traditional chaotic sequence implementation algorithm, and it includes two types. One is to improve the single traditional chaotic sequence, such as piecewise linear chaotic sequence, Chebyshev chaotic sequence, and so on. In order to enhance security, we can consider using multiple chaotic systems. For example, the output of the two chaotic systems can be combined to generate pseudo-random bit stream, such as the combination of logistic map and tent map. In addition, the new algorithm for generating uniform pseudo-random number is studied [135], and spatiotemporal chaotic system is tried to generate pseudo-random numbers [136].

6.3.1.2 New Secure Chaotic System

At present, along with the emergence of the improved algorithm of the traditional chaotic sequences, some new chaotic sequence generators have been put forward. Among them, the TD-ERCS chaotic system has good cryptographic properties due to its discreteness, global chaos, global zero correlation, huge parameter and initial value space, and strong anti-degradation ability [137, 138]. So TD-ERCS chaotic system is a chaotic system with great potential application.

6.3.2 Binary Watermark Image Encryption Algorithm Based on TD-ERCS Map

Binary image processing algorithm is simple and easy to understand and realize, and has fast calculation speed and other features. A lot of digital watermarking system uses binary image as the watermark information, so the watermark is more intuitive. Here, TD-ERCS chaotic map is used to encrypt the watermark image.

6.3.2.1 TD-ERCS Map

TD-ERCS map is a new global two-dimensional discrete chaotic system. It is an ideal model for information encryption with zero correlation, stable probability distribution, and good randomness of the sequence. Set the system parameter $\mu(0 < \mu < 1)$, initial values $x_0(-1 \le x_0 \le 1)$ and $\alpha(0 < \alpha < \pi)$, tangent delay $m(m = 2, 3, 4, 5, 6, \ldots)$. TD-ERCS map is described in Section 3.1.4. The system can generate two independent sequences x_n ($|x_n| < 1$) and k_n ($|k_n| < 1$).

6.3.2.2 Design of Encryption Algorithm

Double-precision floating-point number is used in TD-ERCS iteration, and the data format is shown in Figure 6.11. The data occupies 8 bytes of storage space, and it is composed of the following three parts:

1. Sign bit s. s represents the sign of the number; it is one bit by which 0 means positive, and 1 means negative.
2. Exponent bit e. e represents exponential order code, and it is 11 bits. It is unsigned number ($0 < e < 2047$).
3. Tail bit f. f is the fractional part of the tail number, 52 bits.

If W is a binary watermark image, and the pixel size is $M \times N$, the encryption algorithm is as follows:

1. Taking TD-ERCS system parameters (μ, x_0, α, m) as the key, chaotic sequence is generated by eq. (3.17). Sequence $\{x_n\}$ is composed of $\lceil M \times N/52 \rceil$ numbers generated by the chaotic sequence (Considering security, the initial segment of the sequence is not used).
2. Binary sequence $\{b_i\}$ is obtained from f, $1 \le i \le M \times N$.
3. The binary watermark image is encrypted with the binary sequence $\{b_i\}$; encryption operation is

$$W' = W(i) \oplus b(i) \qquad 1 \le i \le M \times N, \tag{6.8}$$

   where $\oplus$ is XOR operation. At the receiving end, input the same key, and by repeating the encryption algorithm, the image can be decrypted correctly.

6.3.3 Performance Analysis of Encryption Algorithm

The security of encryption algorithm mainly depends on the key, while the key has a great relationship with the key space, the key sensitivity, and the complexity of the algorithm. Here we discuss the security of the algorithms.

6.3.3.1 Key Sensitivity Analysis

To confirm the effectiveness of the algorithm, a watermark image "中大CSU" with the size of 64×64 is selected. The key is set as $\mu = 0.7123$, $x_0 = 0.7654$, $\alpha = 1,000$, and $m = 9$. The encryption effect is shown in Figure 6.12. Figure 6.12(a) is the original

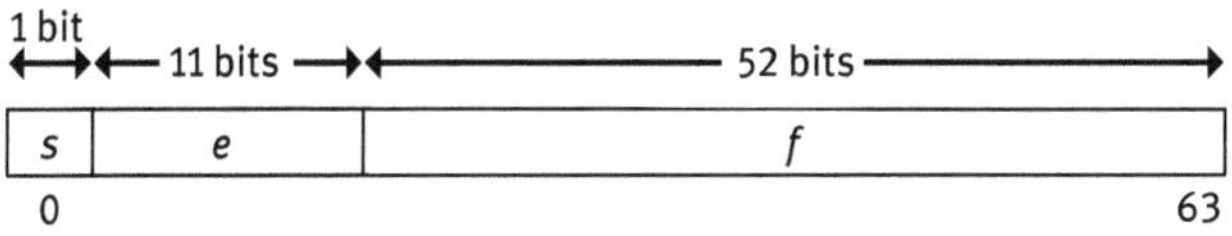

Figure 6.11: IEEE 754 double-precision binary floating-point format.

(a) (b) (c)

Figure 6.12: Image encryption and decryption: (a) original image; (b) encrypted image; and (c) decrypted image with correct key.

(a) (b) (c) (d)

Figure 6.13: Decrypted image with wrong keys: (a) μ deviation with 10^{-16}; (b) x_0 deviation with 10^{-16}; (c) $\alpha = 1,001$; and (d) $m = 10$.

image. Figure 6.12(b) is the encrypted image, and Figure 6.12(c) is the decrypted image with correct key. Figure 6.13 shows the sensitivity to the key. For decryption, the original watermark images are unable to be decrypted with small deviation of the key.

It can be seen from the experiment results that this algorithm is very sensitive to the key. If a slightly modified key is used to decrypt the cipher image, the decryption fails completely. The key contains four dependent parameters: μ, x_0, α, m. μ, and x_0 are in real numbers. So the algorithm is secure with large key space.

6.3.3.2 Scrambling Degree Analysis

The gray difference between pixels and their adjacent pixels is defined as

$$GD(x,y) = \frac{\sum_{i,j} \left| G(x,y) - G(x_i', y_j') \right|}{4}, \tag{6.9}$$

where $G(x, y)$ is the gray value of (x, y). $G(x_i', y_j')$ is the gray value of (x, y) at the four positions: up and down, and left and right.

If the size of the image is $M \times N$, the average adjacent pixel difference of the image is described as

$$E(GD(x,y)) = \frac{\sum\limits_{x=2}^{M-1}\sum\limits_{y=2}^{N-1} GD(x,y)}{(M-2)\times(N-2)}, \tag{6.10}$$

and adjacent pixel scrambling degree is defined by

$$GDD(I,I') = \frac{E'(GD(x,y)) - E(GD(x,y))}{E'(GD(x,y)) + E(GD(x,y))}, \tag{6.11}$$

where E and E' are the average adjacent pixel difference before and after scrambling, respectively. The range of GDD is (−1, 1). The larger the range is, the better the image scrambling is.

Figure 6.14 presents three binary images, and the sizes of them are 64×64. The evaluation of adjacent pixel scrambling is listed in Table 6.2.

$E(GD(x,y))$ is the average adjacent pixel difference of original image I. The value of $E(GD(x,y))$ is about 0.1, which shows that the original image has large pixel values in close smooth area, and the average gray difference of adjacent pixels is small. $E'(GD(x,y))$ is the average adjacent pixel difference of encrypted image I'. In Table 6.2, E' $(GD(x,y))$ is greatly different from E $(GD(x,y))$, which indicates each pixel tends to have random distribution, and the average adjacent pixel difference varies greatly. GDD (I,I') is larger than 0.5, which illustrates that the adjacent pixel scrambling degree is bigger, and the encryption effect is very well.

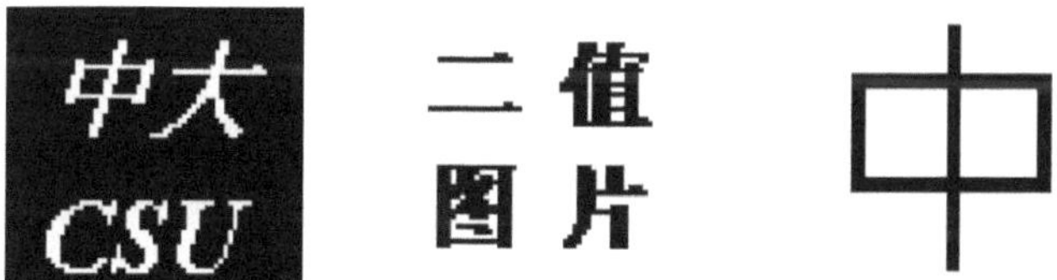

(a) (b) (c)

Figure 6.14: Binary images: (a) image a; (b) image b; and (c) image c.

Table 6.2: Evaluation of adjacent pixel scrambling.

	Image a	Image b	Image c
$E(GD(x,y))$	0.0731	0.0695	0.0419
$E'(GD(x,y))$	0.4750	0.4779	0.4690
$GDD(I,I')$	0.7333	0.7461	0.8360

6.3.3.3 Analysis of Anti-Brute-Force Attack

The user password is binary sequence $K = k_\mu k_{x_0} k_\alpha k_m$. k_μ, k_{x_0}, and k_α have the same data length. They are used to generate floating-point numbers μ, x_0, and α. k_m is used to generate integer m. Let the length of k_m be 12 bits, and then m is generated by

$$m = \begin{cases} 2 & (k_m = 0) \\ 4 & (k_m = 1) \\ k_m + 3 & (2 \le k_m \le 4095) \end{cases} . \quad (6.12)$$

If the computer word is 32 bits, and the precision (floating-point number) is 10^{-9}, then the key space is about $10^{27} \times 2^{12} \approx 10^{30}$, which is large enough to resist brute-force attacks.

The results of experiments and performance analysis show that the binary image watermarking encryption algorithm based on the TD-ERCS map is good with large key space, high key sensitivity, and high security. Therefore, if the encryption algorithm is applied to the watermarking algorithm, it can ensure the security of the watermark.

6.4 Color Image Watermarking Algorithm Based on Chaotic Map and Wavelet Transform

As a new technology of copyright protection and content authentication, digital watermarking has become a hot topic. The application of chaos in digital watermarking technology mainly includes two aspects. One is the pretreatment of watermark information, such as the following: (1) digital watermark information is embedded into the image directly, (2) the watermark information is encrypted by chaotic sequence, and (3) binary image watermarking is encrypted or scrambled. The other is the watermark embedding process encryption, such as the choice of the position, intensity, etc.

In this section, a novel color image watermark algorithm based on discrete chaotic map and 2-D discrete wavelet transform is proposed. The proposed algorithm uses a meaningful binary image as the watermark, which increases the watermark provability. TD-ERCS discrete chaotic system and the commonly generalized Arnold map are used to encrypt the watermark image and determine the embedding position, which improves the security of the watermark image. The watermark is embedded by wavelet transform. Simulation experiments and performance analysis show that the proposed algorithm not only can resist some common image processing and noise interference, but also can resist the image shear, small rotation, histogram modification, and joint attack with good invisibility, robustness, and security.

6.4.1 Wavelet Transformation Algorithm and Its Application

Wavelet transform is an ideal mathematical tool for local analysis, which has been widely used in signal analysis, speech synthesis, image recognition, computer vision, data compression, seismic exploration, and atmosphere and ocean analysis.

6.4.1.1 Continuous Wavelet Transform

If $\psi(x)$ is a real-valued function, and its spectrum $\hat{\psi}(w)$ satisfies the condition

$$C_{\psi} = \int_{-\infty}^{\infty} \frac{\left|\hat{\psi}(w)\right|^2}{|w|} dw < \infty, \tag{6.13}$$

then $\psi(x)$ is called a basic wavelet. $\psi(x)$ is similar to impulse response of a band-pass filter. Assuming that a and b denote expansion and translation operation of $\psi(x)$, respectively, then by the translation and expansion of the basic wavelet, the wavelet basis functions can be generated as

$$\psi_{a,b}(x) = \frac{1}{\sqrt{a}} \psi(\frac{x-b}{a}), \tag{6.14}$$

where $a \in R$, $b \in R$, $a > 0$.

If $f(x)$ is a square integrable function, and $f(x) \in L^2(R)$, then the continuous wavelet transform of $f(x)$ is defined as

$$W_f(a, b) = \langle f, \psi_{a,b}(x) \rangle = \int_{-\infty}^{\infty} f(x)\psi_{a,b}(x)dx. \tag{6.15}$$

It indicates the decomposition of $f(x)$ on function $W_f(a, b)$, and it also can be considered as $f(x)$, which is filtered by using a band-pass filter. The wavelet inverse transform definition is

$$f(x) = \frac{1}{C_{\psi}} \int_{0}^{\infty} \int_{-\infty}^{\infty} w_f(a, b)\psi_{a,b}(x)db\frac{da}{a^2}. \tag{6.16}$$

It should be noted that $\psi_{a,b}(x)$ is not orthogonal, and it has very large redundancy. In addition, the basic wavelet is not unique.

6.4.1.2 Two-Dimensional Discrete Wavelet Transform of Image

Digital image is a discrete two-dimensional signal. The common method for two-dimensional discrete wavelet transform is to let the two-dimensional scaling function be separated; that is,

$$\phi(x, y) = \phi(x)\phi(y), \tag{6.17}$$

where $\phi(x)$ is one-dimensional scaling function. If $\psi(x)$ is the corresponding wavelet, then three two-dimensional basic wavelets are

$$\begin{cases} \psi^1(x,y) = \phi(y)\psi(x) \\ \psi^2(x,y) = \psi(x)\phi(y) \\ \psi^3(x,y) = \psi(x)\psi(y) \end{cases}. \tag{6.18}$$

The scaling function $\phi(x)$ can be treated as a low-pass filter, while the wavelet function $\psi(x)$ is the high-pass filter of the same layer, and the function set is

$$\left\{\psi^l_{j,m,n}(x,y)\right\} = \left\{2^j\psi^l(x-2^jm, y-2^jn)\right\}, \quad j \geq 0, l = 1,2,3. \tag{6.19}$$

It is the standard orthogonal on the basis of $L^2(R^2)$, and j, l, m, and n are integers.

$f_0(x,y)$ is an image with the size $N \times N$, and N is the power of 2. j is resolution parameter, and the scale is 2^j. When $j = 0$, the scale of the original image is 1, and it is a resolution signal for zero*th* layer. The resolution reduces by halve with the value j increasing by 1.

The two-dimensional discrete wavelet transform of an image is carried out as follows. In each layer of the transformation, the image is decomposed into four small images with same size. Each of them is obtained by the inner product of original image with a basic wavelet and 2 times interval sampling in x and y directions. The first layer ($j = 1$) wavelet transform is

$$\begin{cases} f_1^0(m,n) =< f_0(x,y), \phi(x-2m, y-2n) > \\ f_1^1(m,n) =< f_0(x,y), \psi^1(x-2m, y-2n) > \\ f_1^2(m,n) =< f_0(x,y), \psi^2(x-2m, y-2n) > \\ f_1^3(m,n) =< f_0(x,y), \psi^3(x-2m, y-2n) > \end{cases}. \tag{6.20}$$

For the transformation of subsequent levels ($j > 1$), four same interval samplings and filter operation are performed at each layer. $f_j^0(x,y)$ is decomposed into four smaller images on the scale of 2^{j+1} in the same way. The inner product in the form of convolution is

$$\begin{cases} f_{j+1}^0(m,n) = \left\{ [f_j^0(x,y)\phi(-x,-y)](2m,2n) \right\} \\ f_{j+1}^1(m,n) = \left\{ [f_j^0(x,y)\psi^1(-x,-y)](2m,2n) \right\} \\ f_{j+1}^2(m,n) = \left\{ [f_j^0(x,y)\psi^2(-x,-y)](2m,2n) \right\} \\ f_{j+1}^3(m,n) = \left\{ [f_j^0(x,y)\psi^3(-x,-y)](2m,2n) \right\} \end{cases}. \tag{6.21}$$

Because the scaling function and the wavelet function are separable, each convolution can be decomposed into one-dimensional convolution on the row and column

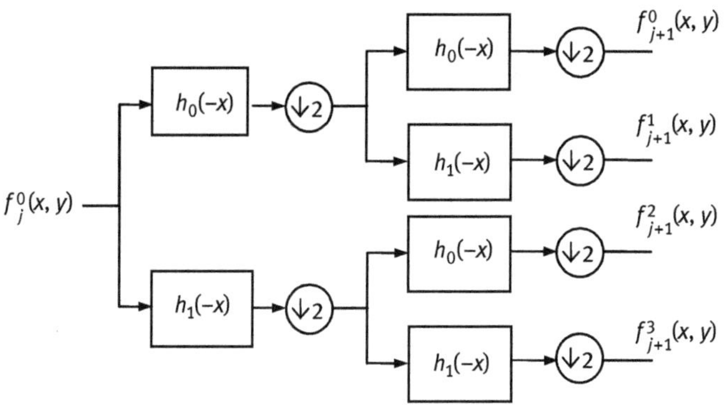

Figure 6.15: Diagram of wavelet decomposition.

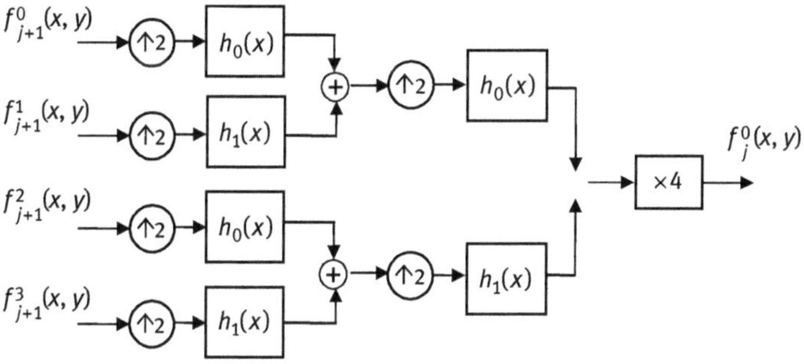

Figure 6.16: Diagram of inverse wavelet transform.

of $f_j^0(x,y)$, as shown in Figure 6.15. h_0 is the low-pass filter, and h_1 is the high-pass filter. ↓2 is two times sampling, and $f_{j+1}^0(x,y)$ is the low-frequency components after wavelet decomposition.$f_{j+1}^1(x,y)$, $f_{j+1}^2(x,y)$, and $f_{j+1}^3(x,y)$ are the high-frequency detail components of horizontal, vertical, and diagonal directions, respectively.

Obviously, the wavelet transform of the image is to filter the image in the horizontal and vertical directions by using the filter module. At each level, $f_j^0(x,y)$ contains the low-frequency approximation components of the previous layer. $f_j^1(m,n)$, $f_j^2(m,n)$, and $f_j^3(m,n)$ contain the high-frequency detail components of horizontal, vertical, and diagonal directions, respectively.

Similarly, the two-dimensional inverse wavelet transform of the image is shown in Figure 6.16, where ⊕ represents convolution.

6.4.1.3 Application of Wavelet Transform for Image Watermarking

The original image, three-level wavelet decomposition diagram, and structure diagram are shown in Figure 6.17. Three high-frequency bands HL_i, LH_i, and $HH_i(i = 1,2,3)$ and a low-frequency band LL_3 are obtained by three-level wavelet decomposition of the original image. According to the characteristics of image wavelet

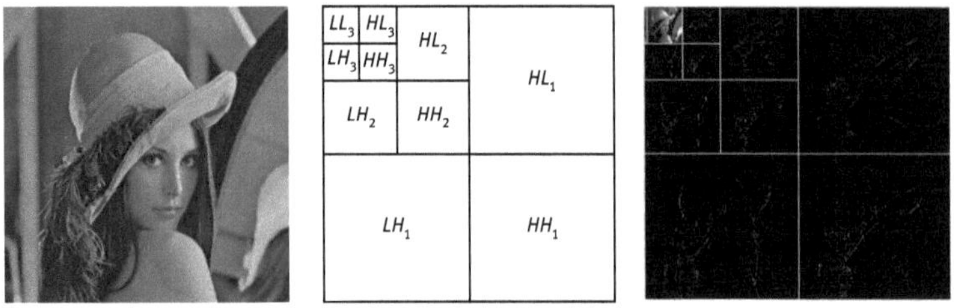

(a) (b) (c)

Figure 6.17: Three-level wavelet decomposition of image; (a) original image; (b) three-level wavelet decomposition; and (c) three-level wavelet decomposition structure.

decomposition, the low-frequency band LL_3 is the best approximation of the original image under the largest scale and minimum resolution decided by wavelet transform decomposition series. Most energy of the image is centered in the low-frequency band. High-frequency band series are the details of image at different scales and different resolution. These bands are not unrelated from the perspective of multiresolution. For each high-frequency band, they are described from being coarse to fine for the same edge, contour and texture information of an image in different directions, different scales, and different resolutions, and there are some relations between them.

Compared with other transforms, the image watermarking algorithm based on wavelet transform has many good characteristics, such as spatial scale localization, multiresolution representation, adaptation to human visual system, different spatial support region, large selectivity, meeting the demand of security, and simple calculation.

6.4.2 Design and Implementation of Watermark Embedding and Extraction Algorithm

6.4.2.1 Principle of Watermarking Algorithm

The principle of watermark algorithm is shown in Figure 6.18. To improve the security of the embedded watermark image, the chaotic sequence generated by TD-ERCS discrete chaotic system is used to encrypt the watermark image. According to the visual characteristics of the human eyes, the human eyes are not sensitive to the blue component of the RGB image, and thus the watermark is embedded in the blue component of the color image.

Based on the advantages of embedded watermark of high- and low-frequency parts in the wavelet domain, the encrypted watermark information frequency is embedded into the four sub-bands of the blue component of the host image. The embedding position is determined by the chaotic sequence generated by the

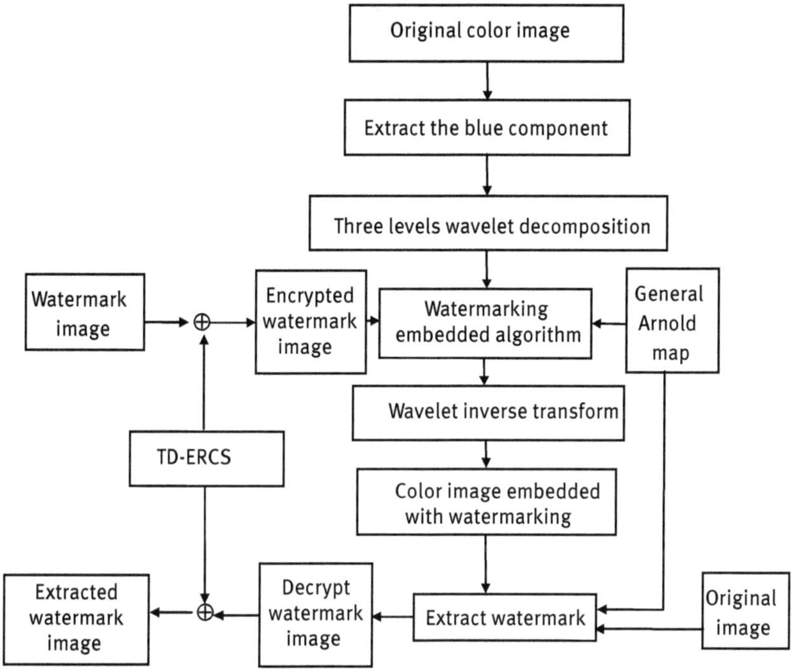

Figure 6.18: Block diagram of watermark embedding and extraction.

generalized Arnold map. Watermark information is spread to the low-frequency and high-frequency coefficients of the three-level decomposition. Because the original image is needed for watermark extraction, it belongs to the nonblind extraction.

6.4.2.2 Choice of Watermark Embedding Position

To embed the watermark into the color image, the first question is to embed the watermark into which component of the color image. There are many color image models, including the RGB model, YUV model, YIQ model, YCbCr model, etc. The RGB model is the most commonly used. A pixel of true color image has three bytes of R, G, and B data structures, which represent the size of red, green, and blue. Bmp image is presented with the RGB method. Based on the brightness equation, gray image transformed from color image is defined by

$$I_{\text{grey}} = 0.299 I_R + 0.587 I_G + 0.114 I_B. \tag{6.22}$$

According to eq. (6.22), the watermark is embedded into the blue component of the color image.

After the blue component of the color image is decomposed by three-level wavelet, a low-frequency band and nine high-frequency bands are obtained. The watermark can be embedded into the low-frequency part or the high-frequency part of the wavelet domain [139]. Here, the encrypted watermark is randomly embedded into the four subbands of three-level wavelet decomposition in the blue component. The watermark is spread to the deepest frequency and high-frequency coefficients of the three level-wavelet decomposition.

6.4.2.3 Selection of Wavelet Base Function

In the design of digital watermarking algorithm based on wavelet transform, the first question is how to choose proper wavelet base function to ensure the robustness and invisibility of the watermarking system. The wavelet base function determines the efficiency and effect of wavelet transform. Different wavelet has different properties, which will lead to the different distribution characteristics and the energy convergence characteristics of wavelet coefficients, and the robustness of the watermark is also different.

At present, biorthogonal wavelet is widely used in the image processing due to the characteristics of regularity, compactness, and symmetry. The research shows that the Harr wavelet has good performance, and it is a good choice in the digital watermarking algorithm. The Harr wavelet is defined as

$$\psi(t) = \begin{cases} 1, & 0 \le t < 0.5 \\ -1, & 0.5 \le t \le 1 \\ 0, & \text{others} \end{cases} . \tag{6.23}$$

Compared to other orthogonal wavelet bases, the Harr wavelet has some advantages as follows:

1. The support length is the shortest.
2. The computation complexity of decomposition and reconstruction is lower than that of other wavelets, and it results in the rapid implementation.
3. When the Mallat algorithm is used, the boundary extension is not required. There is no edge effect due to the limited length of the image signal.
4. When the image is divided into several mutual disjoint image blocks with the same size, each image block has a corresponding independent wavelet block, and these wavelet blocks do not intersect each other.

Based on the discussion above, we select Haar wavelet as the wavelet basis function.

6.4.2.4 General Arnold Map

Arnold map is described as

$$\begin{cases} x_{n+1} = (x_n + y_n) \bmod 1 \\ y_{n+1} = (x_n + 2y_n) \bmod 1 \end{cases} , \tag{6.24}$$

where x mod 1 means only the decimal part of x is obtained. The phase space of (x_n, y_n) is limited in $[0, 1] \times [0, 1]$. The matrix form of eq. (6.24) is

$$\begin{pmatrix} x_{n+1} \\ y_{n+1} \end{pmatrix} = \begin{pmatrix} 1 & 1 \\ 1 & 2 \end{pmatrix} \begin{pmatrix} x_n \\ y_n \end{pmatrix} \mathrm{mod}\, 1 = C \begin{pmatrix} x_n \\ y_n \end{pmatrix} \mathrm{mod}\, 1, \tag{6.25}$$

where C is a matrix, and $|C| = 1$. So Arnold map is a guaranteed area map. At the same time, the Arnold map is the one-to-one map. That is, each point in the unit matrix uniquely transforms to the other point within the unit matrix.

Now, we generalize the phase space of the Arnold map to be $\{0, 1, 2, \ldots, M-1\} \times \{0, 1, 2, \ldots, M-1\}$ and generalize the equation to be two-dimensional reversible area preserving equation, which is described as

$$\begin{pmatrix} x_{n+1} \\ y_{n+1} \end{pmatrix} = \begin{pmatrix} a & b \\ c & d \end{pmatrix} \begin{pmatrix} x_n \\ y_n \end{pmatrix} = C \begin{pmatrix} x_n \\ y_n \end{pmatrix} \mathrm{mod}\ M, \tag{6.26}$$

where a, b, c, and d are positive integers, and the area preserving requires $|C| = ad-bc = 1$. Based on the condition, only three parameters of a, b, c, and d are independent.

The generalized Arnold map is still chaotic. We use the generalized Arnold map to generate the embedding position of the watermark. Taking the pixel coordinates (x, y) of the watermark image as initial values, and taking the three independent parameters and iteration number k of the coefficient matrix C as the key, the generated iterative results (x', y') are used as the embedding position of the watermark in (x, y). When iteration number is large enough, the corresponding embedding position will be greatly separated for any two adjacent watermark pixels.

6.4.2.5 Watermark Embedding and Extracting Algorithm

H is the original host image with the size $N \times N$, and W is binary watermark image with the size $M \times M$. The watermark embedding algorithm is described as follows.

1. Blue component is extracted from the original color image. Wavelet decomposition coefficient $d_l^\theta(i,j)$ is obtained by the three-level wavelet decomposition of the blue component. For $d_l^\theta(i,j)$, l is wavelet decomposition level ($l = 1, 2, 3$), and $\theta = 0, 1, 2, 3$ represent the wavelet coefficients of the low-frequency, horizontal, vertical, and diagonal directions, respectively. A new matrix with size $(N/4) \times (N/4)$ is got by merging the third-level coefficient matrix of the low-frequency, horizontal, vertical, and diagonal directions, and it is marked as $F = \{f(x,y), 1 \le x \le N/4, 1 \le y \le N/4\}$.
2. The watermark image W is encrypted by encryption algorithm based on TD-ERCS map. The encrypted watermark image is denoted by $W' = \{w'(i,j),\ 1 \le i \le M, 1 \le j \le M\}$.
3. Set the three independent parameters of generalized Arnold map and iteration numbers k.
4. For the watermark pixel $w'(i,j)$, let $x_0 = i$, $y_0 = j$. The embedding position (p, q) in the matrix F is obtained by iteration k times of eq. (6.26).

5. Embed watermark. The watermark pixels are embedded into the selected wavelet coefficients according to

$$f'(p,q) = \begin{cases} f(p,q) + \alpha \cdot w'(i,j), & \theta = 0 \\ f(p,q) + \beta \cdot w'(i,j), & \theta = 1,2,3 \end{cases}, \tag{6.27}$$

   where $f(p,\ q)$ is the original wavelet coefficient. $f'(p,q)$ is the wavelet coefficient after embedding watermark. $\theta = 0,1,2,3$ represent wavelet coefficients of low-frequency, horizontal, vertical, and diagonal directions, respectively. α and β indicate the intensity of embedding watermark, which is determined by the specific watermark and the original image, and it needs trade-off between invisibility and robustness of the watermark.
6. Repeat steps (4) and (5) until all the watermark pixels are embedded. After the three-level wavelet inverse transform is carried out by employing all the wavelet coefficients, the blue component image with embedded watermark is obtained. Finally the color image embedded with watermark is obtained.

The watermark extraction algorithm is described as follows.

H^* is the image waiting for detecting. First, we extract blue component of H^*, and then the blue component is decomposed by three-level wavelet. A new matrix marked as $F^* = \{f(x,y)\}$ (where $1 \le x \le N/4$, $1 \le y \le N/4$) is obtained by merging the coefficient matrix of the low-frequency, horizontal, vertical, and diagonal directions in the third level. Refer to steps (3)–(5) of the watermark embedding algorithm; the embedding position of the watermark is obtained by using the same key. The formula for extracting watermark is

$$w'_m = \begin{cases} 1 & \theta = 0\ ,\ f^*(p,q) - f(p,q) \ge \alpha/2 \\ 0 & \theta = 0\ ,\ f^*(p,q) - f(p,q) < \alpha/2 \\ 1 & \theta = 1,2,3\ ,\ f^*(p,q) - f(p,q) \ge \beta/2 \\ 0 & \theta = 1,2,3\ ,\ f^*(p,q) - f(p,q) < \beta/2 \end{cases}, \tag{6.28}$$

where w'_m is the mth extraction watermark pixel. $f(p,\ q)$ is the wavelet coefficients of the original image. $\theta = 0,1,2,3$ represent wavelet coefficients of low-frequency, horizontal, vertical, and diagonal directions, respectively. α, β indicate the intensity of embedding watermark. After extracting the watermark information, it is decrypted by the TD-ERCS map and the correct watermark is recovered.

6.4.3 Simulation and Performance Analysis

Based on Matlab platform, the watermark embedding and extraction is carried out on 24 bits true color Lena image with the size 512×512. The watermark image is the binary image "*中大CSU*" with the size 64 × 64. The embedding strength is $\alpha = 41$ and $\beta = 59$.

The performances of the algorithm are analyzed, including invisibility, robustness, and security. To evaluate the concealment of the algorithm, we employ the peak signal-noise ratio (PSNR). The PSNR of the color images is defined as

$$\mathrm{PSNR} = 10\log_{10}\left(\frac{255^2}{(\mathrm{MSE}(R)+\mathrm{MSE}(G)+\mathrm{MSE}(B))/3}\right), \tag{6.29}$$

where

$$MSE = \frac{1}{N\times N}\sum_{i=1}^{N}\sum_{j=1}^{N}(X'(i,j)-X(i,j))^2, \tag{6.30}$$

where $X'(i,j)$ is the image with embedded watermark, and $X(i,j)$ is the original image, $0 < i, j \le N$. The bigger the PSNR is, the better the invisibility is.

The robustness evaluation is between the original watermark W and the extracted watermark W'. Here, the methods of visibility and normalized correlation coefficient (NC) are employed.

Visibility is evaluated by observing the difference between the original watermark and the extracted watermark. Normalized correlation coefficient is defined as

$$\mathrm{NC} = \frac{\sum_{i,j} W'(i,j)W(i,j)}{\sum_{i,j} W^2(i,j)}, \tag{6.31}$$

where $W'(i,j)$ is the extracted watermark, and $W(i,j)$ is the original image. Obviously, $0 \le \mathrm{NC} \le 1$. The greater the NC is, the better the similarity between the original watermark and the extracted watermark is.

6.4.3.1 Invisibility

First, we test the invisibility of the watermark. Figure 6.19(a) is the original image, and Figure 6.19(c) is the embedded watermark image. Obviously, human eyes cannot distinguish the difference between them. The embedded watermark image has good subjective quality. The PSNR of the embedded watermark image is 38.8074 dB with good objective quality. Figure 6.19(b) is the original watermark, and Figure 6.19(d) is the extracted watermark. The normalized correlation coefficient NC = 1, which illustrates that the two watermarks are identical.

6.4.3.2 Robustness

The robustness of watermarking algorithm is to test the anti-attack ability. The sharpening, distortion, histogram equalization, contrast enhancement, and mosaic effect are completed by Photoshop CS3 Adobe.

(1) Anti-JPEG compression attack

JPEG lossy compression is applied to Figure 6.19(c). The test results of anti-JPEG compression attack capability are shown in Table 6.3. Q is the quality factor, and

(a) (b)

(c) (d)

Figure 6.19: Watermark invisibility experiment: (a) original image; (b) original watermark; (c) embedded watermark image; and (d) extracted watermark.

Table 6.3: Test of anti-JPEG compression attack capability.

Parameter	**Experimental results**								
Q	100	90	80	70	60	50	40	30	20
NC	1	1	0.9489	0.9221	0.8759	0.8005	0.7178	0.6350	0.4915

its range is $0 \sim 100$. The larger the Q is, the less compression distortion the image is. NC is the normalized correlation coefficient. The watermark images with $Q = 70$, 50, and 30 are extracted as shown in Figure 6.20. It can be seen that when Q varies from 100 to 50, NC is bigger than 0.8, which means the watermark image and the original watermark image are basically the same. The extracted watermark is still clearly distinguished when $Q = 30$. So the algorithm has strong ability of resisting JPEG compression.

(2) Antinoise attack

Figure 6.21 shows the noise attack test results. Figure 6.21(a) illustrates the watermark image and the extracted watermark from the 2% Gauss noise attack.

(a) $Q = 70$ (b) $Q = 50$ (C) $Q = 30$

Figure 6.20: Watermark image extracted from JPEG compression: (a) Q = 70; (b) Q = 50; and (c) Q = 30.

(a)

(b)

Figure 6.21: Experiments of antinoise attack: (a) watermark image and the extracted watermark from the 2% Gauss noise attack and (b) watermark image and the extracted watermark from the 4% salt-and-pepper noise attack.

The watermark image and the extracted watermark from the 4% salt-and-pepper noise attack are shown in Figure 6.21(b). The quality of the image has been seriously degraded, but we can still extract a clear watermark image. So the algorithm has a strong ability to resist noise attack.

(a) (b) (c)

Figure 6.22: Anti-filtering attack experiment: (a) extracted watermark after low-pass filtering experiment; (b) extracted watermark after median filtering experiment; and (c) extracted watermark after high-pass filtering experiment.

(3) Anti-filtering attack

The result of an anti-filtering attack is shown in Figure 6.22. When low-pass filtering, median filtering, and high-pass filtering are finished, the NC values are 1, 0.9976, and 0.9951, respectively.

(4) Anti-cutting attack

This algorithm is strongly robust to cutting. Figure 6.23 is the watermark image and the extracted watermark after cutting 1/4 or central 15% cutting. The NCs of Figure 6.23(a) and Figure 6.23(b) are 0.7956 and 0.8637, respectively.

(5) Anti-rotation attack

Rotation attack is the most representative geometric attack. As long as attacker makes a small geometric transform on the original image, such as the rotation of 0.1 degree, the pixel position almost changed completely. The owners cannot extract the watermark in the image, while people cannot see the difference between the attacked image and the original image. When the rotation angle is greater than 1, the human eyes can recognize the difference between the attacked image and the original image. At this situation, we can identify that the image has been tampered. The anti-rotation attack experiment results are shown in Figure 6.24. This algorithm can resist rotation attack with 0 to 0.6 degrees.

Besides the JPEG compression, noise attack, filtering attack, shear attack, rotation attack, and other common noise processing and attack, this algorithm can also resist histogram equalization, contrast enhancement, USM sharpening, distortion, mosaic effect, and some joint attacks. The experiment results are shown in Table 6.4.

6.4.3.3 Security Analysis

The proposed watermarking algorithm has double security. On one hand, the attackers do not know the 3 independent parameters (a, b, and c) and iteration number k of the generalized Cat map. On the other hand, even though the correct watermark

(a)

(b)

Figure 6.23: Anti-cutting attack experiment: (a) watermark image and the extracted watermark with 1/4 cutting of upper left corner and (b) watermark image and the extracted watermark with central 15% cutting.

(a) (b) (c)

Figure 6.24: Anti-rotation attack experiments: (a) watermark extracted from rotation with 0.2 degree; (b) watermark extracted from rotation with 0.4 degree; and (c) watermark extracted from rotation with 0.6 degree.

information is extracted, the original watermark image cannot be recovered without the TD-ERCS parameters (μ, x_0, α, and m). The key is composed of (a, b, and c) and (μ, x_0, α, and m), so the key space is large enough to resist the brute-force attack, and the algorithm has high security.

Table 6.4: Tests of anti-attack ability.

Attack method	PSNR (dB)	NC	Attack method	PSNR (dB)	NC
Gauss low-pass filter	35.8808	1	15% central cutting	33.6596	0.8637
Median filter	35.6995	0.9976	1/4 upper left corner cutting	30.0935	0.7956
High-pass filter	28.3325	0.9951	0.1 degree rotation	35.8816	1
Histogram equalization	27.7572	0.8662	0.3 degree rotation	34.3736	0.9781
Contrast enhancement	29.0262	0.9051	0.5 degree rotation	32.9006	0.8783
2% Gauss noise	28.7673	0.9440	0.6 degree rotation	32.4629	0.7762
4% salt-and-pepper noise	35.5972	0.9805	JPEG compression ($Q = 70$)	35.4051	0.9221
USM sharpening	28.8351	0.9854	JPEG compression ($Q = 50$)	34.6336	0.8005
Distortion (ripple)	34.3541	0.9684	JPEG compression ($Q = 30$)	33.9761	0.6350
Mosaic effect	32.8215	0.7786	Gauss low-pass filtering, then zoom	33.9952	0.9127
Narrow 1/4, then 4 times magnification	33.3291	0.9659	1% Gauss noise and median filter	31.7662	0.9202

6.5 Chaotic Image Encryption Algorithm

From the view of the data-processing methods, the image encryption technology can be divided into three categories: position replacement, value transformation, and its combination. From the view of signal processing, image encryption algorithm can be divided into spatial domain algorithm and frequency domain algorithm. Image encryption based on chaos is a kind of transformation of image data by employing chaotic sequence, which makes the visual difference between the transformed image and the original image. The visual difference here refers to the qualitative or quantitative differences in color, brightness, or outline. In this section, we focus on the chaos-based image encryption technique.

6.5.1 Chaos-Based Image Encryption Technology

6.5.1.1 Image Scrambling Based on Chaotic Map

Digital image scrambling transformation includes one-dimensional, two-dimensional, three-dimensional, and high-dimensional transformations. Objectively they should be one-to-one map, which means they are reversible transformations. The encryption can be achieved through the transformation, and the decryption can be achieved through the inverse transform. In order to provide a larger key space and a more efficient way to change the correlation between pixels, the encryption method based on high-dimensional chaotic systems has become a new research hotspot in

recent years. For designing a chaotic scrambling scheme, we should choose a suitable chaotic map. We think the following factors should be considered:
1. Lyapunov exponent should be as large as possible. If the Lyapunov exponent is larger, the system is more sensitive to the initial value, and it is more suitable for image encryption.
2. Control parameters should be as much as possible. For the chaotic cipher system, parameters are often used as key. Large parameter space can ensure a large key space to resist the brute-force attack.
3. Chaos map should be a reversible one-to-one map to ensure the scrambling transformation is reversible.
4. High-dimensional map should be given priority. If the system structure of high-dimensional chaotic system is more complex, then it can change the correlation between pixels more effectively.

When choosing the encryption parameters, the structure and characteristics of the chaotic system need to be fully investigated. For example, the weak key should be removed, and the parameters range should be determined, and the key space should be as large as possible. The chaotic maps usually applied in image encryption include Kolmogorov flow, two-dimensional Cat map, and two-dimensional Baker map. Some scholars extend the two-dimensional map to the three-dimensional map. When the three-dimensional map is used to scramble pixel, the scrambling speed becomes faster.

In the chaotic scrambling, the sensitivity to initial conditions corresponds to chaotic plaintext sensitivity. The larger the location of the plaintext varies, the smaller the correlation between the two adjacent pixels is. So it is more difficult to attack by analyzing the correlation of pixels. Besides chaos-based pixel scrambling or spatial scrambling, we can also use chaotic sequences to scramble rows and columns of the image. Most image scrambling is carried out in the space domains. Scrambling damages the relationship between pixels in the image encryption, but it increases the encoding difficulty and reduces the compression rate. So there is a problem: How to combine image scrambling encryption with image compression effectively at present?

6.5.1.2 Image Diffusion Based on Chaotic Sequences

Pixel position transformation does not change the pixel values. The statistics of the encryption image is poor, so it is not suitable to be used individually. Traditional encryption theory proposed a scheme of changing the image statistics by diffusion operation and combining the diffusion operation with the pixel scrambling. The diffusion operation can increase the ability of resisting differential attack and the chosen plaintext attack.

Diffusion is to transform change of a pixel to the other pixels, and it leads to the high sensitivity of the encryption system. There are two common methods about diffusion. One is the diffusion of 4 adjacent gray values in the image. The diffusion range increases by the power exponent of 4 with the increase of scrambling cycles. The other is the raster scanning. The diffusion speed is fast. It is realized by "mod and add" operation. The "mod operation" is designed to control the results in the range of pixel values, while the "add operation" is to achieve the transmission function by taking the former change to the next operation. The diffusion process can be expressed as

$$C_i = (P_i + C_{i-1}) \bmod L, \tag{6.32}$$

where P_i is the pixel value before diffusion. C_i is the pixel value after diffusion. L is the pixel gray value. To enhance the randomness of bit distribution, the XOR operation and chaotic sequence are generally introduced in diffusion algorithm, and then eq. (6.32) is rewritten as

$$C_i = l_i \oplus (P_i + l_i) \bmod L \oplus C_{i-1}, \tag{6.33}$$

where $l_i(i = 1, 2, \ldots)$ is the processed chaotic sequence, which limits the range of pixel gray values. In recent years, the encryption scheme based on chaotic sequence is also developed rapidly. One-dimensional chaotic system is used widely, such as logistic map and PLCM map. But research results show that the security of the encryption based on low-dimensional chaotic systems is not high enough [140, 141]. Researchers have begun to focus on the high-dimensional chaotic systems.

Decryption is the symmetric inverse transform of encryption. The random distribution of the diffused image makes statistics attacks difficult. Plaintext sensitivity corresponds to the diffusion rate, and key sensitivity corresponds to the initial value sensitivity of the diffusion function.

6.5.1.3 Comprehensive Selection of Chaotic System in Image Encryption

When we design a chaotic image encryption system, the selected chaotic system should be better than other chaotic systems in the following aspects:

1. Initial value sensitivity. It is beneficial to realize the key sensitivity of the security system, the sensitivity of the original text, and the sensitivity of the cipher text.
2. The uniform distribution of the chaotic states. Chaotic systems can traverse all possible states randomly. It can enhance the security of the secret key and the cipher text and can effectively improve the ability of anti-statistics attack.
3. Non periodicity. After the chaotic system is made discrete, it will not be repeated in the case of limited number of iterations, thus increasing the difficulty of statistical attack.
4. Nonlinearity. The linear structure of the chaotic system increases the information correlation. The higher the nonlinearity of chaotic system is, the more random the state vibration is, and the more secure the encryption system is.

After constructing a chaotic cipher system, it is necessary to carry out the cipher analysis and use cipher attacks to test the security of the system. Even passing all the known cipher attacks, the cipher system is not completely secure.

6.5.2 A New Chaotic Image Encryption Scheme

In this section, a new image encryption scheme based on one-way coupled map lattice (OCML) and TD-ERCS chaotic system is studied. It combines pixel position scrambling with gray value substitution.

6.5.2.1 Spatiotemporal Chaotic System

One-way coupled map lattice is a spatiotemporal chaos system, and it can be described as

$$x_{n+1}(i) = (1 - \varepsilon_i)f(x_n(i)) + 2\varepsilon_i\left\{f[x_n(i-1)]\right\}, \tag{6.34}$$

where n is discrete time coordinate, $n = 1, 2, 3, \ldots, T$ (T is the length of time series). i is discrete space coordinate, $i = 1, 2, 3, \ldots, L$ (L is OCML lattice number). f is local chaotic map. ε_i is the coupling coefficient, $0 < \varepsilon_i < 1$. In this section, we choose logistic map as the local chaotic map, and it is expressed as

$$f(x) = (3.75 + \lambda/4)x(1 - x), \tag{6.35}$$

where $0 \leq \lambda \leq 1$, the initial value is a random number at (0, 1). If the entire OCML system satisfies the periodic boundary conditions $x_n(0) = x_n(L)$, the space sequences and time sequences of the whole system are chaotic.

6.5.2.2 Scrambling Image by Spatiotemporal Chaotic Sequences

Ref. [142]. proposed a chaotic image scrambling algorithm based on rank transform. The replacement address code is generated by the chaotic analog sequence without quantification of the chaotic sequence. It provides a new way for chaotic scrambling.

If the pixel is the basic unit, the scrambling is carried out by the mapping relationship between the pixel and $x(i, n)$. The replacement address code requires to traverse all the addresses, so row-and-column scrambling method is applied to scramble the image. Suppose an image with size $M \times N$, and let $L = M$, $T = N$, then a spatiotemporal matrix x with size $M \times N$ is obtained. Steps of row scrambling image matrix by x are as follows:

1. A chaotic sequence $\{a_1, a_2, \ldots, a_N\}$ is obtained from one row of the spatiotemporal matrix x. $\{\tilde{a}_1, \tilde{a}_2 \cdots \tilde{a}_N\}$ is obtained by ascending (descending) of $\{a_1, a_2, \ldots, a_N\}$.
2. Permutation address set $\{p_1, p_2, \ldots, p_N\}$ is formed by determining the position of each element of the sequence $\{a_1, a_2, \ldots, a_N\}$ in $\{\tilde{a}_1, \tilde{a}_2 \cdots \tilde{a}_N\}$.
3. The same row of the image matrix is replaced according to the permutation address set $\{p_1, p_2, \ldots, p_N\}$.

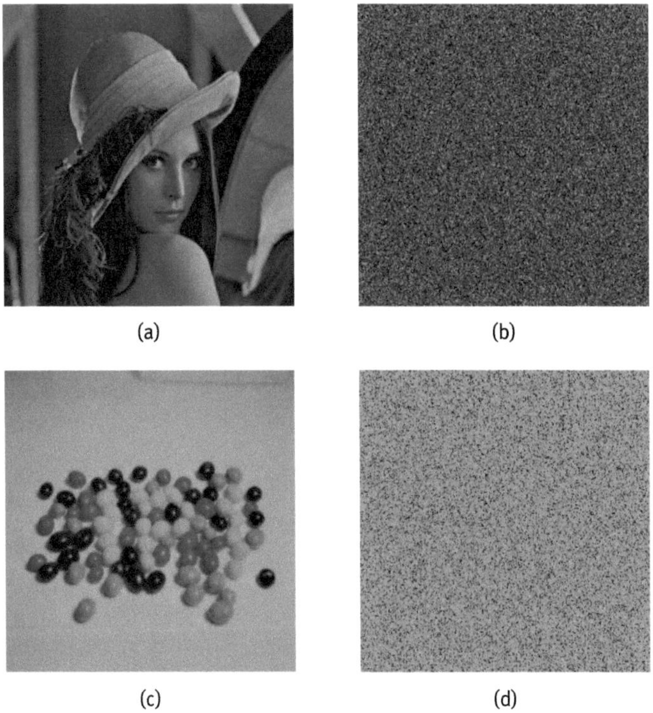

Figure 6.25: Image scrambling experiments: (a) original image; (b) scrambled image; (c) original image; and (d) scrambled image.

The column scrambling of the image matrix is finished by the same way. Because spatiotemporal chaos is chaotic in time and space, each scrambled row and column are satisfied with a good random. Figure 6.25 shows the scrambling results of two different images. It can be seen that the scrambled image does not have any outline of the original image by a round row-and-column scrambling.

6.5.2.3 Initial Conditions and the Key Scheme

Chaotic system is sensitive to initial values and parameters, which is consistent with key sensitivity of the traditional encryption algorithm. The initial values and parameters are often used as keys of the chaotic encryption system. In this scheme, the TD-ERCS chaotic system, with large parameter space and good statistics, is applied to generating two independent random sequences, which are used as the initial value sequence and coupling coefficient ε_i of OCML, respectively.

For the TD-ERCS map, given $\mu(0 < \mu \leq 1)$, initial value $x_0(-1 \leq x_0 \leq 1)$ and $\alpha(0 < \alpha < \pi)$, tangent delay m ($m = 2, 4, 5, 6$), two independent real sequences x_n and k_n are generated. (μ, x_0, α, m) are seed parameters. The two real sequences x_n and k_n are normalized, and then they are assigned to the initial value and coupling coefficient of the OCML system, respectively. The whole encryption system has five keys λ, μ, x_0, α, and m, so the key space is large.

6.5.2.4 Chaotic Diffusion Algorithm

To improve the security of the image, the gray value diffusion is very important. Spatiotemporal chaotic sequence is high-dimensional pseudo-random sequence with good cryptographic properties, so the OCML chaotic system is directly used in the diffusion operation. The security and the encryption speed are greatly improved at the same time.

In this diffusion algorithm, two real sequences with length $M \times N$ are got from spatiotemporal matrix x. Integer sequences L_1 and L_2 in $[0, N_1 - 1]$ are obtained by dealing with the two values sequences. N_1 is the image gray level. Of course, we can do the appropriate transformation for the space-time matrix x and then select the L_1 and L_2 from the time and space, respectively. The selection methods are as follows. L_1 is obtained by quantification after progressive scanning for spatiotemporal matrix x. L_2, first by scanning all the even columns and then scanning all the odd columns, is obtained by quantification and rounding finally.

If $L_1(i) \oplus L_2(i)$ is an even number, the encryption process is

$$C(i) = (I(i) + L_2(i)) \bmod N_1 \oplus C(i-1) \oplus L_1(i). \tag{6.36}$$

If $L_1(i) \oplus L_2(i)$ is an odd number, the encryption process is

$$C(i) = (I(i) + L_1(i)) \bmod N_1 \oplus C(i-1) \oplus L_2(i), \tag{6.37}$$

where $I(i)$ is pixel gray value of the image. $C(i-1)$ is the former encrypted ciphertext pixel value, and $C(i)$ the later encrypted ciphertext pixel value.

If $L_1(i) \oplus L_2(i)$is an even number, the decryption process is

$$I(i) = (C(i) \oplus L_1(i) \oplus C(i-1) + N - L_2(i)) \bmod N_1. \tag{6.38}$$

If $L_1(i) \oplus L_2(i)$ is an odd number, the decryption process is

$$I(i) = (C(i) \oplus L_2(i) \oplus C(i-1) + N - L_1(i)) \bmod N_1. \tag{6.39}$$

6.5.2.5 Image Encryption Scheme

The image encryption scheme is shown in Figure 6.26.

Step 1 Input seed parameters and two normalized sequences are generated by M iterations of the TD-ERCS system (pre iteration 100 times).

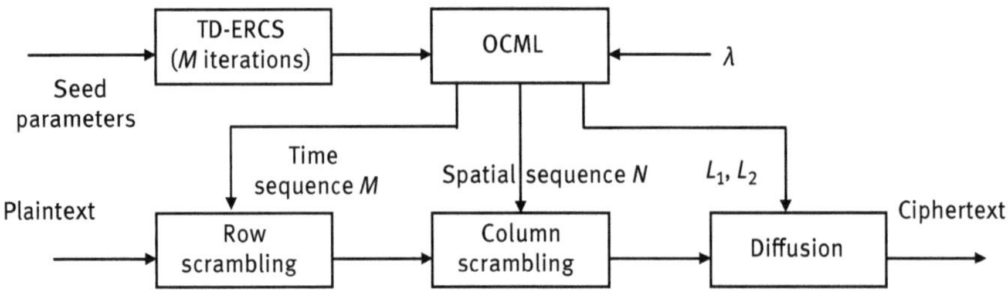

Figure 6.26: Block diagram of image encryption.

Step 2 Set λ, and assign the two normalized sequences generated in Step 1 to initial values and coupling coefficients of OCML system with M order. The OCML system iterates N times.

Step 3 Row scrambling by time series with the length M.

Step 4 Column scrambling by space series with the length N.

Step 5 Get L_1 and L_2 from the OCML system. Do diffusion by employing the algorithm above.

Steps 3 and 5 can be repeated several times to get higher security. In fact, it can get a good effect with one-time encryption.

Decryption is the inverse process of the steps above.

6.5.2.6 System Security Analysis

(1) Statistical analysis. In order to resist statistical attack, the encrypted image needs to be kept in certain diffusion and disturbance. In this section, we take Lena image encryption as an example to analyze the security of the algorithm.

(a) Histogram analysis. Histogram analysis can be used to measure the disturbing of encryption algorithm. The histograms of Lena and its encryption image are shown in Figure 6.27. The histogram of the encrypted image is uniform, so it does not retain any information about the plaintext.

(b) Correlation analysis. Correlation is generally measured by the correlation coefficient r as follows:

$$r = \frac{\text{cov}(p, q)}{\sqrt{D(p)D(q)}}, \tag{6.40}$$

where p and q are gray values of the two adjacent pixels in the image, and cov $(p, q) = E(p-E(p)(q-E(q))$ is the covariance, where $D(p)$ and $D(q)$ are the variances. The correlations are calculated by randomly selecting 1,000 pair adjacent pixels from horizontal, vertical, and diagonal directions of the image, respectively. Table 6.5 lists the correlation coefficients of the original

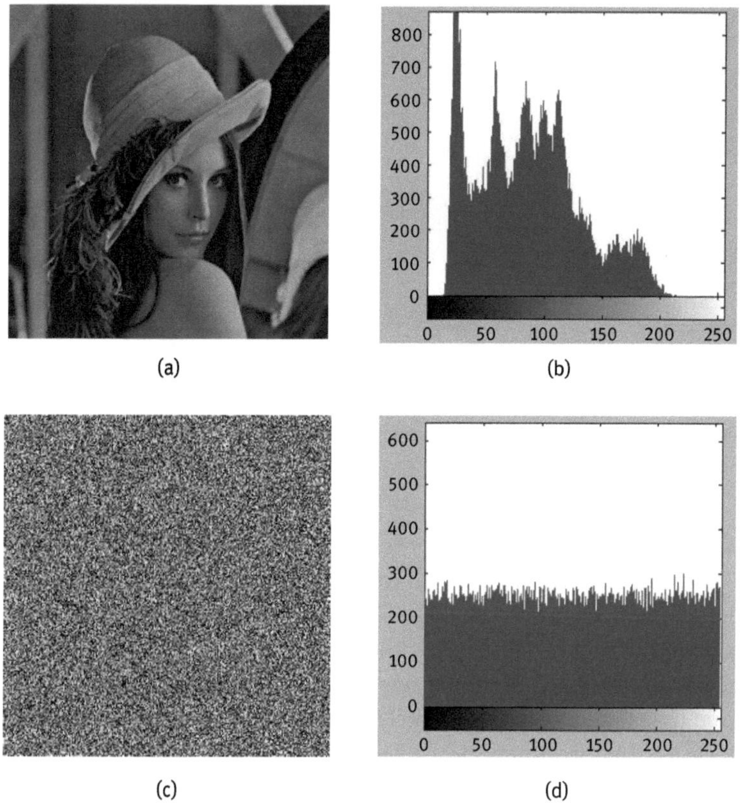

Figure 6.27: (a) Original image; (b) histogram of the original image; (c) encrypted image; and (d) Histogram of the encrypted image.

Table 6.5: Correlation coefficients of two adjacent pixels.

	Original image	Encrypted image
Horizontal	0.9487	0.0207
Vertical	0.9218	0.0106
Diagonal	0.9492	0.0400

image and its encrypted image. The correlation coefficients of the encrypted images are very small, which indicates that no detectable correlations exist between the original image and its corresponding cipher image. The diffusion of the algorithm is good.

(2) Key sensitivity analysis. A good image encryption algorithm should be sufficiently sensitive to the encryption key and the plaintext. We test several images

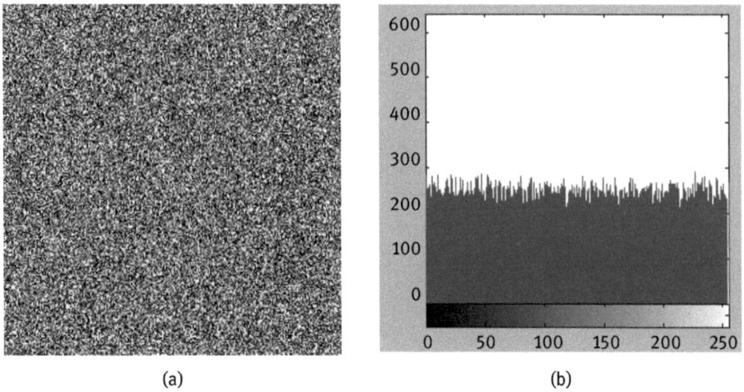

Figure 6.28: Key sensitivity analysis results: (a) encrypted image with $\mu = 0.73120000001$ and (b) histogram of the decrypted image.

to prove the sensitivity of the algorithm to the key. In this encryption experiment, the Lena image is encrypted with the initial values $\lambda = 4$, $\mu = 0.7312$, $x_0 = 0.5654$, $\alpha = 0.9876$, and $m = 9$. Figure 6.28(a) is the decrypted image with $\mu = 0.73120000001$, the other keys being unchanged. Figure 6.28(b) is the histogram of the decrypted image. It can be seen that the decrypted image is totally different from the original image with the key μ varying by 10^{-11}. The histogram is still uniform. So this algorithm is sensitive enough to the key.

For a given key, the encryption of different images will generate different ciphertext. The diffusion operation ensures that even the change of a single pixel will cause a dramatic effect on the whole text, and the encrypted image is completely different. So the algorithm is also sensitive enough to the plaintext.

(3) Analysis of anti-brute-force attack. The password is binary sequence K, $K = k_\lambda k_u k_{x_0} k_\alpha k_m$, where the k_λ, k_u, k_{x_0}, k_α, and k_m are used to generate floating-point number λ, μ, x0, and α and integer m. If the length of k_m is 12 bits, then m is obtained by

$$m = \begin{cases} 2 & (k_m = 0) \\ 4 & (k_m = 1) \\ k_m + 3 & (2 \le k_m \le 4095) \end{cases} . \tag{6.41}$$

If the word "length" in the computer is 32 bytes, and the precision is 10^{-14}, then the key space is $10^{56} \times 2^{12} \approx 10^{59}$. The key space is large enough to resist the brute-force attack.

The experiments above show that the new image encryption scheme based on OCML and TD-ERCS chaotic system has excellent diffusion and disturbance characteristics. In this scheme, the security of the algorithm can be ensured by the combination of the random row–column scrambling with changing pixel gray value. The key space is large enough to resist the brute-force attack, and the encrypted image is greatly sensitive to the keys and the plaintext. The algorithm is able to encrypt the images of any size quickly.

Questions

1. What is chaotic cryptography? What are the differences between chaotic cryptography and traditional cryptography?
2. Analyze the principle of chaotic encryption/decryption.
3. Design a chaotic pseudo-random sequence generator.
4. What are the performance-testing methods of chaotic pseudo-random sequence?

7 Audio and Video Chaotic Encryption and Communication Technology

With the development of network communication, Internet provides an amount of information services to users every day. Because the foundation protocol of Internet is not completely secure, when unencrypted information transmits on the network, it will be directly exposed to the entire network, which is a great threat to information security, especially for audio and video communications, such as eavesdropping, call tracking, phone hijacking, and denial of service. Therefore, we should improve network design and also take a variety of technologies, such as voice encryption technology, to ensure the safety of audio and video communications. With the deepening of chaotic applied research, promising progresses have been made on chaotic cryptosystem [143–146], and a new path has been opened to improve the information security. Thus, it is significant to combine the traditional data encryption algorithm, video image standard MPEG and H.264 with chaos and to apply it in real-time communication. This chapter focuses on audio and video real-time encryption and communication technology based on chaos.

7.1 Principle of Real-Time Voice Encryption Communication

In Chapter 6, we studied the chaotic encryption principle of static data file, and it mainly used the idea of computer process synchronization to realize synchronization of chaotic iterative system. The starting parameters and synchronization strategy themselves are passwords, so it need not use complex physical chaotic synchronization or control method. But if you want to achieve real-time transmission of encrypted data stream, chaotic synchronization is an unavoidable problem. Here, we adopt the network adaptive communication environment (ACE), which is a cross-platform, to implement the real-time transport protocol (RTP)/RTP control protocol (RTCP) and combine the chaotic encryption algorithm with the extended Data Encryption Standard (DES) encryption algorithm. A voice encryption communication system, which is adaptable for network real-time transmission, is developed by equal interval drive synchronous method.

7.1.1 Principle of Digital Encryption Communication System

Although the sequence cipher system, block cipher system, and public key cryptosystem can be used for encryption of digital signal, researchers have certain requirements for the real time and quality (measured by bit error rate) of information transmission in practice. Therefore, an appropriate cryptosystem must be selected according to the specific digital communication system and the corresponding transmission quality of channel [147].

Generally speaking, because there are error diffusion and delay for block cipher system, block cipher system is applied in the occasions of good channel transmission quality or containing data retransmission function. Due to its huge amount of computation for public key system, it is difficult to meet the real-time requirements of voice communication system in the present technology level. Therefore, the public key system is usually not directly applied to real-time data stream encryption. Furthermore, because of the characteristics of small delay and no error diffusion, sequence cipher system is widely used in digital communication system, which is suitable for the quality of all kinds of channels [148]. The principle of digital encryption and communication system based on the sequence cipher system is shown in Figure 7.1.

Digital voice encryption communication system mainly includes source coding/decoding, encryption/decryption operator, cipher sequence generation, cipher synchronization control, and synchronization insertion/detection.

Through sampling, quantization, and encoding (including compressed encoding), source coding/decoding module at the sender transforms the original analog signal into binary digital sequence and inputs them to the encryption device. This module recovers the binary sequence to analog signals at the receiver.

Encryption/decryption operation generally uses the XOR operation. At the sender, ciphertext is obtained by the XOR operation of the binary digital sequence from source encoding module with secret key sequence. At the receiver, the binary plaintext sequence is recovered by the XOR operation to secrete key sequence.

The main function of the cipher sequence generation module is to generate secrete key sequences for encryption or decryption, which is the key part of reliability in digital encryption and communication system, including the encryption sequence generation module (it is used to encrypt the plaintext on the sender) and the decryption sequence generation module (it is used to decrypt the ciphertext at the receiver).

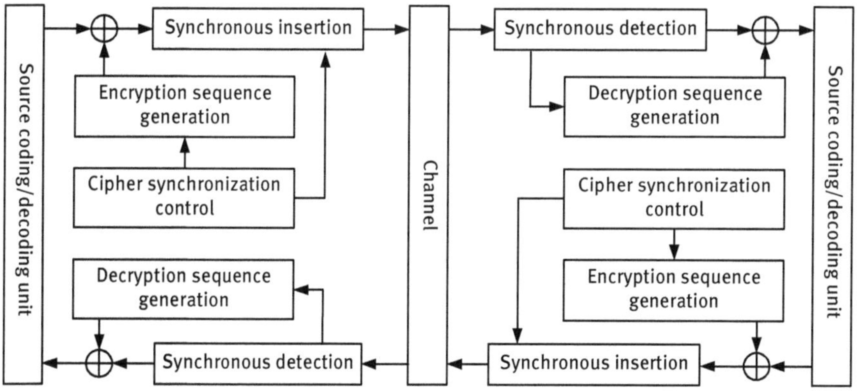

Figure 7.1: Principle block diagram of digital encryption and communication system.

The cipher synchronization control module mainly generates a cipher synchronization signal to update the status of secrete key sequence at both the receiver and the sender. To ensure the cipher synchronization in the encryption and decryption processes, the secrete key (initial conditions or parameters of chaotic system) must be the same.

At the sender, synchronous insertion/detection module inserts the cipher synchronization signal into the corresponding position of ciphertext stream. At the receiver, digital information stream is detected in real time. Once the cipher synchronization signal is found, it is extracted from the information stream, and its message key is sent to the decryption sequence generation module to update its status and the noncipher synchronization information is transmitted transparently at the same time.

7.1.2 Real-Time Voice Encryption and Communication System

Point-to-point (P2P) real-time voice chaotic encryption and communication system is shown in Figure 7.2. At the transmitter, first, analog signal is transformed into binary plaintext data stream through the A/D converter. Then, chaotic sequence generator generates a chaotic pseudorandom sequence to XOR the plaintext data stream, and a ciphertext data stream is produced. Subsequently, ciphertext is sent to the network channel according to the network protocol. At the receiver, a chaotic sequence, which is synchronous with the sender, is used to decrypt the ciphertext sequence by the XOR operation to restore the binary voice stream. Finally, through D/A converter, the binary voice stream is returned to the analog voice signal and it is broadcasted.

The encryption process at the sender is expressed as

$$S(n) = M(n) \oplus E(n), \tag{7.1}$$

where $M(n)$ is the voice sequence, $E(n)$ is the chaotic sequence, $S(n)$ is the ciphertext sequence, and $\oplus$ is the XOR operator.

The decryption process at the receiver is the inverse process of encryption, i.e.

$$M'(n) = R(n) \oplus D(n), \tag{7.2}$$

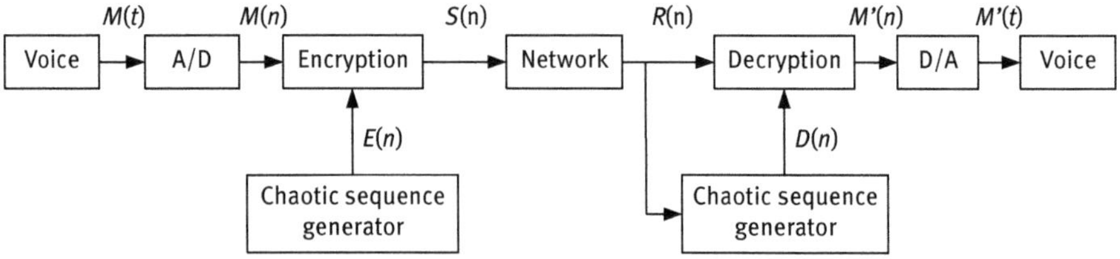

Figure 7.2: Principle block diagram of the voice chaotic encryption system.

where $R(n)$ is the ciphertext on the receiving side and $R(n) = S(n)$ for an ideal channel. $D(n)$ is the chaotic sequence on the receiving side. $M'(n)$ is the decryption output. Obviously, when both sides of the system are synchronized, i.e. $D(n) = E(n)$, then $M'(n) = M(n)$. So the voice encryption communication is achieved.

Obviously, the key of realizing the encryption communication is to realize the synchronization of chaotic sequence generators between the receiver and the sender. Because the object of computer processing is digital signal, the existing analog synchronization method cannot meet the requirements of network transmission. Combined physical synchronization with computer programming techniques, the equal interval drive synchronization method is proposed, which is suitable for computer network communication. Theory analysis and experimental results show that this method has strong fault tolerance capability.

7.1.3 Cipher Synchronization Technology

Cipher synchronization is the key technology in the digital real-time encryption and communication system. In practice, to ensure the strict consistency of the starting points of system communication on the encryption and decryption sides, the cipher synchronization must be established before encryption and communication. The order of cipher synchronization is shown in Figure 7.3.

First, bit synchronization is set up for the system. It can be realized by the clock provided by the source terminal or channel transmission device, and also can be achieved by extracting the clock from the source terminal or the digital stream provided by the transmission device. Second, cipher synchronization is established. After the establishment of bit synchronization of communication system, the sender sends the cipher synchronization signal (including the synchronization head and the message key) to the other side to make the cipher generator of two sides to work synchronously. For some real-time encryption communication systems, it needs to establish frame synchronization (group synchronization) before the establishment of cipher synchronization. Finally, ciphertext sequence is transmitted. After the code synchronization signal of the sending side is transmitted, the ciphertext sequence is transmitted immediately. The encrypted sequence is obtained by the corresponding encryption algorithm, which is based on the same message key as the receiving side. Receiver extracts the message key from the cipher synchronization signal and injects it into the cipher generator. Simultaneously, the cipher generator generates the cipher

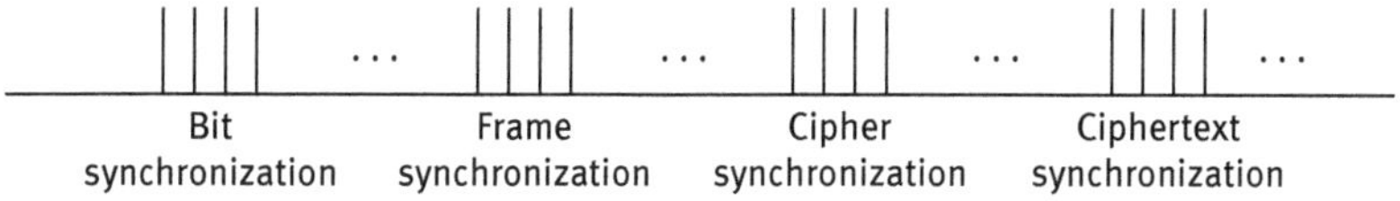

Figure 7.3: Order of the cipher synchronization.

sequence, which makes the decryption cipher sequence and the encryption cipher sequence to work in the synchronized state.

Cipher synchronization methods include continuous synchronization, unique initial synchronization, self-synchronization, and special synchronization. In this section, a continuous synchronization method is adopted to realize the cipher synchronization of encryption communication. The continuous cipher synchronization is to send a cipher synchronization signal with a fixed interval, and it has advantages, such as allowing the lag to enter, plaintext beyond function and multiple communication ways. It is worth noting that the transmission cipher synchronization information may reduce the transmission rate, but the security and reliability of system are improved.

7.2 Real-Time Data Stream Encryption Algorithm

In real-time voice encryption algorithm, the traditional DES cryptography is no longer safe [149]. Researchers proposed various improved algorithms, such as 3DES based on secrete key expansion, but its key space is still limited. Therefore, 3DES encryption method based on chaotic map is adopted to guarantee the security of data stream. Specifically, synchronized chaotic map systems are set on the encryption and decryption sides, the secrete key of each data packet is generated by chaotic system to realize real-time encryption and transmission of voice data.

7.2.1 3DES Encryption Algorithm

3DES is defined by the National Institute of Standards of Technology (NIST) in 1999 [150]. 3DES defines two kinds of work modes. In the first mode, secrete key k_1 is used to encrypt the 64-bit plaintext. Then secrete key k_2 is used for the second encryption. Finally, secrete key k_3 is used for the third encryption, and the ciphertext is outputted. This pattern is called DES-EEE, which is expressed as

$$C = E_{k3}(E_{k2}(E_{k1}(M))), \tag{7.3}$$

where M is the plaintext and C is the ciphertext. This encryption mode extends the key space into 168 bits. In the second mode, k_1 is used to encrypt the 64-bit plaintext. Then the result is decrypted by k_2. Finally, k_3 is used to encrypt the decryption result. This pattern is called DES-EDE, which is expressed as

$$C = E_{k3}(D_{k2}(E_{k1}(M))) \tag{7.4}$$

where M is the plaintext and C is the ciphertext. This encryption mode extends the key space into 112 bits. In the encryption process, secrete keys k_1, k_2, and k_3 have three options as follows.

1. k_1, k_2, and k_3 are independent of each other, and they are different from each other.
2. k_1 and k_2 are different, but k_3 is same with k_1.
3. k_1, k_2, and k_3 are same, and this way is equivalent to the standard DES.

Here, we adopt the second mode of 3DES, and the second way of secrete key. The encryption model is shown in Figure 7.4. 64-bit data is encrypted with k_1 to obtain the ciphertext first. Then k_2 is used to decrypt the result to obtain "plaintext," and then k_3 (k_1) is used to encrypt the decryption results. Obviously, after removing the bits of eight times from the 128-bit secrete key, there are 112-bit secrete key, and the key space is 2^{112}. The only disadvantage of this method is that three times ordinary DES encryption of the algorithm spends about three times implementation time. But it is acceptable for very cheap computing resource at present.

7.2.2 Encryption Algorithm Based on 3DES and Chaotic Map

Chaotic dynamics system can be used in many application fields, such as digital communication and multimedia data security [151].

Similarly, logistic map studied in the sixth chapter is defined as

$$x_{n+1} = \mu \times x_n \times (1 - x_n), \tag{7.5}$$

where $\mu \in (0, 4]$ and $x_n \in (0, 1)$. When $3.5699\ldots < \mu \leq 4$, it is chaotic. According to the statistical analysis in Section 3.1.1, the chaotic sequence of logistic map $\{x_n,\ n = 0, 1, 2, 3, \ldots\}$ is nonperiodic and nonconvergent, and it is very sensitive to initial conditions.

DES algorithm with key expansion (3DES) is the core of real-time voice encryption algorithm, and the secrete keys of encryption and decryption are generated by two synchronized chaotic systems (logistic map). The encryption model is shown in Figure 7.5. K_i is the secrete key of encryption and decryption of each data packet, which is generated by the logistic map. x_0 and μ are the initial condition and parameter of logistic map, and that is the secrete key transmitted in the secret channel.

When the logistic map is used to generate secrete key, two problems need to be solved. One is how to transform the generated sequence into a 64-bit secrete key, and another is how to synchronize two chaotic maps to ensure that each data packet to

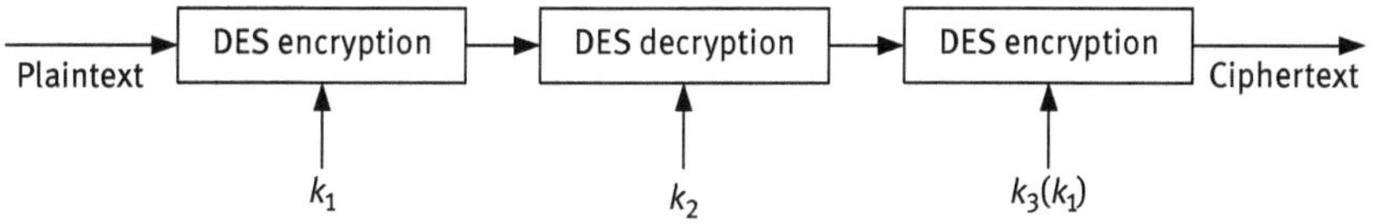

Figure 7.4: 3DES encryption model.

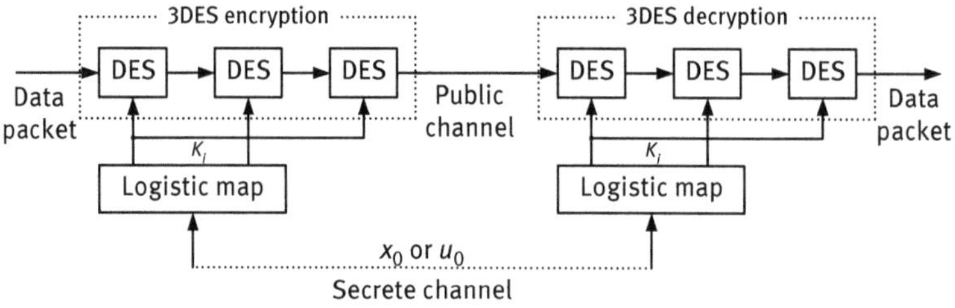

Figure 7.5: Encryption model based on 3DES and Logistic map.

i	11-bit order code		52-bit tail code	
0	1	11	12	63

Figure 7.6: Storage format for double type data.

obtain the right secrete key for decryption. Here, the double precision type is used to store the initial condition x_0 and parameter μ of logistic map. The double type is 8 bytes (64 bits) in Win32. So it can be used as the secrete key of DES. The secrete key of 3DES is obtained by iterating the logistic map two times to produce 64-bit double data. The storage format of double data is shown in Figure 7.6.

In this figure, the 0th bit is sign bit. The first bit to the 11th bit is 11-bit order code, and the 12th bit to the 63rd bit is 52-bit mantissa. The calculation method of double type data is

$$x_0 = (-1)^i \times 2^{e-1023} \times 1.f, \tag{7.6}$$

where i is the sign bit, e is the order code, and f is the mantissa.

The range of iteration value is (0, 1), and thus $e \in [0, 1022]$. The number of its valid bits is still 11, and its sign bit is unchanged. So the total number of valid bits in double type is 63. The iteration space is $2 \times (2^{63} - 2^{52})$, in which multiplying by 2 represents two 64-bit iterations, and minus 2^{52} means to subtract it at $e = 1,023$. Because the number of 8 times in DES algorithm is only used as check bit, the key space of an actual data packet is about 2^{112}.

To realize synchronization between two chaotic systems, the iteration number of system is saved in the header of each encrypted data packet as plaintext. According to the iteration number in the header, the receiver iterates the logistic map to produce decryption key and uses it to decrypt. In practice, the former packet of secrete key can be memorized to reduce iterations. Since the number of iterations is transmitted as plaintext in the header data, it may be intercepted by third party. But the decryption key cannot be obtained without the initial condition and parameter of this map. As long as the synchronous data in the header is received correctly, the synchronization can be achieved.

Shannon proves that one-time pad is unable to be deciphered, and 3DES has high security level. So, the encryption algorithm has a high security. Corresponding to real-time voice communication, if a single data packet is deciphered, it will not affect the entire transmission process. Because the pseudorandomness of chaos leads to the condition that secrete keys are relatively independent, so that each data packet has the same crack strength to ensure the security of data communication to the maximum extent.

7.2.3 Network Synchronization Method

Synchronization is the key to achieve real-time data stream encryption. If transmission errors of the sequence generated by the logistic map occur, it will cause the failure of synchronization of encryption and decryption. Here, a synchronization scheme is designed, which fully depends on the network to achieve self-synchronization as shown in Figure 7.7.

It can be seen that the processes of encryption and decryption are symmetrical, and the synchronization between them is completely dependent on the error-free transmission on the network. Tests show that the synchronization of encryption and decryption is good in the beginning of a few minutes. However, when the time is longer (usually between 5 and 10 min), the phenomenon of out of step will suddenly happen, so that the entire decryption is destroyed, and it cannot be restored. On one hand, the transmission control protocol (TCP) cannot be fully realized error-free transmission. On the other hand, both the encryption and decryption sides do not have fault-tolerant ability, and a little mistake could waste all the previous efforts. After further analysis, it is found that the above scheme is not without merit, because it is synchronized at the beginning of a few minutes. That is to say, we could establish an error tolerance and error correction mechanism to make the less probability of data error. But how do you know when the data is wrong, unless the server sends the voice data and error correction code at the same time. Otherwise, it is not known. However, considering

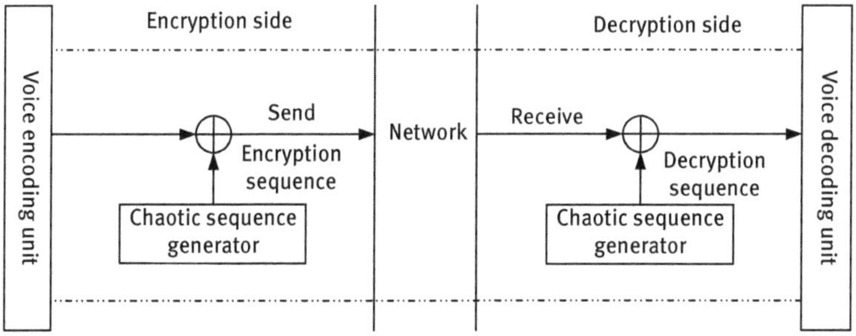

Figure 7.7: Network self-synchronization scheme.

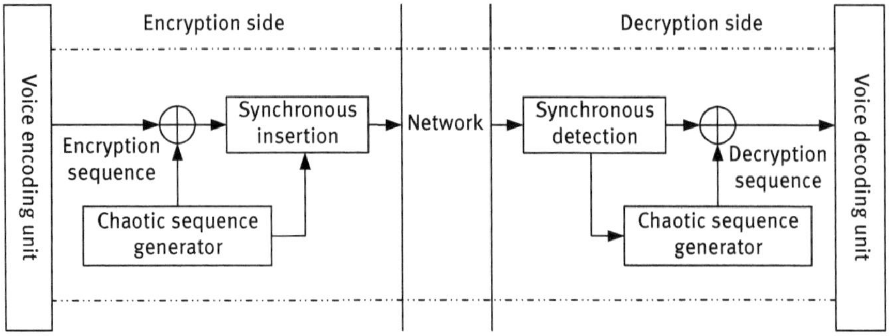

Figure 7.8: Equal interval drive network synchronization scheme.

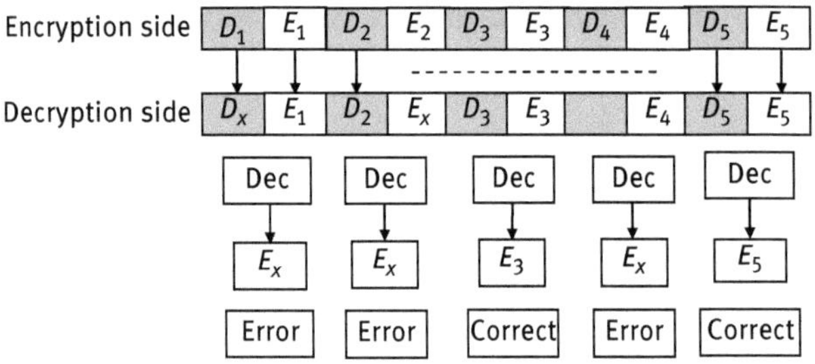

Figure 7.9: Block diagram of fault tolerance capacity analysis.

the particularity of voice information, it does not necessarily have the ability of error correction as long as the fault is tolerant. In real life, when a sentence does not listen to the other side, he can ask the other side to say it again. Based on the above analysis, an equal interval drive synchronization scheme is designed in Figure 7.8.

Compared with Figure 7.7, the synchronous insertion and synchronous detection module are added to the system in Figure 7.8. In the encryption side, the drive code of chaotic sequence (number of iterations) is inserted into the header of ciphertext data packet, and it is sent with the ciphertext data packet. In decryption side, the driver code is taken out by the synchronous detection module, and it is used to drive the chaotic sequence generator to generate secrete key to decrypt the cipher sequence, and the secrete key is consistent with that of the sender.

The fault tolerance capability of the above synchronization scheme is analyzed. We will consider all errors that may occur in the network transmission. The transmission process is shown in Figure 7.9.

The upper part of Figure 7.9 is the data sent to the network, where D_i is the drive code and E_i is encrypted voice stream. The middle data is read from the Internet on the decryption side, where D_i is the drive code and E_i is the encrypted voice stream.

The bottom data is the decrypted voice data stream. Obviously, transmission errors are classified into the following categories:
1. Drive code is transmitted in error. For instance, D_1 is transmitted to D_X.
2. Encrypted voice data is transmitted in error. For instance, E_2 is transmitted to E_X.
3. Drive code is lost. For instance, D_4 is lost.

As shown in Figure 7.9, the first and second errors only affect one data packet of voice stream. When D_2 or D_3 is received on the decryption side, all contacts between E_X and D_X will be cut off. D_3 is used to iterate two values to generate two double precision decryption sequences to complete decryption, while the encryption side also use D_3 to generate two double-precision encryption sequence to accomplish encryption. So the encryption and decryption sequences are same, and the synchronization of encryption and decryption is realized. For the third case, it only affects one voice data packet too. When the correct driver code is received, it will return to the state of synchronization. After a long period of testing, it is proved that this scheme is feasible and can keep synchronization of encryption and decryption in infinite time. Although transmission errors occur accidentally in the middle, they will not affect the whole process of call. There is no security risk to expose the drive code to the outside because the following n-bit sequence does not have the possibility to be predicted without knowing the values of parameter μ and initial condition x_0.

In general, the main idea of this synchronization scheme is to make use of the tiny bit error rate provided by network transmission protocol. Regardless of whether there is a bit error, the decryption side is forced to make a new synchronization at set intervals (it is very small). That is to say, the synchronization of the entire call time is changed to the synchronization of many times. Even if a period of time is not synchronized, it will only affect the data in this segment, and the entire call is not affected.

7.2.4 Security Analysis of Real-Time Voice Encryption System

Dachselt et al. [152] studied the cryptanalysis method of an encryption system based on discrete chaotic map. Yang et al. [153] thought that analog chaotic encryption system is difficult to resist the return mapping attack. Zhou et al. [154] reported that simple chaotic encryption system did not have a high security level and proposed an improved algorithm by increasing the number of iterations. Here, we design an encryption system, which combines the block encryption algorithm with chaotic encryption algorithm, and the security of system is analyzed by the following cryptanalysis methods:
1. The system has the ability to resist the brute-force attack. Because the encryption system is not dependent on the encryption scheme or algorithm, but relies on the secrete key. It is worth noting that secrete key signal generator could not be open, and some literatures about cryptanalysis often adopted this design principle. This

is particularly important for the advantages of chaotic encryption. The previous analysis indicates that the system adopts the improved DES algorithm, i.e. 3DES, to encrypt data packets, and its key space is 2^{112}. Since the secrete key of system is generated by chaotic map, the secrete key of each data packet is alterative. Therefore, the key space of each data packet is 2^{112}, which is large enough to resist brute-force attack. Even if the breaker acquires the equation of chaotic map, it is difficult to determine its parameters and initial conditions.

2. The system has the ability to resist the chosen plaintext and chosen ciphertext attacks. As the secrete key of system is generated by the chaotic map, and the secrete key sequence is unpredictable, there is no one corresponding to the plaintext and the ciphertext. The ciphertexts corresponding to the same text are different and not related. So, the chaotic encryption system is equivalent to one-time pad mechanism, and Shannon has proved that it is unable to be deciphered.
3. The system has the ability to resist spectrum analysis. Spectrum analysis is one of cipher cracking methods often adopted by cryptanalysts. However, because chaotic signal with wide frequency spectrum is similar to the noise, attackers cannot get any useful information about the encrypted signals through the spectrum analysis method. Therefore, this method has no effect on the output signal of chaotic encryption system. In addition, there is no correlation between the chaotic encrypted signal and the original signal, which ensures the security of encryption system. Any deciphered error to the ciphertext is amplified by the chaotic system, and it also improves the encryption intensity.

7.3 Realization of Real-Time Voice Encryption System

7.3.1 Structure of Real-Time Voice Encryption System

Based on ACE, RTP, and real-time encryption scheme, a real-time network voice encryption system is developed. Its structure is shown in Figure 7.10. The whole system consists of the following modules: voice acquisition/broadcast module, encoding/decoding module, encryption/decryption module, and transmission module. As a real-time system, it is necessary to have a real-time voice acquisition and broadcast module. Encryption operation is achieved by using real-time encryption scheme. The communication protocol is based on RTP, which is transmitted on the Internet through the network layer and physical layer.

Figure 7.10 is a P2P communication model. Multicast function is realized by using RTP in voice transmission module. So, it is not only used in P2P communication, but also can be used in many-to-many communication. The bidirectional arrows in the figure represent the bidirectional data stream.

The realization of each module is independent of each other. It communicates with other module by interfaces. Each module is realized by the static library (.lib file), such as VoiceIO.lib of voice capture/broadcast module, Codec.lib of voice

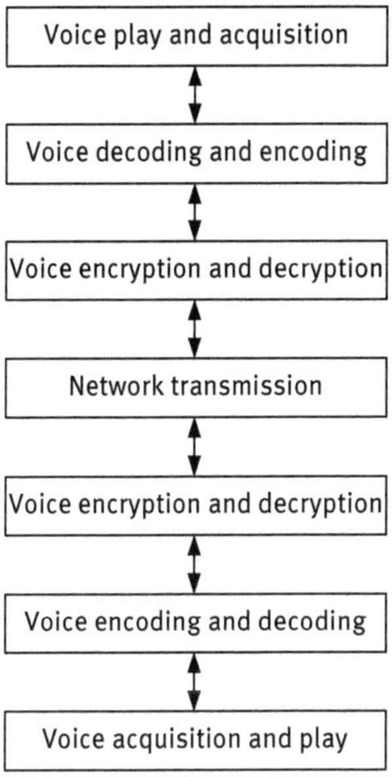

Figure 7.10: Structure of real-time voice encryption system.

encoding/decoding module, Crypto.lib of voice encryption/decryption module, and RTPLib.lib of voice transmission module. These static libraries can be used independently or in combination with external applications, which greatly enhances the reusability of each module.

Speech acquisition module records the voice data collected by microphone through sound card, while speech broadcast outputs the speech data by sound card and speaker. The signals via microphone and speaker are analog signals, and signals of computer processing and transmission are digital signals. So the role of sound card is to achieve the mutual conversion of the two kinds of signals. The conversion process is related to the sampling frequency, encoding bits, and channels number. Human voice frequency is about 10 kHz. According to the sampling theorem, the sampling frequency is chosen as 22.050 kHz. Here, the system uses 16-bit memory Pulse Code Modulation (PCM) encoding, single channel.

Encoding/decoding module employs A-law compression encoding. Through this method, 16-bit linear code obtained by sampling can be converted to 8-bit nonlinear encoding. In this way, the transmission rate can be reduced by half.

Encryption/decryption module uses the encryption scheme based on the 3DES (extended DES) and chaotic system (logistic map), and block encryption algorithm is used in the stream communication. Logistic map, synchronization process, and 3DES encryption algorithm are packaged together, and only two interfaces of encryption/decryption are exposed to external. As described in Section 7.2.3, chaotic synchronization is accomplished by the header of data packets, which include the

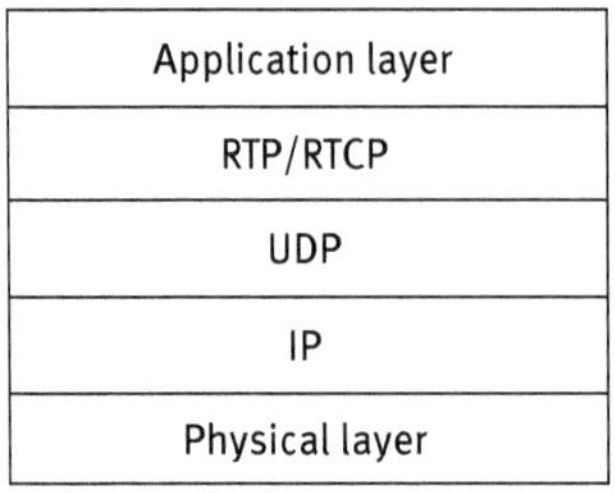

Figure 7.11: Application model of network voice stream based on RTP.

iteration number of logistic map, and the iteration number is directly used to generate the secrete key for encryption/decryption.

Network transmission module is based on RTP, and the network application model is shown in Figure 7.11. User Datagram Protocol (UDP) protocol is used in the RTP transport layer to achieve simple multicast function. The speech transmission module exposes the RTP_Session class to outside, and inside of this module is the specific implementation of RTP.

The RTP is a transport protocol for multimedia data stream on Internet. RTP is defined in the transmission conditions of P2P or one-to-many, and its purpose is to provide time information and achieve flow synchronization. RTP usually uses UDP protocol to transmit data, but RTP can also work on other protocols such as TCP or Asynchronous Transfer Mode (ATM). When an application starts an RTP session, two ports are used, one for RTP and the other for RTCP. RTP itself does not provide flow control or congestion control, but it provides these services by RTCP. RTP algorithm is usually not implemented as an independent network layer, but as a part of the application code. During the RTP session, the participants periodically transmit RTCP packets. The RTCP packet contains some statistical information such as the number of sending packets and lost packets, so that the server can use these information to dynamically change the transmission rate, and even change the effective load type. With the use of RTP and RTCP, they can optimize the transmission efficiency by effective feedback and minimum overhead, which is especially suitable for the transmission of real-time data on Internet.

The software running platform of voice encryption system based on RTP is Windows 2000 professional or Red Hat Linux 9.2. The development tools are Rose Rational 2000, C++ Visual 6, g++, and emacs. The third library is ACE 5.4 and Standard Template Library. The software framework is shown in Figure 7.12.

7.3.2 Real-Time Transmission Protocol and Its Data Structure

The RTP data packet is composed of two parts, the RTP header and the continuous media data with indefinite length, and its format is shown in Figure 7.13, where V represents the version number of RTP specified as 2. P is the padding bit. When $P = 1$,

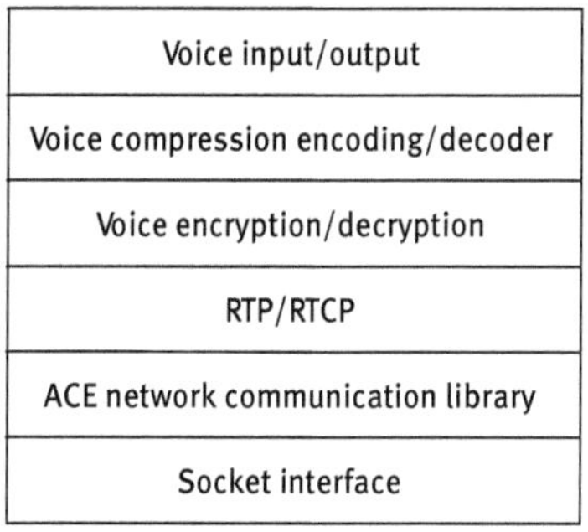

Figure 7.12: Software architecture diagram.

it indicates that the data packet tail is loaded with a nonload information, which can be used to encrypt or notify the bottom layer protocol. A data unit has several RTP packages. X is an extension bit, and when $X = 1$, it indicates that the RTP header is attached to an extended header with variable length. M is used to mark the important event in the data stream, such as the frame boundary. Load type is used to identify the type of media in the receiver. If it is the encoding data, it determines the decoder of the receiver. The serial number is used to identify the order of the RTP packets sent by the sender, and the serial number increases by 1 when an RTP packet is sent. The receiver detects the loss situation of data packet by serial number, and then the packet is recovered after reordering. The time stamp reflects the sampling interval of RTP packet in the first byte. The sampling frequency must be derived from a clock with linear growth, and the clock frequency usually depends on the data transmission format. If the RTP packet is periodically generated, then the sampling frequency is determined by the sampling clock, rather than the system clock. The synchronization source identifier (SSRC) identifies the synchronization source, which is randomly generated, and the SSRC identifier of two synchronization sources in the same RTP session is different. Contributing source identifier (CSRC) can contain 0–15 items. Contributing source refers to a new combination of RTP packets after the procession of mixer when one or more of RTP data packets are received from the synchronous source. The mixer is used as the SSRC of the combination RTP packet, and all the SSRC are transmitted to the receiver as a special source identifier, so that the receiver knows all the SSRC of RTP packets.

Another protocol is RTCP, which is usually used with RTP, and it can be used to carry out the feedback about transmission quality (such as the jitter and average loss rate) and transfers some identity information of participants. RTCP packet contains the following formats:

SR: source report package; it is used to send and receive the statistics of active source.

RR: receiver report package; it is used to receive the statistics of nonactive station.

<table>
<tr><td>V</td><td>P</td><td>X</td><td>CC</td><td>M</td><td>Load type</td><td>Serial number</td></tr>
<tr><td colspan="7">Timestamp</td></tr>
<tr><td colspan="7">SSRC</td></tr>
<tr><td colspan="7">CSRC</td></tr>
<tr><td colspan="7">Data payload
----------------</td></tr>
</table>

Figure 7.13: RTP packet format.

SDES: source site description information packet; it is used to report the related information of site, including the specification name CNAME.
BYE: send this packet when you want to leave.
APP: application package; it contains custom functions.

An RTCP composite package usually contains several RTCP packages. To perform this protocol, the following two points must be observed. First, in the case of the bandwidth allowed, receiving statistics (RR or SR) must be transmitted as soon as possible, so that the accuracy of statistics is improved. Therefore, each transmitted RTCP composite package must contain a report package. Second, the new source site must receive CNAME as soon as possible to verify the other source sites. So each RTCP composite package must contain a CNAME SDES too.

With the aid of the above control packet, RTCP can complete four control functions, including quality of service monitoring control, congestion control, flagging media synchronization, providing identification information, and estimation and planning of session size.

7.3.3 Implementation of RTP Based on Adaptive Communication Environment

ACE is an object-oriented (OO) toolkit, which implements many basic design patterns of communication software. The target user of ACE is the developer who develops high-performance communication services and applications on UNIX or Win32 platform. ACE simplifies the development of communication between processes, event channel separation, display dynamic link, and concurrent OO network applications and services. By dynamically linking services and applications at run time and executing these services in one or more processes or threads, ACE enables the system's configuration and reconfiguration to be automated.

As shown in Figure 7.14, the hierarchical architecture of ACE consists of three basic layers: uses layer of operating system (OS), C++ packet layer, and frame component layer [155]. ACE contains a lot of reusable packaging, framework, and classes,

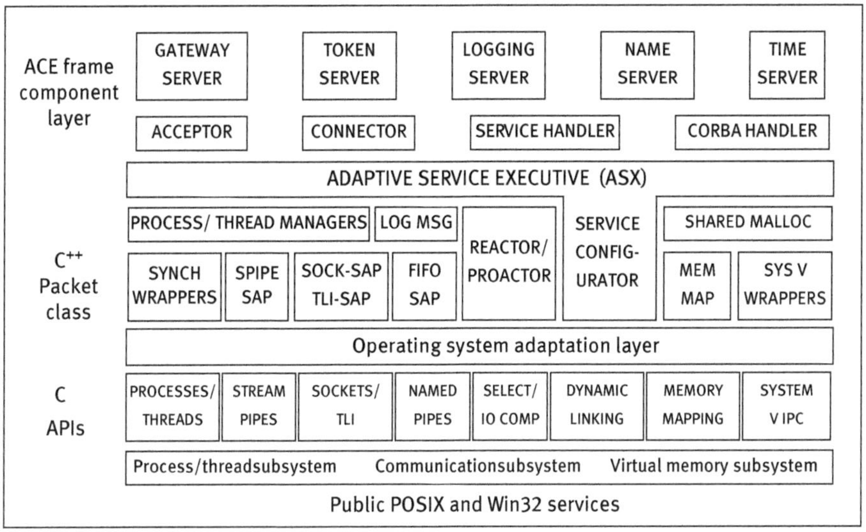

Figure 7.14: Structure of ACE and components.

which can be used to perform common network programming tasks across many OSs. ACE components can be used to achieve the following objectives: event demultiplexing and event processor allocation, connection establishment and service initialization, inter-process communication and shared memory management, dynamic configuration of distributed communication services, concurrency/concurrent, and synchronization.

The important ACE modules in this system are presented as follows:

1. SOCK_SAP. It is inherited from the IPC SAP, and it is the OO package of the bottom layer network programming interface.
2. ACE reactor. It is the OO framework for event demultiplexing.
3. ACE_Task class. It is the center of the active object and the passive object, which is used to create a user definition and process application messages in ACE.

The structure of RTP library is shown in Figure 7.15. Among them, RTP_Session module is the core module of the whole protocol library and is also the only interface exposed to the upper-level users. Users can control this module to start real-time data communication, and it is responsible for creating the RTCP_Recv_Handler module and RTCP_Timer_Handler modules. RTP_Sender is used to send RTP data packets. RTP_Reactor_Task is a thread object. It is the central hub for the protocol part, and it is also an I/O event dispatcher. RTP_Reactor_Task is used to create three classes, including RTCP_Recv_Handler, RTCP_Timer_Handler, and RTP_Recv_Handler. Then it records the arrival events of RTP and RTCP packets and the timing events of RTCP transmission. Finally, it calls these modules when these events come.

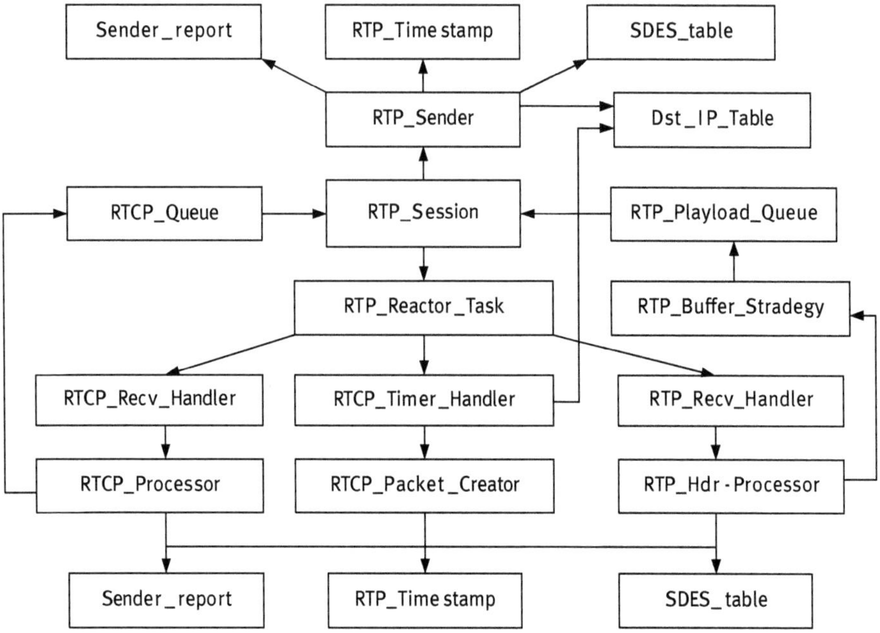

Figure 7.15: Structure of RTP library based on ACE.

RTP_Recv_Handler module is used to process the arrival event of RTP data packets. It will receive the RTP message when the event arrives. RTP_Hdr_Processor module is used to process the RTP message and extracts the valid data. RTCP_Recv_Handler module is used to receive RTCP message, which has a similar function with RTP_Recv_Handler. RTCP_Processor module is used to process the RTCP message and obtain its statistical information. RTCP_Timer_Handler module is used to send RTCP message. RTCP_Packet_Creator module is used to generate RTCP message. The Sender_Report module is used to set up and save some statistical information in the RTCP message. SDES_Table module is used to save and set SDES information. DST_IP_Table module is used to save the IP information of source site that is involved in the communication. RTP_Timestamp module provides some operations related with time, such as accessing system time or RTP timestamp. RTP_Payload_Queue module is used to save the received RTP data. RTCP_Queue module is used to save the received statistical information of RTCP.

7.3.4 System Debugging and Performance Testing

Test project RTVC (real time) is applied to debug each module of the system. CContoller class is used to combine and coordinate each module, which is responsible for receiving user's parameter settings (audio parameters, secrete key, and communication members) and coordinating the transmission of data stream among each module.

The problems encountered during the debugging process as well as their solution methods are illustrated as follows.

During the test of speech acquisition/playback module, when you stop recording, you are prompted that recording buffer still has data, and the stop recording operation failed, which results in the failure of recording operation again. The error reason is that the recording buffer is released in the message processing function WM_MIM_DONE, and it is not fully released when the recording is stopped. Therefore, the recording buffer memory is managed automatically. When you stop recording, the recording thread is hanged up (also the execution of message processing function WM_MIM_ DONE is suspended), and then the recording buffer is released explicitly. The recording thread is restored when you restart recording.

In addition, communication often automatically terminates when the system works for a period of time. Through testing and tracking, it is found that the output rate of data queue is faster than that of input, which results in that the playback module occasionally does not get data, and then the playback module will be shut down automatically. So communication is terminated. To solve this problem, the whole process is controlled manually. When the play module cannot obtain the data, this module is still working, but it does not fill data to the playing buffer.

The communication system is tested by the release version. The results are shown in Table 7.1. In the conditions of 22.050 kHz sampling frequency, 16-bit quantization data, single channel, and 1,024 bytes recording buffer (sampling time is approximately 22.68 ms), the effect of common mode (no encryption) is best, and the delay cannot be felt. The effect of encryption mode is second, and there is a little noise, but still can be understood. For the use of equal interval drive of the continuous chaotic network synchronization method, the system works in synchronized state. Even if a data packet header is damaged, it has no effect on communication process.

In the conditions of 22.050 kHz sampling frequency, 8-bit quantization data, single channel, and 1,024 bytes recording buffer (the sampling time is 45 ms), the communication effect of common mode is apparently worse than the above mode, and there are some time delay. The reason is that its sampling time is too long. The encryption mode of communication is not as good as the common mode, but its security is superior to the common mode.

Table 7.1: Results of system test using the published version (release).

Mode	Quantization bits	Sampling delay	Processing delay	Transmission delay	Voice quality	Safety performance
Common	16	22.68	0.125	0	4.5	Poor
Encryption	16	22.68	≈ 8	0	4.2	Excellent
Common	8	45	0	0	4.0	Poor
Encryption	8	45	≈ 8	0	3.9	Excellent

In this section, we study and develop a real-time voice encryption system based on chaotic map and 3DES encryption algorithm. The system has the characteristics of fast encryption speed, less encryption time, and strong ability of anti-breaking. In the implementation of the whole system, the modular design and multi-thread technology are used to minimize the coupling of each packaging module and improve their reusability. In the implementation of RTP/RTCP, the use of cross-platform ACE greatly improves the cross-platform characteristics of RTP/RTCP module (RTPLib.lib), and it can be ported to Windows 2000 Professional, Linux RedHad 9.2, and other OSs. Through the network test, the encryption system can realize the real-time encryption and transmission of voice signals, and it has good communication quality and high security. However, it is necessary to improve and optimize the encryption speed and signal delay for commercial use.

7.4 MPEG Video Encryption Algorithm Based on Chaotic Sequence

Video image data has a strong correlation, i.e. there is a large amount of redundant information. The redundant information can be divided into spatial redundant information and temporal redundant information. Compression technology is used to remove the redundant information in data stream (to remove the correlation between data), and it includes the intra-image data compression technology, inter-image data compression technology, and entropy encoding compression technology. At present, the commonly used video compression algorithms are MPEG, H.26X based on discrete cosine transform (DCT), and compression encoding based on discrete wavelet transform. H.26X standard is mainly used for real-time video communication, such as video conference and video phone. MPEG standard is mainly used for video storage (DVD), video broadcast and video streaming media, such as Digital Subscriberloop (DSL), Internet video, and wireless video. Here, we focus on MPEG video encryption algorithm based on chaos.

In recent years, the encryption algorithm combined with the compression algorithm is applied to MPEG video stream, which has received extensive attention. Bhargava et al. proposed an encryption algorithm, which the secrete key is used to vary the coefficients of DCT and sign bit of motion vector [156]. Based on this method, researchers proposed an algorithm to further encrypt the intra-DC coefficients. A perceptual video encryption algorithm (PVEA) is proposed by encrypting the fixed length code data of all compressed video stream. Due to the smaller amount of encrypted data, the compression algorithm has little effect on video coding and decoding and will not affect the compressed format. But there is still a defect about the security of MPEG. To maintain a fast encryption speed, the above algorithms often use a single stream cipher to carry out data encryption. But the stream cipher cannot resist known plaintext attack. Other optimization measures, such as the scrambling and the plaintext or ciphertext feedback mechanism, are gradually adopted. Considering that both the DC and AC coefficients have a very important effect on encryption, researchers advised to shuffle all DCT coefficients of the same frequency location [157]. To maintain the size of data, some researchers proposed a scheme to shuffle the variable length code or run-length encoded (RLE) event [158, 159] before the entropy encoding, and the same or better results can be obtained from these cryptographic algorithms.

7.4.1 Encryption Principle and Procedure of MPEG Video Data

A practical MPEG video encryption algorithm should meet the requirements of the invariance of compression format and ratio, the less computation quantity, and the widely adjustable security level. Encryption of fixed length code data can meet these requirements at utmost extreme, but the encryption degree and security need to be improved further. The analysis and experiment results show that the effective texture information is distributed in all the DCT coefficients. Scrambling for internal macro-block of each video frame is an effective means to hidden information, which is added to the fixed length data encryption algorithm. In general, the entire MPEG video encryption scheme consists of two parts as follows [160].

Step 1: The data are encrypted including the intra-DC coefficients, non-intra-DC coefficients, sign bit of AC coefficients, ESCAPE DCT coefficients, motion vector symbol, and residual fixed length code.

Step 2: The bit stream of macro-block is scrambled, and each frame is assigned a different scrambling table.

As all the operations are done on the video stream, the entire encryption algorithm has no effect on the compression ratio. Stream cipher and scrambling table are generated by the chaotic map. The principle block diagram of encryption algorithm is shown in Figure 7.16. It shows the complete process of chaotic MPEG video encryption. To reduce the chaotic iterations, we uses the same Cat map (with the same initial conditions and parameters) to generate the stream cipher and permutation table. So, the sequence cipher and block cipher (permutation) will share the same set of secrete key parameters $\{x_1(0), y_1(0), u, v\}$. Of course, to enhance the security, we can use chaotic map with different initial conditions and parameters to generate chaotic sequences for different encryption processes. In this case, the stream encryption and scrambling have independent secrete key, but it requires more chaos iteration and the amount of computation.

From Figure 7.16, the key space of whole encryption system is composed of $\{x_1(0), y_1(0), u, v, p, x_2(0)\}$. In the experiments, secrete key is reset for each image group. You can change the secrete key to improve the security of system by setting the step size vector $d = \{d_u, d_v, d_p, d_{x1}, d_{y1}, d_{x2}\}$. The decryption process is the symmetric operation of the encryption process.

The details of encryption process will be presented in the following sections.

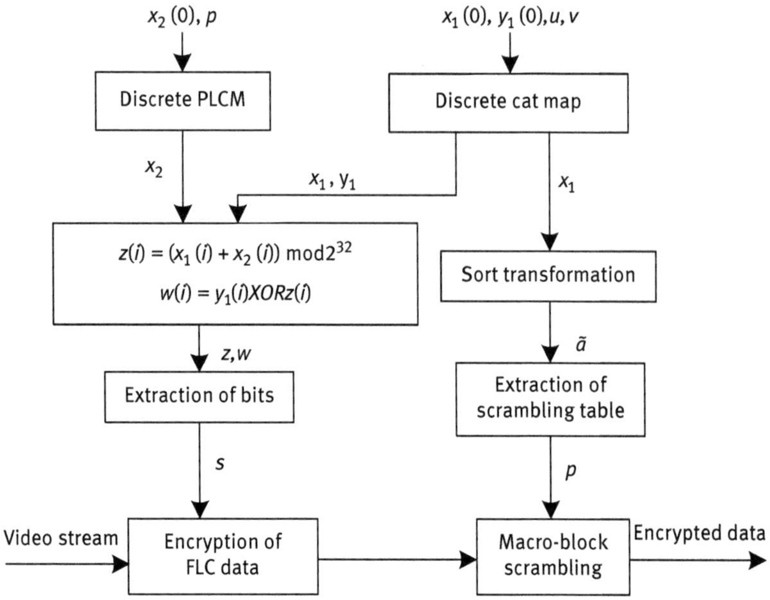

Figure 7.16: Principle block diagram of MPEG encryption based on chaos.

7.4.2 Design of Stream Cipher Based on Chaos

Chaos-based cryptography has been widely used in video encryption. To design a fast and secure chaotic cipher, we should using fixed point operation instead of floating point operation, using the simple chaotic system, a large encryption unit, and so on. So two simple chaotic maps are selected to design the stream cipher: they are cat map and piecewise linear chaotic map.

Cat map is defined by

$$\begin{bmatrix} x_1(k+1) \\ y_1(k+1) \end{bmatrix} = \begin{bmatrix} 1 & u \\ v & uv+1 \end{bmatrix} \begin{bmatrix} x_1(k) \\ y_1(k) \end{bmatrix} \bmod 1, \tag{7.7}$$

where u and v are control parameters.

Piecewise linear chaotic map is defined by

$$x_2(k+1) = \begin{cases} x_2(k)/p, & 0 \le x_2(k) < p \\ (x_2(k) - p)/(0.5 - p), & p \le x_2(k) < 0.5 \\ (1 - p - x_2(k))/(0.5 - p), & 0.5 \le x_2(k) < 1 - p \\ (1 - x_2(k))/p, & 1 - p \le x_2(k) < 1 \end{cases}, \tag{7.8}$$

where p is the control parameter.

To avoid floating point operations, the above two maps are discretized to operate in the integer domain. The discretized cat map is described by

$$\begin{bmatrix} x_1(k+1) \\ y_1(k+1) \end{bmatrix} = \begin{bmatrix} 1 & u \\ v & uv+1 \end{bmatrix} \begin{bmatrix} x_1(k) \\ y_1(k) \end{bmatrix} \bmod 2^{32}, \tag{7.9}$$

where u and v are unsigned integers, and initial conditions $x_1(0)$ and $y_1(0)$ are restricted to unsigned integers. Obviously, after the discretization, $x_1(k), y_1(k) \in [0, 2^{32})$, and they are integers, which ensure that all operations are fixed point operation. The discretized piecewise linear chaotic map is described by

$$x_2(k+1) = \begin{cases} \lfloor 2^{32} \cdot x_2(k)/p \rfloor, & 0 \le x_2(k) < p \\ \lfloor 2^{32} \cdot (x_2(k) - p)/(2^{31} - p) \rfloor, & p \le x_2(k) < 2^{31} \\ \lfloor 2^{32} \cdot (2^{32} - p - x_2(k))/(2^{31} - p) \rfloor, & 2^{31} \le x_2(k) < 2^{32} - p \\ \lfloor 2^{32} \cdot (2^{32} - x_2(k))/p \rfloor, & 2^{32} - p \le x_2(k) < 2^{32} \end{cases} \tag{7.10}$$

where $\lfloor x \rfloor$ represents an integer that is no more than x, and integer $p \in (0, 2^{31})$. Correspondingly, $x_2(k)$ is an integer that belongs to $[0, 2^{32})$.

Thus, x_1, x_2, and y_1 are three different chaotic integer sequences generated by two chaotic maps. To avoid the degradation of chaotic characteristics after discretization, the following improvement measures are taken. First, two chaotic maps are iterated for 50 times as pretreatment. Then, starting from the 51th iteration, the generated data is processed to obtain two new integer sequences z and w as follows:

$$z(i) = (x_1(i) + x_2(i)) \bmod 2^{32}, \tag{7.11a}$$

$$w(i) = y_1(i) \mathrm{XOR} z(i), \tag{7.11b}$$

where XOR is the XOR operation between bits, and mod is modulo operation. This operation not only reduces the degradation of original chaotic systems but also improves the complexity of the generated chaotic sequence. The new sequences, $z = [z(0), z(1), \ldots, z(n-1)]$ and $w = [w(0), w(1), \ldots, w(n-1)]$, have better random properties. Then, each of them is transformed into the corresponding binary form, i.e. $z(i) = b_0 b_1 \ldots b_{31}$ and $w(i) = b'_0 b'_1 \ldots b'_{31}$ (b'_j and b_j are binary numbers, $j = 0, 1, \ldots, 31$). A binary random sequence s is obtained by extracting the bits of integer random sequence z and w, and alternately permuting them. Therefore, when the Cat and Piecewise Linear Chaotic Map (PLCM) maps iterate one time, a 64-bit secrete key stream $b_0 b'_0 b_1 b'_1 \ldots b_{31} b'_{31}$ is produced.

Through a series of random tests, it is proved that the sequence s has good randomness. First, set the initial conditions of $u = 20$, $v = 2$, $p = 2^{20}$, $x_1(0) = 600$, $y_1(0) = 100$, and $x_2(0) = 1,000$. Then, for each of iteration, bits of $z(i)$ and $w(i)$ are alternates extracted to obtain three binary random sequences s_1, s_2, and s_3 with length of 200, 2,000, and 20,000, respectively. These binary random sequences are tested

Table 7.2: Random test results for different length sequences.

Sequence length	Balance test	Runs test				
		1-Run	2-Run	3-Run	4-Run	5-Run
200	0.0552	0.4754	0.2452	0.1321	0.0632	0.0352
2,000	0.0215	0.4895	0.2485	0.1239	0.0624	0.0315
20,000	0.0047	0.4977	0.2496	0.1248	0.0627	0.0313

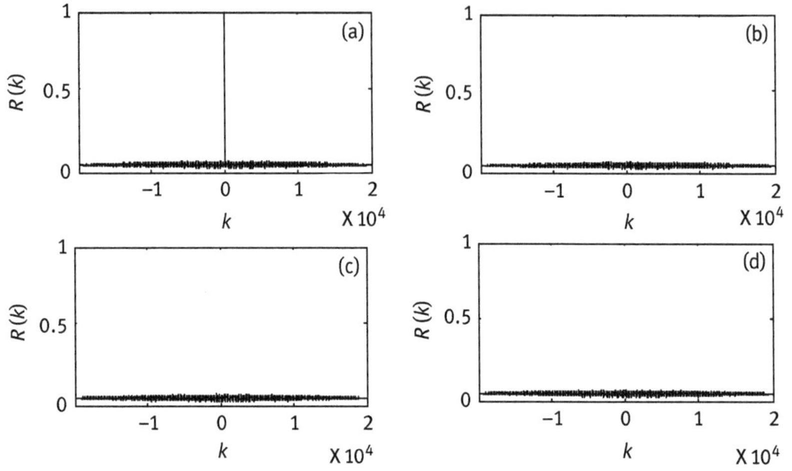

Figure 7.17: Autocorrelation and cross-correlation tests of chaotic sequence: (a) autocorrelation; (b) cross-correlation of two sequences with different parameter p; (c) cross-correlation of two sequences with different $x_1(0)$; and (d) cross-correlation of two sequences with different parameter u.

randomly, and the results are shown in Table 7.2. As shown, all of them pass the test and show excellent performance.

To ensure the security of the cipher, it is necessary to carry out the correlation test. The sequences with different lengths are chosen to carry out the autocorrelation and cross-correlation tests. Test results are shown in Figure 7.17.
Obviously, the results of autocorrelation tests are ideal as shown in Figure 7.17(a). Using two different values of p, $x_1(0)$, and u, respectively, two different chaotic sequences with the same length are obtained to make the cross-correlation test, and the results are shown in Figure 7.17(b–d). As shown, the test results are also satisfied, which means that the chaotic binary sequence has high sensitivity to parameters of chaotic system.

In this scheme, the sequence s as the stream cipher is used to encrypt the fixed length data of video stream, and the ciphertext feedback mode is adopted to resist the known plaintext attack in the encryption process.

7.4.3 Design of Scrambling Algorithm Based on Chaos

The algorithm requires that each frame takes different scrambling tables to scramble macro-block. Here, the sorting-based method is adopted, and it has advantages on simple operation and fast running speed. We suppose that each frame is made up of N macro-blocks. The specific scrambling process for each frame is as follows.

1. A sequence with length N, $a = \{a_1, a_2, \ldots, a_N\}$, is extracted from the chaotic integer sequence. Here, the chaotic sequence is generated by the same chaotic map with the upper stream cipher. That is to say, a sequence with length N can be obtained from the chaotic sequence x_1, y_1 or x_2, and x_1 is selected in this experiment.
2. Rearrange the sequence in ascending or descending order, and then a new sequence is generated as $\tilde{a} = \{\tilde{a}_1, \tilde{a}_2, \ldots, \tilde{a}_N\}$. This method is feasible because there do not exist the same elements of chaotic sequence.
3. Record the location in new sequence $\{\tilde{a}_1, \tilde{a}_2, \ldots, \tilde{a}_N\}$ of each element in the original sequence $a = \{a_1, a_2, \ldots, a_N\}$ to get the corresponding address sequence $p = \{p_1, p_2, \ldots, p_N\}$.
4. Use the scrambling addresses $p = \{p_1, p_2, \ldots, p_N\}$ to scramble the macro-block after the current frame is scanned.

The above scrambling method does not increase the redundancy iteration, which makes the whole process simple and quick.

7.4.4 Encryption Experiment and Encryption Effect Analysis

The experimental environment are the open-source codec of MPEG-4, xvidcore-1.1.2. XVID has high encoding speed, and it is considered to be the fastest MPEG-4 encoder at present. Moreover, the XVID encoder is designed for Windows users. The source code is completely open, and the quality of image compression is close to DVD quality. But the data quantity is only one-eighth of DVD. So XVID encoder is the best choice for Windows users.

XVID encoder is developed in the UNIX OS and C++6.0 Visual environment. When it is used in the Windows system, you must remove the UNIX format first. The specific method is to use the writing board to open the *.dsw, *.dsp, and Makefile files, and then save them directly. So it does not appear that there is no phenomenon of engineering loading when opening the xvidcore.dsw file.

This encryption algorithm is combined with compression algorithm. The basic idea of encoding includes initializing the encoder, creating an encoding instance handle, circular encoding for each frame data, and destroying the existing encoding instance handle to stop encoding. Encryption algorithm is embedded in compressed encoding. For each frame of the image data, the process is shown in Figure 7.18. First,

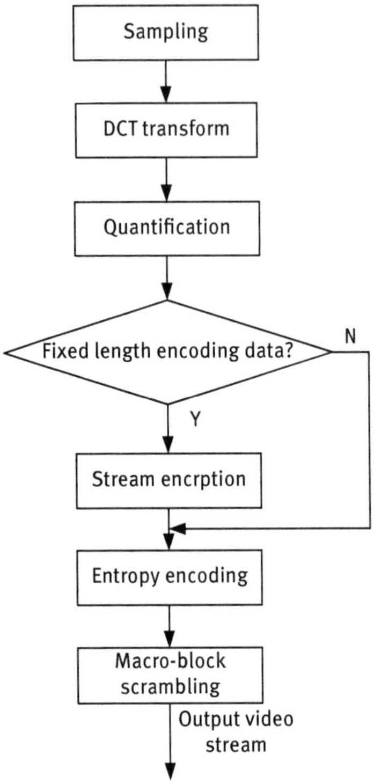

Figure 7.18: Flowchart of the encryption algorithm combined with compression encoding.

the encoder samples the pixels, and then do DCT. The encryption of the fixed length data is carried out after quantification, and the scrambling of macro-block is done after the entropy encoding. To do so, it will not affect encoding.
According to this procedure, a series of standard test sequences are tested. Figure 7.19 is the encryption effect diagram for two different standard test sequences. Figure 7.19(a) and 7.19(c) is the first frame image of foreman and akiyo, respectively. Figure 7.19(b) is the first frame decoding image of foreman only using the macro-block scrambling method, and Figure 7.19(d) is the first frame decoding image of akiyo using the complete encryption algorithm.

Therefore, attackers cannot get any valuable information from the scrambled and encrypted image. This scheme has good property of format invariance and its control parameters can be further set up to achieve different security levels.

7.4.5 Performance Analysis of MPEG Video Encryption System

The performances of encryption system are analyzed mainly from the following two aspects.

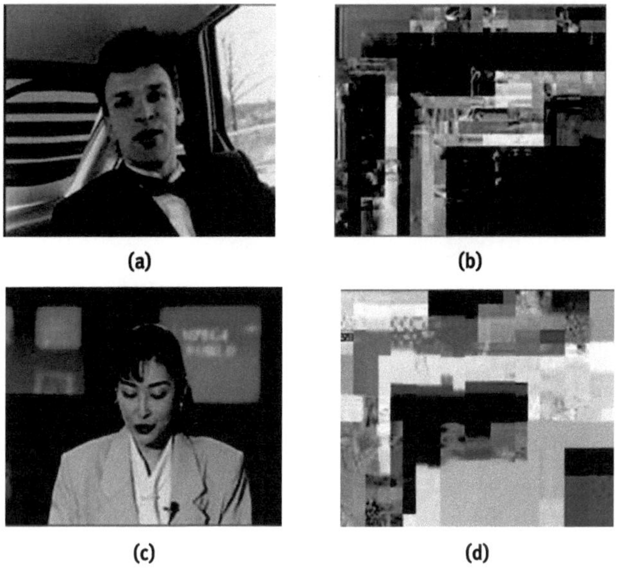

Figure 7.19: Video encryption effects: (a) foreman plain image; (b) effect of macro-block scrambling; (c) akiyo plain image; and (d) full encryption effect of this scheme.

7.4.5.1 Analysis of the System Security

As discussed previously, for a group of images, key space of the chaotic encryption system consists of $\{u, v, p, x_1(0), y_1(0), x_2(0)\}$, which is large enough to resist a brute-force attack. Considering the scrambling process with the size N, for frame p, the majority of macro-blocks come from the corresponding prediction of the previous frame. Because the previous frame has been scrambled, attackers are unable to restore the current image. The same results can be obtained for frame B. For frame I, even in the *qicf* video clips, the size of scrambling is $N = 176 \times 144/16^2 = 99$. So the search space is 99! which is large enough to resist the brute-force attack too.

In the fixed length code data encryption, the feedback operation makes secrete key stream to depend on the whole plaintext, and thus can resist the known/chosen plaintext attack. In the scrambling process, if an attacker obtains a pair of plaintext and ciphertext, it is clear that the scrambling table can be obtained by comparing. However, in this system, the scrambling table changes according to the chaotic sequence, and the evolution of chaotic sequence is difficult to be predicted. Therefore, it is not possible to infer the original chaotic sequence from the scrambling algorithm. In addition, the secrete key is changed in real time with the change of images, resulting in that the attack is more difficult.

Experiments show that this algorithm can also resist error concealment attack (ECA). In our experiments, different encryption algorithms are used to encrypt a 176×144 akiyo video fragment, and then they are attacked by ECA as shown in

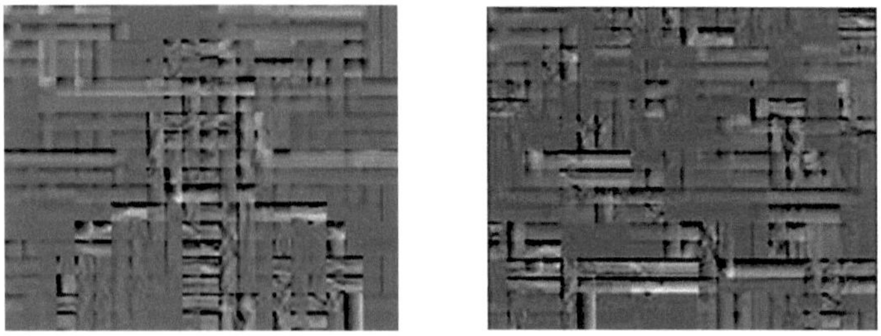

(a) (b)

Figure 7.20: Results of ECA attack on the first frame of "akiyo": (a) encryption algorithm with fixed length data and (b) new algorithm.

Table 7.3: Results of encryption speed test.

Video clips	Size	Ratio (%)
Akiyo	176 × 144	2.5
Carphone	176 × 144	1.9
Foreman	176 × 144	2.1
Miss America	176 × 144	1.6
Dancer	352 × 288	3.1
Singer	352 × 288	2.4

Figure 7.20. As can be seen, if a single compressed video stream is encrypted by using fixed length data, a basic outline can be seen after the ECA attack. But it is completely unable to be identified by using our encryption scheme.

7.4.5.2 Analysis of Encryption Speed

Experiments show that the new encryption algorithm has a fast speed. By employing two simple chaotic maps and iterating the chaotic maps in the integer domain, the computation time of encryption system is greatly reduced. On the other hand, because of the selected finite fixed length code data to be encrypted and simple scrambling scheme, the average number of iterations is greatly reduced. Taking several 90 frames video fragments as example, the ratios of encryption time and compression time are tested and recorded. Results are shown in Table 7.3. As can be seen, the time cost of encryption process has little effect on the MPEG4 compression, and this scheme can meet the real-time communication requirements.

7.5 H.264 Video Encryption Algorithm Based on Chaos

H.264 is a video encoding standard with extremely low bit rate, which is adopted by the International Telecommunication Union. It is the tenth part of MPEG-4 and is also called advanced video coding by the International Organization for Standardization (ISO). H.264 standard has better image quality, higher compression ratio, and better network adaptability than that of the previous video encoding standard. However, its good encoding effect is at the cost of a huge amount of computing. Therefore, the encryption of H.264 should be considered as a simplified encryption algorithm, which can reduce the computational complexity and the cost of secrete key, and simultaneously the compression effect does not be affected.

In this section, a new chaotic video encryption scheme is designed.

7.5.1 Chaotic Stream Cipher Based on Tangent Delay-Ellipse Reflecting Cavity System

Chaotic stream cipher is constructed by tangent delay-ellipse reflecting cavity system (TD-ERCS) chaotic system. In fact, the TD-ERCS chaotic system is an iterative map, which is discussed in Section 3.1.4. Iterate the chaotic map for one round at each time, and then the tail numbers of results are extracted to construct pseudorandom binary sequence. The generated sequences x and k are denoted as $\{x_j|j = 1, 2, 3, \ldots\}$ and $\{k_j|j = 1, 2, 3, \ldots\}$, where x_j and k_j represent the iteration values x and k in round $m + j - 1$. The binary codes of k_j and x_j mantissa are extracted from low to high and are denoted as $\{K_j^P|P = 1, 2, \ldots, 51, 52\}$ and $\{X_j^P|P = 1, 2, \ldots, 51, 52\}$, respectively. The binary pseudorandom sequence s of TD-ERCS is obtained by alternant arranging K_j^P and X_j^P. As it can be seen, when the TD-ERCS goes into normal state, each round of iteration can generate 104-bit binary sequence, which is 13 bytes.

According to the characteristics of H.264 standard, the selective encryption algorithm is investigated based on chaotic pseudorandom sequence. The main feature of this algorithm is that chaos is introduced to enhance the security of system and the encryption speed is improved. The encryption algorithm has low complexity and good performance on security and real time. It has little influence on the compression performance of encoding and has certain data operability, which is consistent with the basic requirements of video data encryption.

7.5.2 H.264 Video Encryption Scheme

To reduce the amount of data to be encrypted, the most effective data must be selected for encryption. Because the human eye is more sensitive to the brightness information than the color information, and the brightness information and color information are contained in the compressed file separately. Here a few key data are selected to encrypt. The basic encryption framework is shown in Figure 7.21. The encryption content of this scheme includes all sign bits of DC and AC coefficients in 4×4 luminance block, the sign bits of DC coefficients in chroma block, and the sign bits in motion vector of P/SP frame and B frame. The reason for choosing these data to encrypt is

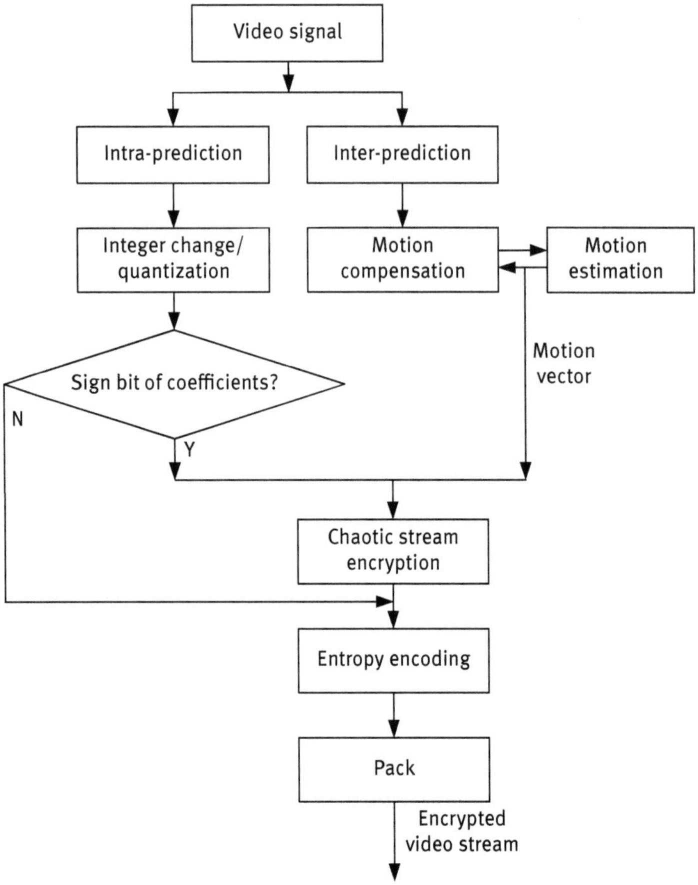

Figure 7.21: Flowchart of H.264 encryption process.

because those data can maximize the impact of visual effects and maintain the basic compression ratio and signal-to-noise ratio (SNR) of encoding at the minimum cost.

The chaotic stream cipher is designed by using the above pseudorandom sequence *s*. To improve the security, the following ciphertext feedback mode is adopted. Encryption process is described by

$$C(i) = \mathrm{XOR}(s(i), C(i-1-s(i))), \tag{7.12}$$

where XOR is the bit XOR operation; *s* is the secrete key stream; and $C(i)$ is the ciphertext.

Experiments are carried out on the H.264 open-source codec JM. JM is one of the three official encoders of H.264. Although the algorithm is complex and its speed is slow, it is suitable for the second development. To run JM in the VC6.0 compiler, you

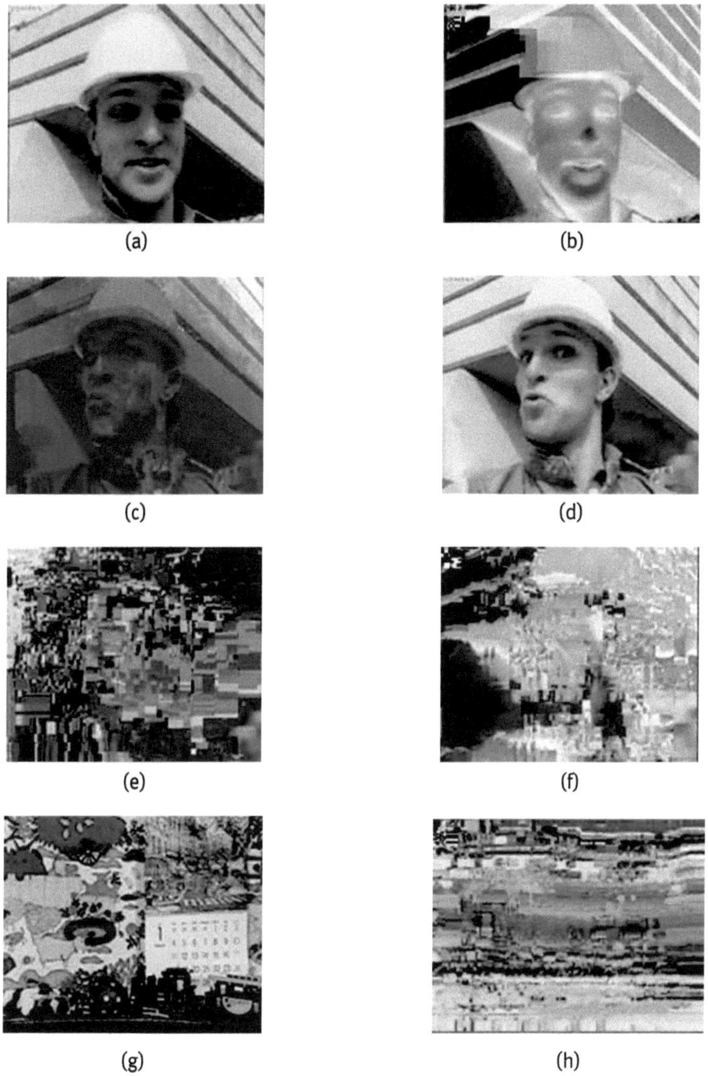

Figure 7.22: H.264 video encryption effects. (a) First frame of foreman. (b) First frame full encryption of foreman. (c) Encryption symbol bit of brightness of DCT coefficient. (d) Encryption symbol bit of chroma DC coefficient. (e) mv encryption for frame 89 of foreman. (f) Full encryption for frame 89 of foreman. (g) Mobile plain image. (h) Full encryption of mobile image.

have to set up configuration file before the coding and decoding. For example, the encoding configuration file can be set to encoder_baseline.cfg.

In this experiment, the encryption effect of different data in the standard video sequence is recorded as shown in Figure 7.22. As it can be seen, different data can be encrypted to achieve different results. For the encryption of motion vector, an inevitable result is that the encryption effect is not obvious for a set of video sequences at the beginning as shown in Figure 7.22(b). With the accumulation of motions, the ciphertext information of encrypted motion vector will continue to be accumulated, and the effect is more and more obvious as shown in Figure 7.22(e) and 7.22(f). After full encryption, the image has become very blurred as shown in Figure 7.22(f) and 7.22(h). To improve the security of encryption further, we can scramble the frequency coefficient, 4× 4 change block or Flexible Macroblock Ordering (FMO)-based macro-block. Of course, those algorithms are based on the cost of compression ratio and encryption delay.

7.5.3 Performance Analysis of Video Encryption

7.5.3.1 Security Analysis of the Encryption System

On one hand, for the chaotic pseudorandom sequence generator based on the TD-ERCS map, the total secrete key parameters are (x_0, $tg\alpha$, μ, m), and small changes in parameters will generate a completely different sequence. Obviously, the key space is large enough to resist the brute-force attack. At the same time, because of the real-time variation of the chaotic iteration parameters and initial conditions in each image, the algorithm has the ability to resist the chosen plaintext attack.

On the other hand, for the selective encryption scheme, encrypted data includes the DCT coefficients, which represent the texture information, and the motion vector, which represents the motion information. To do so, the readability of video data can be greatly damaged.

7.5.3.2 Real-Time Analysis of the Encryption System

The encryption speed depends on the producing speed of the chaotic stream cipher and the amount of encrypted video data. For TD-ERCS, the system can generate 104-bit secrete key stream for each round of iteration. For encrypted data, the proposed encryption scheme only selects the key data of video stream and only involves the sign bit. For example, considering that the majority of AC coefficients of chroma are zero, the encryption effect is not affected. So the DC coefficients of chroma are chosen to be encrypted. Experiments are carried out on several standard sequences with Quarter Common Intermediate Format (QCIF) format, and the real-time performance is shown

Table 7.4: Results of encryption speed.

Video clips	Size	Frame number	Ratio (%)
Carphone	176 × 144	90	1.26
Foreman	176 × 144	90	1.82
Mobile	176 × 144	300	2.76
Dancer	352 × 288	150	3.05

in Table 7.4. The percentage of encryption time accounting for encoding time shows that the encryption system has good real-time performance.

Any encryption method may have an impact on the encoding effect of video sequences. Practical encryption algorithm should reduce the influence of compression ratio and SNR of video sequences. Because only the sign bits of video data are encrypted, the encryption algorithm has little impact on the SNR and compression ratio of video sequence. This algorithm takes the macro-block, the frame, and the image group as the encoding unit, so it can keep the frame synchronization information of the video data to be invariable, and it has certain data operability, such as supporting the operations of position index, pasting, and cutting. For the encryption to DCT and quantization coefficients, it does not change any information about data format. The experimental results show that the SNR and compression ratio are not affected.

Questions

1. What are the technical difficulties of real-time data stream encryption?
2. How to evaluate the security of real-time data stream encryption?
3. How to deal with the relationship between encryption and compression in video transmission?
4. Design a wireless transmission scheme based on chaos.

8 Analysis and Simulation of Fractional-Order Chaotic System

With the further research of nonlinear chaotic theory, fractional-order chaotic system has become a hot topic. To apply the fractional-order chaotic system in the field of information security, it is necessary to reveal its characteristics. Moreover, key technical problems are required to be solved quickly, such as accurate solution of fractional differential equations, digital signal processing (DSP) circuit realization, and so on. In this chapter, we focus on solution algorithm and simulation technology. It is the theoretical and experimental basis for applications of fractional-order chaotic system.

8.1 Development of Fractional-Order Chaotic System

Theoretical research of fractional calculus has more than 300 years history, and it has the same long history as the integer-order calculus. However, due to the lack of application background, its development is relatively slow. Actually, fractional dynamic behavior can be observed in many physical systems, such as viscous system, dielectric polarization, colored noise, electromagnetic waves, and so on. Compared to integer-order calculus equation, fractional-order equation can describe nature phenomenon more accurately in applied science areas [161] such as memory of various materials, mechanics and electrical characterization, viscoelastic dampers, power fractal network, fractional-order sinusoidal oscillator, robots, electronic circuits, electrolytic chemical, fractional capacitance theory, fractal theory, fractional proportional–integral–derivative controller, viscoelastic vibration control system, flexible configuration object, and so on. Since Mandelbort first reported that there are numerous fractal phenomena in nature in 1983 [162], fractional calculus has become a hot topic and has developed quickly.

At the beginning, the investigation of fractional-order chaotic system was based on the research of integer-order chaotic system. When fractional differential operator is introduced to nonlinear dynamic system, the fractional-order system still exhibits chaotic behavior, such as Chua circuit [163], a nonautonomous Duffing system [164], improved Duffing system [165], fractional Jerk system [166], Rössler system [167], and so on. Although dynamic characteristics of some fractional-order chaotic system have been investigated, there still remain many other fractional-order chaotic systems to be studied.

Up to now, there are many definitions about fractional-order calculus [168]. Li and Deng studied three definitions of fractional-order calculus deeply [169]. Among these definitions, Riemann–Liouville (R-L) definition and Caputo definition are usually used in practice. Although the R-L definition can simplify the calculations of fractional differential, its physical meaning is not clear and its value is difficult to be measured. However, when using Caputo definition, its physical meaning of fractional differential is clear and its value is easy to be measured. Based on these two definitions, people proposed two different approximate methods for solving the fractional differential equation. One is time–frequency method based on R-L definition [170]. First, the fractional integral operator in the time domain is transformed into transfer function in frequency domain. Then piecewise linear approximation method is used to realize approximation calculation in frequency domain. The other is predictor–corrector method based on Caputo definition [171–173]. It is a time domain method. The time series of the fractional-order system are obtained by Adams–Bashforth prediction equation and Adams–Moulton correction equation. However, frequency domain method cannot reflect the dynamic characteristic of fractional-order chaotic system accurately. As for the time domain method, the calculation step can be set small enough for more accuracy, but its calculation is more complex relatively.

By far, the chaotic characteristic study of fractional-order system is mainly based on existing integer-order chaotic system. The main idea is to observe whether the system is still chaotic when the system order changed from integer to fractional. However, the system order when system entering into chaotic still needs further research, as well as dynamic characteristic of fractional-order chaotic system. For example, Ahmad reported that when the order of system is $2 + \varepsilon$ $(0 < \varepsilon < 1)$, third-order chaotic systems still perform chaotic behavior [6]. And fourth-order chaotic systems still perform hyperchaotic behavior when the system order is $3 + \varepsilon$ $(0 < \varepsilon < 1)$ [174]. Obviously, the range of ε is too big in this case. Is it possible to narrow? For example, fractional-order Chua system performs chaotic behavior when the order is 2.7. People get the conclusion only through observing system behavior when the system order is 2.7 and 2.6. So is it chaotic when the order is 2.65? As a result, smaller step size should be chosen and smaller error fractional differential approximation algorithm model should be designed. Besides, dynamic characteristics such as Lyapunov exponent spectrum, bifurcation, and dimension remain further research.

Synchronization technology is the key of application of chaotic system to secure communication. In 1990, Pecora and Corroll first proposed the concept of chaotic synchronization [10]. They also realized synchronization of two mutually coupled chaotic systems in the laboratory by using the same signal circuits driven [11]. In the next 20 years, chaotic synchronization developed a lot. Even though synchronization of integral chaotic system has achieved much research results, research of fractional-order system synchronization is still in the beginning because it is too complex and the effective mathematical tool is deficient. If we apply integral-order control method to fractional-order system, true characteristic will be ignored. As a result, studying control method of fractional-order chaotic system has practical significance.

By far, more effective synchronization control method has not been proposed for fractional-order chaotic system. In most cases, we develop integral chaotic synchronization method to fractional-order chaotic, so its synchronization control mechanism and performance need further study. Studies show that the mutual coupling synchronization, projective synchronization, and adaptive method have unique advantages on improve synchronization performance. The target of fractional-order system synchronization research is to improve system performance and finding general law when changing order and parameter. The most suitable synchronization method should be chosen through comparing synchronization establishment time and synchronization accuracy. It is significantly important to lay technical foundation for the application of fractional-order chaotic system in secure communication.

Application of fractional-order chaotic systems is becoming a new direction of nonlinear applied science research area [175], especially in the aspects of information secure communication. Complexity of fractional-order chaotic sequence is a key point of the system safety performance. It is directly related to the performance of cryptography of fractional-order chaotic encryption system. For example, to ensure maximum communication capacity of spread spectrum communication, practical chaotic pseudorandom code sequence should have the biggest complexity. Complexity of fractional-order chaotic system contains two meanings. One is for chaotic system, which is always related to nonlinear term of system equation. As for fractional-order system, it may also be associated with order. This kind of complexity is mainly measured by calculating system fractal. The other meaning is for complexity of time series generated by fractional-order chaotic system. It is mainly used to measure the similarity degree of time signal and random signal. Some parameters, for example, information entropy, Lyapunov exponent, and fractal dimension, are adopted to measure the degree of system disorder. But these methods are not suitable to describe the complexity of time sequence generated by chaotic system. Therefore, the phase space observation method and structural complexity algorithm should be combined to analyze the fractional-order chaotic system and corresponding chaotic pseudorandom sequence. It will provide basis for the application of fractional-order chaotic system to cryptography and secure communication.

Circuit implementation of fractional-order chaotic system is another key issue of chaotic secure communication. In 2006, Bohannan et al. [176] applied for a patent to design fractional devices. Liu led his team to achieve circuit simulation for a number of fractional-order chaotic system through designing equivalent circuit of fractional integral [177–179]. But these methods are all based on frequency domain method. Even though the time domain method has significant advantages in studying fractional-order chaotic system, its circuit implementation is quite difficult. However, as the development of microelectronics, it is possible to realize time domain approximate method based on Field Programmable Gate Array (FPGA) or DSP. This method is a new way to realize the application of fractional-order chaotic system.

In all, research of fractional-order chaotic system is still at the initial stage. There still remain a lot of theoretical and practical issues that need further study. In this chapter, we will analyze the relationship of dynamic characteristics and fractional order. Synchronization of fractional-order chaotic system will also be studied.

8.2 Definition and Physical Meaning of Fractional Calculus

Generally, fractional calculus is denoted by the operator ${}_aD_t^q$, which indicates both of the fractional differential and the fractional integration. When q is positive, the operator represents fractional differential. When q is negative, it indicates fractional integration. Here, we introduce several common definitions about fractional calculus.

Definition 8.1: Gamma function $\Gamma(x)$ is a basic function for fractional calculus, and its definition is

$$\Gamma(x) = \int_0^{\infty} t^{x-1} e^{-t} dt. \tag{8.1}$$

When $x = n$, $\Gamma(n) = (n-1)!$

Definition 8.2: Grünwald–Letnikov fractional calculus is defined as

$${}_aD_t^q f(t) = \frac{d^q f(t)}{d(t-a)^q} = \lim_{N\to\infty} \left[\frac{t-a}{N}\right]^{-q} \sum_{j=0}^{N-1} (-1)^j \binom{q}{j} f\left(t - j\left[\frac{t-a}{N}\right]\right). \tag{8.2}$$

Definition 8.3: Riemann–Liouville fractional calculus is defined as

$${}_aD_t^q f(t) = \begin{cases} \frac{1}{\Gamma(-q)} \int_a^t (t-\tau)^{-q-1} f(\tau) d\tau, & q < 0 \\ f(t) & q = 0, \\ D^n [{}_aD_t^{q-n} f(t)] & q > 0 \end{cases} \tag{8.3}$$

where $q \in R$, n is the first integer greater than q, which means $n-1 \leq q < n$, $\Gamma(\cdot)$ is Gamma function. According to the above definition, q-order differential of power function and constant is

$$D_{t_0}^q t^r = \frac{\Gamma(r+1)}{\Gamma(r+1-q)} (t-t_0)^{r-q}, \tag{8.4}$$

$$D_{t_0}^q C = \frac{C}{\Gamma(1-q)} (t-t_0)^{-q}. \tag{8.5}$$

The Laplace transform of R-L fractional calculus is defined as

$$L\{{}_0D_t^q f(t)\} = s^q F(s), \quad q \leq 0, \tag{8.6}$$

$$L\{{}_0D_t^q f(t)\} = s^q F(s) - \sum_{k=0}^{n-1} s^k {}_0D_t^{q-k-1} f(0), \quad n-1 < q \leq n \in N. \tag{8.7}$$

As the Laplace transform needs the value of noninteger-order differential at $t = 0$, it is difficult to solve Laplace transform of R-L fractional-order differential. However, it can be solved according to the Caputo definition of fractional differential

Definition 8.4: Caputo fractional differential is defined as

$$
{}_0D_t^q f(t) = \begin{cases} \frac{1}{\Gamma(m-q)} \int_0^t \frac{f^{(m)}(\tau)}{(t-\tau)^{q+1-m}}, & m-1 < q < m \\ \frac{d^m}{dt^m} f(t), & q = m \end{cases}, \tag{8.8}
$$

where m is the first integer greater than q. The Laplace transform of Caputo fractional differential is

$$
L\{{}_0D_t^q f(t)\} = s^q F(s) - \sum_{k=0}^{n-1} s^{q-1-k} f^{(k)}(0), \quad n-1 < q \le n \in N. \tag{8.9}
$$

When compared with Laplace transform of R-L fractional differential, for Caputo fractional differential, only integer-order differential is needed in Laplace transform. When the initial value of operator is zero, its corresponding Laplace transform is

$$
L\{{}_0D_t^q f(t)\} = s^q F(s). \tag{8.10}
$$

Two basic properties of Caputo fractional differential definition are

$$
{}^*D_{t_0}^q J_{t_0}^q f(t) = f(t), \tag{8.11}
$$

$$
J_{t_0}^q ({}^*D_{t_0}^q) f(t) = J_{t_0}^q J_{t_0}^{m-q} D_{t_0}^m f(t) = J_{t_0}^m D_{t_0}^m f(t) = x(t) - \sum_{k=0}^{m-1} x^{(k)}(t_0^+) \frac{(t-t_0)^k}{k!}. \tag{8.12}
$$

Thus, the q-order differential of the power function and constant are defined as

$$
{}^*D_{t_0}^q t^r = \frac{\Gamma(r+1)}{\Gamma(r+1-q)} (t-t_0)^{r-q}, \tag{8.13}
$$

$$
{}^*D_{t_0}^q C = 0. \tag{8.14}
$$

Integer-order calculus has clear geometric interpretation and physical meaning. However, because of the complexity of fractional calculus, it is difficult to understand the concept of fractional calculus. Therefore, there exist some obstacles in practical applications. At present, fractional calculus does not have universal and unified physical meaning and geometric interpretation yet [180].

It is easy to get the physical meaning of fractional integral according to its definition. If we suppose integral as storage, then the fractional integral can be taken as memory storage. The fractional integral only store near information and abandon the past information gradually. Professor Liu et al. in Peking University described the fractional differential as a "bridge between weather and climate," which means the fractional differential of climate is weather. It is the fractional differential that makes climate has greater memory than weather [181].

We can get some conclusions according to the definition of fractional differential. (1) The initial value of input function is added to the output as attenuate form. (2) When the initial value is zero, the fractional differential is formed as convolution. Thus, the fractional differential is actually an integral and it forgets characteristic gradually (or time decay of memory). (3) Introduction of the definition of R-L fractional differential can simplify the calculation of fractional derivative. (4) Introduction of the definition of Caputo fractional differential makes Laplace transform more concise, which benefits the calculation of fractional differential equation [182].

8.3 Solution Methods of Fractional-Order Chaotic System

8.3.1 Frequency Domain Method for Solving Fractional-Order Chaotic System

There are several methods and algorithms for solving fractional differential equations and fractional-order systems, for example, Laplace transform, Fourier transform algorithm, and numerical algorithms. However, these methods have some drawbacks. At present, people are most interested in using integer-order model to approximate the fractional-order system.

Bode graphics approximating method of the fractional-order system is realized by the time–frequency conversion. The transfer functions are calculated by piecewise linear approximation in frequency domain. This method is in favor of the project implementation. The idea is to obtain frequency domain expansion of fractional integral operator by solving $1/s^{\alpha}$ in frequency domain. And then the frequency domain function is transformed to the integer-order state equation [170]. This approximation error is within the permissible range and can fully meet the needs of engineering.

If the initial condition is set to zero, then the Laplace transform of R-L in formula (8.7) becomes to

$$L\left\{\frac{d^q f(t)}{dt^q}\right\} = s^q L\{f(t)\}. \tag{8.15}$$

Therefore, the fractional integral operator in frequency domain with q-order can be described by the transfer function $F(s) = 1/s^q$. To analyze dynamics of fractional-order chaotic system effectively, standard integral-order operator is required to approximate fractional-order operators (Bode plot approximation). In this case, approximation error will definitely be within the permissible range and can fully meet engineering requirement. Fractional integral operators are defined as

$$H(s) = 1/s^{\alpha}, \tag{8.16}$$

where $s = j\omega$ is complex frequency, and α is a positive fraction which is also called fractional power. Since the magnitude of this function is relatively big at low frequency and the system is of multifrequency, a fractional-order system with multifrequency can be described as

$$H(s) = \frac{1}{\prod_{i=1}^{n}\left(1+\frac{s}{p_{T_i}}\right)^{\alpha_i}}, \quad 0 < \alpha_i < 1; \; i = 1, 2, \ldots, n, \tag{8.17}$$

where $1/p_{T_i}$ is relaxation time and α_i is fractional power factor. Formula (8.16) is called single structural model of single fractional-order system. Through a fractional extreme representation of single frequency, the transfer function in frequency domain can be represented as

$$H(s) = \frac{1}{\left(1+\frac{s}{p_T}\right)^{\alpha}}, \tag{8.18}$$

where $1/p_T$ is relaxation time and α is fractional power factor ($0 < \alpha < 1$).

By using −20 dB segments with "zigzag" type to approximate, formula (8.18) can be approximated as [170]

$$H(s) = \frac{1}{\left(1+\frac{s}{p_T}\right)^{\alpha}} \approx \lim_{N\to\infty} \frac{\prod_{i=0}^{N-1}\left(1+\frac{s}{z_i}\right)}{\prod_{i=0}^{N}\left(1+\frac{s}{p_i}\right)}, \tag{8.19}$$

where N represents the number of poles. Since working frequency in engineering practice is limited, the number of the above formula can be limited to N. Then the above formula can be further approximated as

$$H(s) = \frac{1}{\left(1+\frac{s}{p_T}\right)^{\alpha}} \approx \frac{\prod_{i=0}^{N-1}\left(1+\frac{s}{z_i}\right)}{\prod_{i=0}^{N}\left(1+\frac{s}{p_i}\right)}. \tag{8.20}$$

As a result, the problem of getting the fractional function expression is transformed into the problem of finding zero-pole pair. If the approximation error is controlled within y dB (y is positive), the pole-zero pair is determined as follows:

The first pole: $p_0 = p_T 10^{y/(20\alpha)}$

The first zero: $z_0 = p_0 10^{y/(10(1-\alpha))}$.

The second pole: $p_1 = z_0 10^{y/(10\alpha)}$.

The second zero: $z_1 = p_1 10^{y/(10(1-\alpha))}$.

⋮

The Nth zero: $z_{N+1} = p_{N+1} 10^{y/(10(1-\alpha))}$.

The (N+ 1)th pole: $p_N = z_{N+1} 10^{y/(10(\alpha))}$.

Here P_T is the angular frequency and p_0 is the first singular value which is determined by the setting error y. P_N is the last singular value which is determined by N.

Now, let $a = 10^{y/(10(1-\alpha))}$, $b = 10^{y/(10\alpha)}$, then we have

$$ab = 10^{y/(10\alpha(1-\alpha))}. \tag{8.21}$$

So the distribution relationship of zeros and poles can be written as

$$\frac{z_0}{p_0} = \frac{z_1}{p_1} = \cdots = \frac{z_{N-1}}{p_{N-1}} = a, \tag{8.22}$$

$$\frac{p_1}{z_0} = \frac{p_2}{z_1} = \cdots = \frac{p_{N-1}}{z_{N-1}} = b, \tag{8.23}$$

$$\frac{z_1}{z_0} = \frac{z_2}{z_1} = \cdots = \frac{z_{N-1}}{z_{N-2}} = ab, \tag{8.24}$$

$$\frac{p_1}{z_0} = \frac{p_2}{p_1} = \cdots = \frac{p_N}{p_{N-1}} = ab. \tag{8.25}$$

According to the above relationship and p_0, we can obtain all zero-pole values:

$$p_i = (ab)^i p_0, \quad i = 1, 2, 3, \ldots, \tag{8.26}$$

$$z_i = (ab)^i a p_0, \quad i = 1, 2, 3, \ldots. \tag{8.27}$$

N of formula (8.19) is deduced by the following procedure. The constraint relationship of system frequency bandwidth is

$$p_{N-1} < \omega_{\max} < p_N, \tag{8.28}$$

$$(ab)^{N-1} p_0 < \omega_{\max} < (ab)^N p_0, \tag{8.29}$$

$$(ab)^{N-1} < \frac{\omega_{\max}}{p_0} < (ab)^N, \tag{8.30}$$

$$(N-1)\log(ab) < \log\left(\frac{\omega_{\max}}{p_0}\right) < N\log(ab). \tag{8.31}$$

Therefore,

$$N - 1 < \frac{\log\left(\frac{\omega_{\max}}{p_0}\right)}{\log(ab)} < N, \tag{8.32}$$

$$N = \operatorname{int}\left[\frac{\log\left(\frac{\omega_{\max}}{p_0}\right)}{\log(ab)}\right] + 1. \tag{8.33}$$

Generally, we set $y = 2$ dB or $y = 3$ dB, $\omega_{\max} = 0.01$–100 rad/s. According to the above formula, $1/s^\alpha$ in frequency domain is obtained. The expansion form of $1/s^\alpha$ is shown in Table 8.1 [166], where α changes from 0.1 to 0.9 and approximation error is 2 dB. When the approximation error is 3 dB, the expansion form of $1/s^\alpha$ is shown in Table 8.2 [163].

Table 8.1: α-Order fractional integral linearity conversion function approximation. (Approximation error is 2 dB.)

α	$1/s^{\alpha}$	Transfer functions
0.1	$\frac{1}{s^{0.1}}$	$\frac{220.4s^4+5004s^3+5038s^2+234.5s+0.4840}{s^5+359.8s^4+5742s^3+4247s^2+147.7s+0.2099}$
0.2	$\frac{1}{s^{0.2}}$	$\frac{60.95s^4+816.9s^3+582.8s^2+23.24s+0.04934}{s^5+134.0s^4+956.5s^3+383.5s^2+8.953s+0.01821}$
0.3	$\frac{1}{s^{0.3}}$	$\frac{23.76s^4+224.9s^3+129.1s^2+4.733s+0.01052}{s^5+64.51s^4+252.2s^3+63.61s^2+1.104s+0.002267}$
0.4	$\frac{1}{s^{0.4}}$	$\frac{25.00s^4+558.5s^3+664.2s^2+44.15s+0.1562}{s^5+125.6s^4+840.6s^3+317.2s^2+7.428s+0.02343}$
0.5	$\frac{1}{s^{0.5}}$	$\frac{15.97s^4+593.2s^3+1080s^2+135.4s+1}{s^5+134.3s^4+1072s^3+543.4s^2+20.10s+0.1259}$
0.6	$\frac{1}{s^{0.6}}$	$\frac{8.579s^4+255.6s^3+405.3s^2+35.93s+0.1696}{s^5+94.22s^4+472.9s^3+134.8s^2+2.639s+0.009882}$
0.7	$\frac{1}{s^{0.7}}$	$\frac{5.406s^4+177.6s^3+209.6s^2+9.197s+0.01450}{s^5+88.12s^4+279.2s^3+33.30s^2+1.927s+0.0002276}$
0.8	$\frac{1}{s^{0.8}}$	$\frac{5.235s^3+1453s^2+5306s+254.9}{s^4+658.1s^3+5700s^2+658.2s+1}$
0.9	$\frac{1}{s^{0.9}}$	$\frac{1.766s^3+38.27s^2+4.914}{s^3+36.15s^2+7.789s+0.01000}$

Table 8.2: α-Order fractional integral linearity conversion function approximation. (Approximation error is 3 dB.)

α	$1/s^{\alpha}$	Transfer functions
0.1	$\frac{1}{s^{0.1}}$	$\frac{501.14(s+0.16811)}{(s+0.3162)(s+681.1)}$
0.2	$\frac{1}{s^{0.2}}$	$\frac{141.2538(s+0.1334)(s+10)}{(s+0.035623)(s+4.217)(s+316.2)}$
0.3	$\frac{1}{s^{0.3}}$	$\frac{125.8925(s+0.08483)(s+2.276)(s+61.05)}{(s+0.03162)(s+0.8483)(s+22.76)(s+610.5)}$
0.4	$\frac{1}{s^{0.4}}$	$\frac{26.6073(s+0.07499)(s+1.334)(s+23.71)}{(s+0.02371)(s+0.4217)(s+7.499)(s+133.4)}$
0.5	$\frac{1}{s^{0.5}}$	$\frac{50.1187(s+0.07943)(s+1.259)(s+19.95)(s+316.2)}{(s+0.01995)(s+0.3162)(s+5.021)(s+79.43)(s+12.59)}$
0.6	$\frac{1}{s^{0.6}}$	$\frac{28.1838(s+0.1)(s+1.778)(s+31.62)(s+562.3)}{(s+0.01778)(s+0.3162)(s+5.623)(s+100)(s+1778)}$
0.7	$\frac{1}{s^{0.7}}$	$\frac{7.9433(s+0.1638)(s+4.394)(s+117.9)}{(s+0.016389)(s+0.4393)(s+11.79)(s+100)(s+316.2)}$
0.8	$\frac{1}{s^{0.8}}$	$\frac{8.1752(s+0.487)(s+36.52)(s+2,738)}{(s+0.0154)(s+1.155)(s+86.6)(s+6,494)}$
0.9	$\frac{1}{s^{0.9}}$	$\frac{4.2987(s+14.68)(s+31,620)}{(s+0.01468)(s+31.62)(s+68,130)}$

8.3.2 Time Domain Method for Solving Fractional-Order Chaotic System

8.3.2.1 Adams–Bashforth–Moulton Predictor–Corrector Scheme

A typical fractional-order time domain approximation method is Adams–Bashforth–Moulton predictor–corrector scheme [171–173]. This method is described as follows.

For the differential equation

$$\begin{cases} \frac{d^q x}{dt^q} = f(t,x), & 0 \le t \le T \\ x^k(0) = x_0^{(k)}, & k = 0,1,2,\dots,\lceil q \rceil - 1 \end{cases}. \tag{8.34}$$

It equals Volterra integral equation

$$x(t) = \sum_{k=0}^{n-1} x_0^{(k)} \frac{t^k}{k!} + \frac{1}{\Gamma(q)} \int_0^t (t-\tau)^{q-1} f(\tau, x(t)) d\tau. \tag{8.35}$$

Obviously, the two parts of the right side in the above equation totally determined by the initial value. A typical case is $0 < q < 1$, and the Volterra equation has weak singularity. In Refs [171–173], a predictor–corrector scheme for eq. (8.35) is reported. It can be regarded as a similar algorithm of Adams–Moulton algorithm with one step.

Let $h = T/N$, $t_j = jh$ $(j = 0,1,2,\dots,N)$, where T is the integral upper limit of this equation. Then the corrector formula of eq. (8.35) is given by

$$x_h(t_{n+1}) = \sum_{k=0}^{\lceil q \rceil -1} x_0^{(k)} \frac{t_{n+1}^k}{k!} + \frac{h^q}{\Gamma(q+2)} f(t_{n+1}, x_h^p(t_{n+1})) + \frac{h^q}{\Gamma(q+2)} \sum_{j=0}^{n} a_{j,n+1} f(t_j, x_h(t_j)), \tag{8.36}$$

where

$$a_{j,n+1} = \begin{cases} n^{q+1} - (n-q)(n+1)^q, & j = 0 \\ (n-j-2)^{q+1} + (n-j)^{q+1} - 2(n-j+1)^{q+1}, & 1 \le j \le n \end{cases}. \tag{8.37}$$

Then we use one-step Adams–Bashforth rule to replace one-step Adams–Moulton rule. The predictor term $x_h^p(t_{n+1})$ is given by

$$x_h^p(t_{n+1}) = \sum_{k=0}^{n-1} x_0^{(k)} \frac{t_{n+1}^k}{k!} + \frac{1}{\Gamma(q)} \sum_{j=0}^{n} b_{j,n+1} f(t_j, x_h(t_j)), \tag{8.38}$$

where

$$b_{j,n+1} = \frac{h^q}{q}((n-j+1)^q - (n-j)^q), \quad 0 \le j \le n. \tag{8.39}$$

The basic algorithm of fractional-order Adams–Bashforth–Moulton method is shown in formulas (8.36) and (8.38), where $a_{j,n+1}$ and $b_{j,n+1}$ are presented in formulas (8.37) and (8.39), respectively. The estimate error of this method is $e = \max_{i=0,1,\ldots,N} |x(t_j) - x_h(t_j)| = o(h^p)$, where $p = \min(2, 1+q)$. Although the time domain method is more complex than the frequency domain method, the step size can be smaller, and it is much better for analysis of dynamics of fractional-order system.

8.3.2.2 Adomian Decomposition Algorithm

Adomian decomposition algorithm (Adomian decomposition method, ADM) is a time domain approximation method for solving fractional-order system [183]. This method is used to solve linear and nonlinear problems. The corresponding approximate numerical solution is highly accurate with fast convergence. Moreover, it needn't be discreted and too many computer memories. The numerical solutions of fractional-order Chen system and fractional-order Rössler system based on ADM are studied [184, 185], which proves the effectiveness of the algorithm.

If the fractional-order chaotic system is ${}^*D_{t_0}^q \mathbf{x}(t) = f(\mathbf{x}(t)) + \mathbf{g}(t)$, where $\mathbf{x}(t) = [x_1(t), x_2(t), \ldots, x_n(t)]^{\mathrm{T}}$ is the functional variable and $\mathbf{g}(t) = [g_1(t), g_2(t), \ldots, g_n(t)]^{\mathrm{T}}$ is the constant term for autonomous system and f includes both linear and nonlinear terms, then the system can be separated into three parts by

$$\begin{cases} {}^*D_{t_0}^q \mathbf{x}(t) = L\mathbf{x}(t) + N\mathbf{x}(t) + \mathbf{g}(t) \\ \mathbf{x}^{(k)}(t_0^+) = \mathbf{b}_k, \quad k = 0, \ldots, m-1 \\ m \in N, \quad m-1 < q \leq m \end{cases}, \tag{8.40}$$

where ${}^*D_{t_0}^q$ represents q-order Caputo differential operator. L represents the linear term. N represents the nonlinear term, and $\mathbf{b}_k$ is the initial value. Applying the integral operator to both sides of the equation, we have

$$\mathbf{x} = J_{t_0}^q L\mathbf{x} + J_{t_0}^q N\mathbf{x} + J_{t_0}^q \mathbf{g} + \Phi, \tag{8.41}$$

where $\Phi = \sum_{k=0}^{m-1} b_k \frac{(t-t_0)^k}{k!}$ is the initial value. According to the Adomian decomposition algorithm, the solution of the equation can be represented as $\mathbf{x} = \sum_{i=0}^{\infty} \mathbf{x}^i$. The nonlinear term can be decomposed as

$$\begin{cases} A_j^i = \frac{1}{i!} \left[\frac{d^i}{d\lambda^i} N(v_j^i(\lambda)) \right]_{\lambda=0} \\ v_j^i(\lambda) = \sum\limits_{k=0}^{i} (\lambda)^k x_j^k \end{cases}, \tag{8.42}$$

where $i = 0, 1, \ldots,$ and $j = 1, 2, \ldots, n$. The nonlinear term is

$$N\mathbf{x} = \sum_{i=0}^{\infty} \mathbf{A}^i(\mathbf{x}^0, \mathbf{x}^1, \ldots, \mathbf{x}^i). \tag{8.43}$$

Then, the solution of the equation is

$$\mathbf{x} = \sum_{i=0}^{\infty} \mathbf{x}^i = J_{t_0}^q L \sum_{i=0}^{\infty} \mathbf{x}^i + J_{t_0}^q \sum_{i=0}^{\infty} \mathbf{A}^i + J_{t_0}^q \mathbf{g} + \Phi, \tag{8.44}$$

where its iterative relationship is

$$\begin{cases} \mathbf{x}^0 = J_{t_0}^q \mathbf{g} + \Phi \\ \mathbf{x}^1 = J_{t_0}^q L\mathbf{x}^0 + J_{t_0}^q \mathbf{A}^0(\mathbf{x}^0) \\ \mathbf{x}^2 = J_{t_0}^q L\mathbf{x}^1 + J_{t_0}^q \mathbf{A}^1(\mathbf{x}^0, \mathbf{x}^1) \\ \vdots \\ \mathbf{x}^i = J_{t_0}^q L\mathbf{x}^{i-1} + J_{t_0}^q \mathbf{A}^{i-1}(\mathbf{x}^0, \mathbf{x}^1, \ldots, \mathbf{x}^{i-1}) \\ \vdots \end{cases} . \tag{8.45}$$

Next, we take fractional-order simplified Lorenz system as an example to explain this method further.

The differential equations of fractional-order simplified Lorenz system with a parameter c is [186]

$$\begin{cases} {}^*D_{t_0}^q x_1 = 10(x_2 - x_1) \\ {}^*D_{t_0}^q x_2 = (24 - 4c)x_1 - x_1x_3 + cx_2 \\ {}^*D_{t_0}^q x_3 = x_1x_2 - 8x_3/3 \end{cases} . \tag{8.46}$$

The linear and nonlinear terms of system are decomposed as

$$\begin{bmatrix} L_{x_1} \\ L_{x_2} \\ L_{x_3} \end{bmatrix} = \begin{bmatrix} 10(x_2 - x_1) \\ (24 - 4c)x_1 + cx_2 \\ -8x_3/3 \end{bmatrix}, \begin{bmatrix} N_{x_1} \\ N_{x_2} \\ N_{x_3} \end{bmatrix} = \begin{bmatrix} 0 \\ -x_1x_3 \\ x_1x_2 \end{bmatrix}, \begin{bmatrix} g_1 \\ g_2 \\ g_3 \end{bmatrix} = \begin{bmatrix} 0 \\ 0 \\ 0 \end{bmatrix}. \tag{8.47}$$

According to eq. (8.44), the first seven items of nonlinear term are [187]

$$\begin{cases} A_2^0 = -x_1^0x_3^0 \\ A_2^1 = -x_1^1x_3^0 - x_1^0x_3^1 \\ A_2^3 = -x_1^2x_3^0 - x_1^1x_3^1 - x_1^0x_3^2 \\ A_2^4 = -x_1^3x_3^0 - x_1^2x_3^1 - x_1^2x_3^2 - x_1^0x_3^3 \\ A_2^5 = -x_1^5x_3^0 - x_1^4x_3^1 - x_1^1x_3^4 - x_1^3x_3^2 - x_1^2x_3^3 - x_1^0x_3^5 \end{cases} , \tag{8.48}$$

$$\begin{cases} A_3^0 = x_1^0x_2^0 \\ A_3^1 = x_1^1x_2^0 + x_1^0x_2^1 \\ A_3^3 = x_1^2x_2^0 + x_1^1x_2^1 + x_1^0x_2^2 \\ A_3^4 = x_1^3x_2^0 + x_1^2x_2^1 + x_1^2x_2^2 + x_1^0x_2^3 \\ A_3^5 = x_1^5x_2^0 + x_1^4x_2^1 + x_1^1x_2^4 + x_1^3x_2^2 + x_1^2x_2^3 + x_1^0x_2^5. \end{cases} \tag{8.49}$$

According to the given initial values, we have

$$\begin{cases} x_1^0 = x_1(t_0) \\ x_2^0 = x_2(t_0) \\ x_3^0 = x_3(t_0) \end{cases}, \tag{8.50}$$

Let $c_1^0 = x_1^0$, $c_2^0 = x_2^0$, $c_3^0 = x_3^0$, which means $c^0 = [c_1^0, c_2^0, c_3^0]$. According to $\mathbf{x}^1 = J_{t_0}^q L\mathbf{x}^0 + J_{t_0}^q \mathbf{A}^0$ in formula (8.45) and properties of fractional calculus, we have

$$\begin{cases} x_1^1 = 10(c_2^0 - c_1^0)\frac{(t-t_0)^q}{\Gamma(q+1)} \\ x_2^1 = [(24-4c)c_1^0 + cc_2^0 - c_1^0 c_3^0]\frac{(t-t_0)^q}{\Gamma(q+1)} \\ x_3^1 = \left(-\frac{8}{3}c_1^0 + c_1^0 c_2^0\right)\frac{(t-t_0)}{\Gamma(q+1)} \end{cases}. \tag{8.51}$$

If $c_1^1 = 10(c_2^0 - c_1^0)$, $c_2^1 = (24-4c)c_1^0 + cc_2^0 + c_4^0 - c_1^0 c_3^0$, $c_3^1 = -(8/3)c_1^0 + c_1^0 c_2^0$, then $x^1 = c^1(t-t_0)^q/\Gamma(q+1)$. So we only need to solve the corresponding coefficient of each term. According to formula (8.46), the other five coefficients are obtained as

$$\begin{cases} c_1^2 = 10(c_2^1 - c_1^1) \\ c_2^2 = (24-4c)c_1^1 + cc_2^1 - c_1^1 c_3^0 - c_1^0 c_3^1 \\ c_3^2 = -\frac{8}{3}c_1^1 + c_1^1 c_2^0 + c_1^0 c_2^1 \end{cases}, \tag{8.52}$$

$$\begin{cases} c_1^3 = 10(c_2^2 - c_1^2) \\ c_2^3 = (24-4c)c_1^2 + cc_2^2 - c_1^2 c_3^0 - c_1^1 c_3^1 \frac{\Gamma(2q+1)}{\Gamma^2(q+1)} - c_1^0 c_3^2 \\ c_3^3 = -\frac{8}{3}c_1^2 + c_1^2 c_2^0 + c_1^1 c_2^1 \frac{\Gamma(2q+1)}{\Gamma^2(q+1)} + c_1^0 c_2^2 \end{cases}, \tag{8.53}$$

$$\begin{cases} c_1^4 = 10(c_2^3 - c_1^3) \\ c_2^4 = (24-4c)c_1^3 + cc_2^3 - c_1^3 c_3^0 - (c_1^2 c_3^1 + c_1^1 c_3^2)\frac{\Gamma(3q+1)}{\Gamma(q+1)\Gamma(2q+1)} - c_1^0 c_3^3 \\ c_3^4 = -\frac{8}{3}c_1^3 + c_1^3 c_2^0 + (c_1^1 c_2^2 + c_1^2 c_2^1)\frac{\Gamma(3q+1)}{\Gamma(q+1)\Gamma(2q+1)} + c_1^0 c_2^3 \end{cases}, \tag{8.54}$$

$$\begin{cases} c_1^5 = 10(c_2^4 - c_1^4) \\ c_2^5 = (24-4c)c_1^4 cc_2^4 - c_1^4 c_3^0 - (c_1^3 c_3^1 + c_1^1 c_3^3)\frac{\Gamma(4q+1)}{\Gamma(q+1)\Gamma(3q+1)} - \\ \quad c_1^2 c_3^2 \frac{\Gamma(4q+1)}{\Gamma^2(2q+1)} - c_1^0 c_3^4 \\ c_3^5 = -\frac{8}{3}c_1^4 + c_1^4 c_2^0 + (c_1^1 c_2^3 + c_1^3 c_2^1)\frac{\Gamma(4q+1)}{\Gamma(q+1)\Gamma(3q+1)} + c_1^2 c_2^2 \frac{\Gamma(4q+1)}{\Gamma^2(2q+1)} + c_1^0 c_2^4 \end{cases}, \tag{8.55}$$

$$\begin{cases} c_1^6 = 10(c_2^5 - c_1^5) \\ c_2^6 = (24-4c)c_1^5 + cc_2^5 - c_1^5 c_3^0 - (c_1^4 c_3^1 + c_1^1 c_3^4)\frac{\Gamma(5q+1)}{\Gamma(q+1)\Gamma(4q+1)} - \\ \quad (c_1^2 c_3^3 + c_1^3 c_3^2)\frac{\Gamma(5q+1)}{\Gamma(2q+1)\Gamma(3q+1)} - c_1^0 c_3^5 \\ c_3^6 = -\frac{8}{3}c_1^5 + c_1^5 c_2^0 + (c_1^1 c_2^4 + c_1^4 c_2^1)\frac{\Gamma(5q+1)}{\Gamma(q+1)\Gamma(4q+1)} + \\ \quad (c_1^2 c_2^3 + c_1^3 c_2^2)\frac{\Gamma(5q+1)}{\Gamma(2q+1)\Gamma(3q+1)} + c_1^0 c_2^5 \end{cases}. \tag{8.56}$$

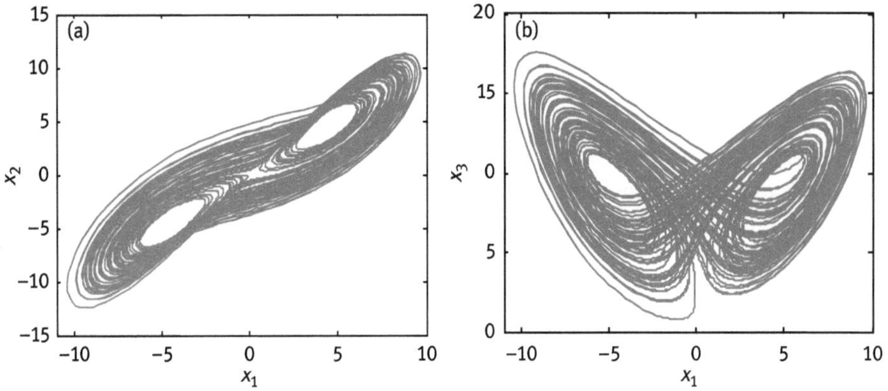

Figure 8.1: Attractor of fractional-order simplified Lorenz system: (a) x_1–x_2 plane and (b) x_1–x_3 plane.

So the solution of system equation is

$$\begin{aligned}\tilde{x}_j(t) = c_j^0 + c_j^1\frac{(t-t_0)^q}{\Gamma(q+1)} + c_j^2\frac{(t-t_0)^{2q}}{\Gamma(2q+1)} + c_j^3\frac{(t-t_0)^{3q}}{\Gamma(3q+1)} \\ + c_j^4\frac{(t-t_0)^{4q}}{\Gamma(4q+1)} + c_j^5\frac{(t-t_0)^{5q}}{\Gamma(5q+1)} + c_j^6\frac{(t-t_0)^{6q}}{\Gamma(6q+1)}\end{aligned}, \tag{8.57}$$

where $j = 1, 2, 3$. Then we can get the analytic solution of system variable. In practical calculation, the time period $[t_0, t]$ should be divided into smaller time interval $[t_k, t_{k+1}]$. In each interval, t_0 is set as t_k, and then the value of corresponding t_{k+1} is obtained. The attractor phase diagram of fractional-order simplified Lorenz system based on formula (8.57) is shown in Figure 8.1. In this figure, the length of small time interval is 0.01 s. The initial values are [0.1, 0.2, 0.3]. Parameter $c = 5$ and fractional order $q = 0.98$.

8.4 Simulation Method for Fractional-Order Chaotic System

8.4.1 Dynamic Simulation Method for Fractional-Order Chaotic System

Dynamic simulation analysis method is to use the dynamic model to represent the state change over time, and the dynamical characteristics of the system are analyzed by observing the real-time evolution of the variables. By employing this method, we can directly observe the simulation results and also can make a quantitative analysis for the simulation output data. It has the characteristics of intuition and flexibility. Based on frequency domain analysis method of fractional calculus operator, fractional-order system simulation model is designed through state-space module in Simulink. State-space module is a state-space description module for input–output variables, which is defined as

$$\begin{cases} \dot{x} = \mathbf{A}x + \mathbf{B}u \\ y = \mathbf{C}x + \mathbf{D}u \end{cases}, \tag{8.58}$$

where x is the state vector, u is the input vector, and y is the output vector. **A**, **B**, **C**, and **D** are coefficient matrices. Obviously, those modules receive inputs and then generate outputs. Parameters of coefficient matrices **A**, **B**, **C**, and **D** are corresponding to that of transform function, respectively, which can be calculated by tf2ss transform function in Matlab.

Steps of the dynamic simulation are as follows:

1. Determine the function modules listed according to the differential equation of fractional-order chaotic system. The simulation system is built by connecting these modules.
2. Set simulation parameters. It includes system operating parameters (e.g., system operating time, integration step size, tolerance error, integration algorithms, etc.) and function modules operating parameters (such as integrator initial value, signal amplitude, frequency, etc.).
3. Set observation module. It is used to output data and observe operation status of simulation system.
4. Analyze the simulation results. Simulation results are analyzed by output data and waveform, and compared with that of theoretical analysis. If it is not consistent with each other, the model and parameters should be revised.

During the simulation, Scope module is used to show real-time output signal dynamically. Phase diagram can be observed by XY graph module. In phase space, convergence motion corresponds to the equilibrium point, periodic motion corresponds to a closed trajectory, and chaotic motion corresponds to random separation of the never closed trajectory in a certain region (strange attractor). So, the output phase diagram of XY graph module is used to preliminarily judge whether the system is chaotic. Out module can save results to workspace automatically, so the saved data can be output by yout() function, and it is used to calculate the power spectrum and the maximum Lyapunov exponent of the system.

How to set the state-space module is the key of realizing dynamic simulation of fractional-order chaotic system. Parameter matrices **A**, **B**, **C**, and **D** in the property dialog box are determined by the order of fractional-order system. To obtain each matrix element, we can call command [**A**,**B**,**C**,**D**] = TF2SS(NUM,DEN) (where NUM represents the molecular coefficient matrix and DEN represents the denominator coefficient matrix) or [**A**,**B**,**C**,**D**] = ZP2SS(**Z**,**P**,**K**) (where **Z** represents the zeros matrix of transform function, **P** represents the poles matrix, and **K** represents the coefficient).

For example, when $q = 0.95$, the transform function of fractional-order integral operator is written as [167]

$$H(s) = \frac{1}{s^{0.95}} \approx \frac{1.2831s^2 + 18.6004s + 2.0833}{s^3 + 18.4738s^2 + 2.6547s + 0.003}. \tag{8.59}$$

For formula (8.59), calling format of the function TF2SS is [*A*,*B*,*C*,*D*] = TF2SS (1.2831,18.6004,2.0833], [1,18.4738,2.6547,0.003]). Its corresponding return value is

$$A = \begin{bmatrix} -18.4738 & -2.6547 & -0.003 \\ 1 & 0 & 0 \\ 0 & 1 & 0 \end{bmatrix}, B = \begin{bmatrix} 1 \\ 0 \\ 0 \end{bmatrix}, C = [1.2831, 18.6004, 2.0833140], D = 0.$$

Next, we take the fractional-order diffusionless Lorenz system as an example to explain further. The diffusionless Lorenz system is

$$\begin{cases} \dot{x} = -x - y \\ \dot{y} = -xz \\ \dot{z} = xy + R \end{cases}, \tag{8.60}$$

where R is the parameter of system. When $R \in (0, 5)$, the system is chaotic. Its corresponding fractional-order diffusionless Lorenz system is written as

$$\begin{cases} \dfrac{d^\alpha x}{dt^\alpha} = -x - y \\ \dfrac{d^\beta y}{dt^\beta} = -xz \\ \dfrac{d^\gamma z}{dt^\gamma} = xy + R \end{cases}, \tag{8.61}$$

where $0 < \alpha, \beta, \gamma \le 1$ are fractional orders. Dynamic simulation model of the system is constructed as shown in Figure 8.2. The initial values of system are set as [0.3782, 0.6748, 2.8831]. Let $\alpha = \beta = \gamma = 0.95$ and simulation time is 200 s. By using ODE45 differential equation solvers, the strange attractor of diffusionless Lorenz system in x–z plane is shown in Figure 8.3.

8.4.2 Circuit Simulation Method of Fractional-Order Chaotic System

Design of fractional integral operator circuit is the key of Electronics Workbench circuit simulation for fractional-order chaotic system. Because the fractional integral operator with q-order is described by transform function $F(s) = 1/s^q$ in frequency domain, the unit circuit as shown in Figure 8.4 is used to achieve the expansion of $1/s^q$ [177]. The nonlinear terms are realized by multipliers. Then the equivalent circuit of the fractional-order system is constructed.

In Figure 8.4, system function of the equivalent circuit between A and B is described as

$$H(s) = \frac{R_1}{sR_1C_1 + 1} + \frac{R_2}{sR_2C_2 + 1} + \frac{R_3}{sR_3C_3 + 1} + \cdots + \frac{R_n}{sR_nC_n + 1}, \tag{8.62}$$

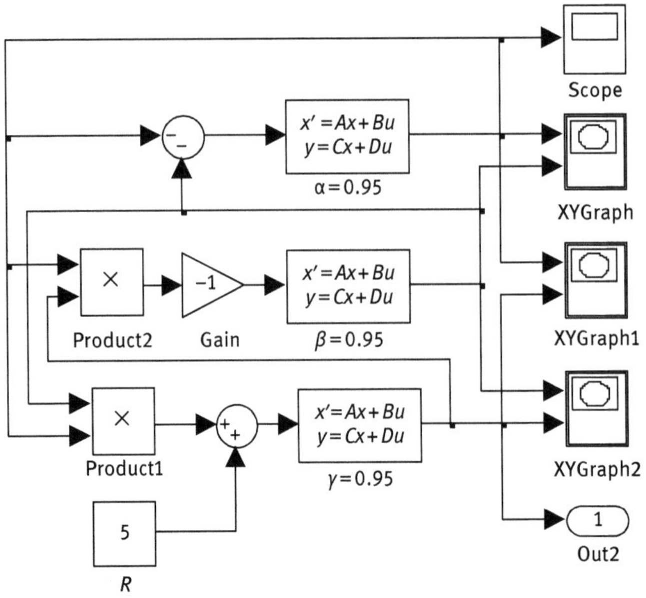

Figure 8.2: Simulation model of the fractional-order diffusionless Lorenz system.

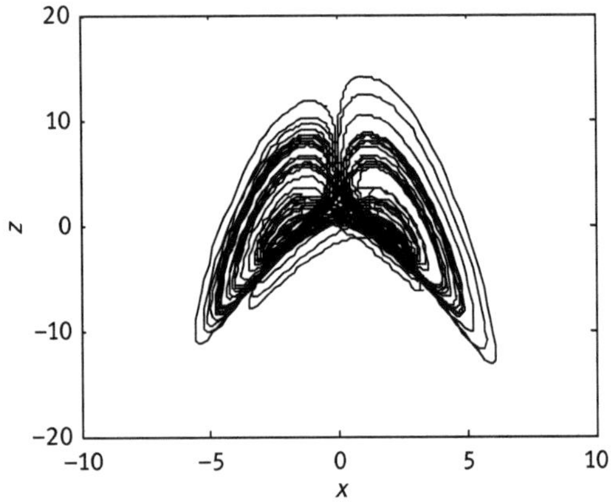

Figure 8.3: Attractor of diffusionless Lorenz system in the x–z plane when $q = 0.95$.

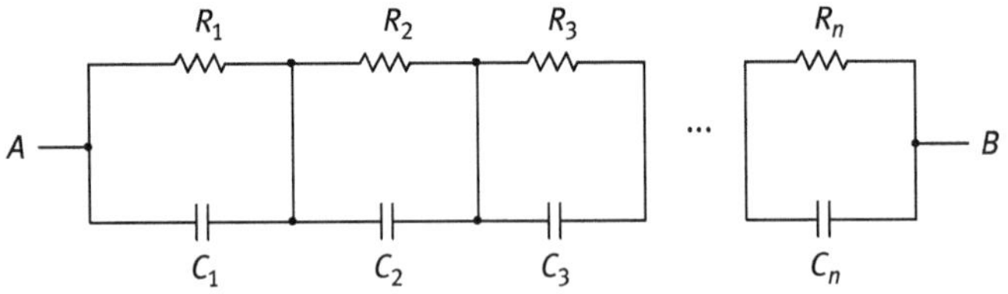

Figure 8.4: Unit circuit of fractional-order integrator.

where n is the highest order of s in denominator of $1/s^{\alpha}$ when q varied from 0.1 to 0.9. For example, when $q = 0.95$, $n = 3$, according to eq. (8.59), we have

$$F(s) = \frac{R_1}{sR_1C_1 + 1} + \frac{R_2}{sR_2C_2 + 1} + \frac{R_3}{sR_3C_3 + 1}. \tag{8.63}$$

Comparing eq. (8.63) with eq. (8.59), we have $R_1 = 15.1\ \text{k}\Omega$, $R_2 = 1.51\ \text{k}\Omega$, $R_3 = 692.9\ \text{M}\Omega$, $C_1 = 3.616\ \mu\text{F}$, $C_2 = 4.602\ \mu\text{F}$, $C_3 = 1.267\ \mu\text{F}$. Resistors and capacitors of $1/s^q$ calculated in Ref. [166] are shown in Tables 8.3 and 8.4, respectively [177].

The circuit of fractional-order diffusionless Lorenz system is shown in Figure 8.5, in which the resistance values are $R_1 = R_2 = R_3 = R_4 = R_7 = 10\ \text{k}\Omega$, $R_6 = R_8 = R_9 = 100\ \text{k}\Omega$,

Table 8.3: Resistance of fractional-order integrator unit circuit with order α.

$\frac{1}{s^{\alpha}}$	$R_1/\text{M}\Omega$	$R_2/\text{M}\Omega$	$R_3/\text{M}\Omega$	$R_4/\text{M}\Omega$	$R_5/\text{M}\Omega$	$R_6/\text{M}\Omega$
$\frac{1}{s^{0.1}}$	0.636	0.3815	0.5672			
$\frac{1}{s^{0.2}}$	1.130	0.6070	0.3500	0.425	0.2498	
$\frac{1}{s^{0.3}}$	2.050	0.9380	0.4830	0.262	0.1430	0.106
$\frac{1}{s^{0.4}}$	3.744	1.3920	0.6310	0.294	0.1230	0.068
$\frac{1}{s^{0.5}}$	6.824	1.9440	0.7440	0.296	0.0754	0.030
$\frac{1}{s^{0.6}}$	12.330	2.4480	0.7380	0.233	0.0754	0.006
$\frac{1}{s^{0.7}}$	21.900	2.6000	0.5260	0.113	0.0246	
$\frac{1}{s^{0.8}}$	37.850	1.7540	0.1700	0.017	0.0018	
$\frac{1}{s^{0.9}}$	62.840	0.2500	0.0025			

Table 8.4: Resistance of fractional-order integrator unit circuit with order α.

$\frac{1}{s^{\alpha}}$	$C_1/\mu\text{F}$	$C_2/\mu\text{F}$	$C_3/\mu\text{F}$	$C_4/\mu\text{F}$	$C_5/\mu\text{F}$	$C_6/\mu\text{F}$
$\frac{1}{s^{0.1}}$	15.720	0.1572	0.0006335			
$\frac{1}{s^{0.2}}$	27.990	2.9300	0.285	0.0132		
$\frac{1}{s^{0.3}}$	22.640	5.5200	1.200	0.2460	0.029	
$\frac{1}{s^{0.4}}$	15.020	5.9260	1.920	0.6050	0.183	0.036
$\frac{1}{s^{0.5}}$	9.246	5.1450	2.129	0.8480	0.324	0.925
$\frac{1}{s^{0.6}}$	5.527	4.0850	1.990	0.9260	0.420	0.156
$\frac{1}{s^{0.7}}$	3.284	3.1390	1.700	0.8860	0.454	0.207
$\frac{1}{s^{0.8}}$	1.980	2.4000	1.390	0.7800	0.420	
$\frac{1}{s^{0.9}}$	1.232	1.8400	1.100			

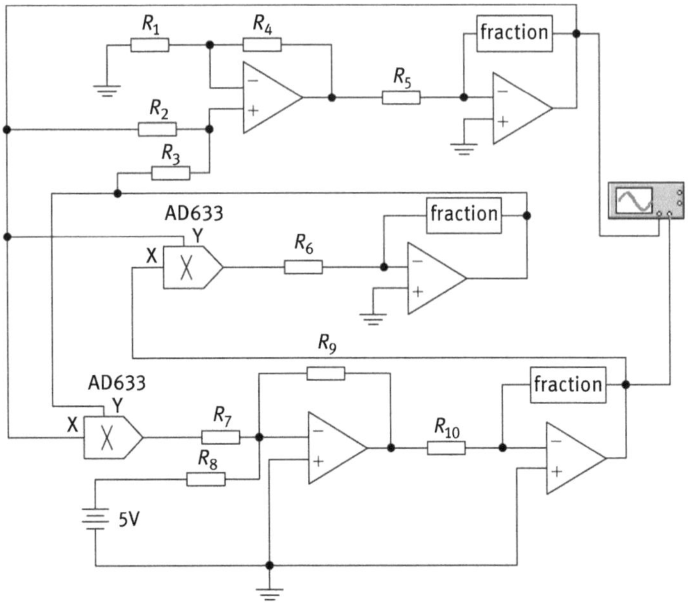

Figure 8.5: Circuit simulation model of diffusionless Lorenz system with q = 0.95.

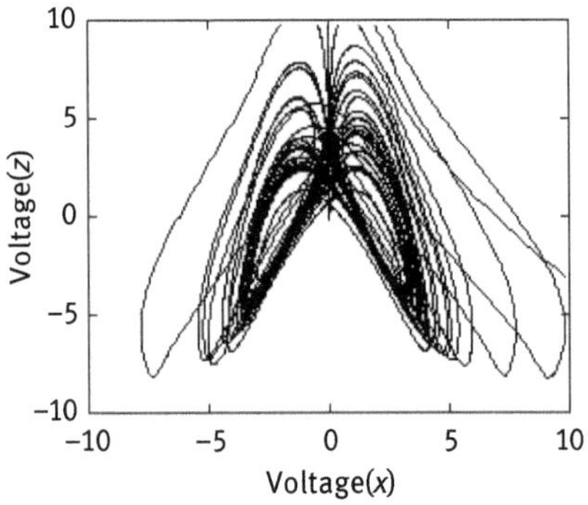

Figure 8.6: Attractor of fractional-order diffusionless Lorenz system based on circuit simulation.

$R_5 = R_{10} = 1\ \mathrm{M\Omega}$. "Fraction" is the integrator unit circuit of 0.95-order after packaged. When $R = 5$ and $q = 2.85$, its attractor of x–z plane is shown in Figure 8.6. Obviously, it is consistent with that of simulation result as shown in Figure 8.3.

8.4.3 Numerical Simulation Method of Fractional-Order Chaotic System

Numerical simulation method is to transform the mathematic model of fractional-order chaotic system into recursive calculation form which is suitable for processing on a computer. The main advantage of this method is that it can solve complex problems analytically. The basic steps are as follows [188]:

1. Establish mathematical model for fractional-order chaotic system.
2. According to the characteristic of this model, choose suitable software as simulation tool and design the computer program according to the mathematical model.
3. Run the program and get the numerical solution of the system.
4. According to the numerical solution, analyze dynamic characteristic such as calculating Lyapunov exponent and fractal dimension.

For example, according to the approximation formula based on Adams–Bashforth–Moulton time domain method, the fractional-order diffusionless Lorenz chaotic system is presented as

$$\begin{cases} x_{n+1} = x_0 + \dfrac{h^\alpha}{\Gamma(\alpha+2)}\left\{[-y_{n+1}^p - x_{n+1}^p] + \sum\limits_{j=0}^{n} a_{1,j,n+1}(-y_j - x_j)\right\} \\ y_{n+1} = y_0 + \dfrac{h^\beta}{\Gamma(\beta+2)}\left[(-x_{n+1}^p z_{n+1}^p) + \sum\limits_{j=0}^{n} a_{2,j,n+1}(-x_j z_j)\right] \\ z_{n+1} = z_0 + \dfrac{h^\gamma}{\Gamma(\gamma+2)}\left\{[x_{n+1}^p y_{n+1}^p + R] + \sum\limits_{j=0}^{n} a_{3,j,n+1}(x_j y_j + R)\right\} \end{cases}, \tag{8.64}$$

where

$$\begin{cases} x_{n+1}^p = x_0 + \dfrac{1}{\Gamma(\alpha)} \sum\limits_{j=0}^{n} b_{1,j,n+1}(-y_j - x_j) \\ y_{n+1}^p = y_0 + \dfrac{1}{\Gamma(\beta)} \sum\limits_{j=0}^{n} b_{2,j,n+1}(-x_j z_j) \\ z_{n+1}^p = z_0 + \dfrac{1}{\Gamma(\gamma)} \sum\limits_{j=0}^{n} b_{3,j,n+1}(x_j y_j + R) \end{cases}, \tag{8.65}$$

$$\begin{cases} b_{1,j,n+1} = \dfrac{h^\alpha}{\alpha}((n-j+1)^\alpha - (n-j)^\alpha), & 0 \le j \le n \\ b_{2,j,n+1} = \dfrac{h^\beta}{\beta}((n-j+1)^\beta - (n-j)^\beta), & 0 \le j \le n, \\ b_{3,j,n+1} = \dfrac{h^\gamma}{\gamma}((n-j+1)^\gamma - (n-j)^\gamma), & 0 \le j \le n \end{cases} \tag{8.66}$$

$$\begin{cases} a_{1,j,n+1} = \begin{cases} n^\alpha - (n-\alpha)(n+1)^\alpha & j=0 \\ (n-j+2)^{\alpha+1} + (n-j)^{\alpha+1} - 2(n-j+1)^{\alpha+1}, & 1 \le j \le n \end{cases} \\ a_{2,j,n+1} = \begin{cases} n^\beta - (n-\beta)(n+1)^\beta & j=0 \\ (n-j+2)^{\beta+1} + (n-j)^{\beta+1} - 2(n-j+1)^{\beta+1}, & 1 \le j \le n \end{cases} \\ a_{3,j,n+1} = \begin{cases} n^\gamma - (n-\gamma)(n+1)^\gamma & j=0 \\ (n-j+2)^{\gamma+1} + (n-j)^{\gamma+1} - 2(n-j+1)^{\gamma+1}, & 1 \le j \le n \end{cases} \end{cases}. \tag{8.67}$$

Based on Matlab, we let $\alpha = \beta = \gamma = 0.95$, $R = 5$, $h = 0.01$, $T = 200$, and the computer program is designed. The chaotic attractor of fractional-order diffusionless Lorenz

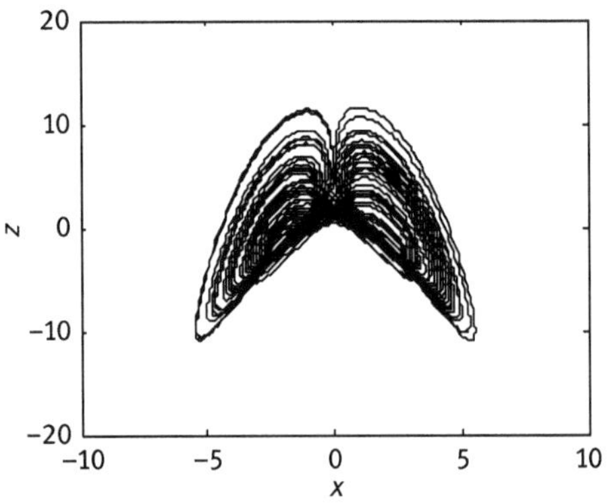

Figure 8.7: Attractor of fractional-order diffusionless Lorenz system based on numerical simulation.

chaotic system with $q = 0.95$ is shown in Figure 8.7. Obviously, it is consistent with the results of dynamic simulation and circuit simulation.

8.4.4 Comparison of Simulation Methods

1. Comparison among dynamic simulation, circuit simulation, and numerical simulation based on predictor–corrector method. The three simulation methods have their own advantages and disadvantages. We summarized their features as shown in Table 8.5 from the aspects of analysis method, fractional order change, result output, and simulation speed. Dynamic simulation method can construct dynamic system model quickly by employing abundant functional blocks in Simulink. You don't need to write a line of codes. When simulating complicated fractional-order nonlinear chaotic system, this method is more simple and intuitive. However, its limitation is that the value of q can only take finite value. Circuit simulation has similar characteristic with dynamic simulation, but because the circuit simulation is more close to the realistic engineering system, it is more important for the application of the fractional-order chaotic system. Programming is required for numerical simulation, and dynamic output process of variable evolution cannot be observed directly. But when the value of fractional-order changes continuously, it is more convenient to analyze the chaotic characteristic of the fractional-order system.

Table 8.5: Comparison of simulation methods.

Simulation methods	Analysis methods	Variation of q	Output results	Speed
Dynamic simulation	Frequency domain	Finite	Dynamic	Fast
Circuit simulation	Frequency domain	Finite	Dynamic	Fast
Numerical simulation	Time domain	Infinity	Static	Slow

Table 8.6: Comparison between numerical simulation methods for fractional-order chaotic system.

	Adomian method	**Predictor–corrector scheme**
Time complexity	$O(n)$, fast	$O(n^2)$, slow
Space complexity	$O(1)$, small	$O(n)$, large
$N = 1,000$	0.9701 s	2.0154 s
$N = 2,000$	1.8403 s	7.7051 s
$N = 5,000$	4.5162 s	51.3995 s

2. Comparison between ADM and predictor–corrector scheme. Both of them are time domain methods. Comparison of time complexity and space complexity of the two methods is shown in Table 8.6. It can be seen that the ADM is better than predictor–corrector method from both of these two aspects. The data in last three lines of Table 8.6 is the requirement time for getting time series with the different lengths (the results are different with different CPU frequencies. Here, the computer parameter is Intel Dual E2180 2.0 GHz). It can be seen that the growth rate of time required for predictor–corrector method is much faster than that of Adomian decomposition algorithm. So, we can get the conclusion that Adomian decomposition algorithm is better than predictor–corrector scheme when solving fractional-order chaotic system.

In this section, we analyzed three simulation methods of fractional-order chaotic system. For different methods, design methods and simulation steps are described. Furthermore, to show the effectiveness of different simulation methods, the fractional-order diffusionless Lorenz chaotic system is simulated based on Matlab, Simulink, and EWB.

8.5 Dynamics of Fractional-Order Unified System Based on Frequency Domain Method

In this section, we use dynamic simulation method to analyze fractional-order unified system based on Matlab/Simulink.

8.5.1 Fractional-Order Unified Chaotic System Mode

In 2002, a new chaotic system is proposed [84], which connects Lorenz attractor and Chen attractor. Meanwhile, Lü system is a special case of this system. So it is called unified chaotic system [189], and its mathematic model is

$$\begin{cases} \dot{x} = (25a + 10)(y - x) \\ \dot{y} = (28 - 35a)x - xz + (29a - 1)y \\ \dot{z} = xy - (a + 8)/3z \end{cases}, \tag{8.68}$$

where a is the system parameter. When $a \in [0,1]$, it is chaotic. This system represents a family of chaotic system. Chaotic attractor in x–z plane of unified chaotic system with different parameters is shown in Figure 3.20.

After replacement of integer-order differential operator with fractional-order differential operator, fractional-order unified system becomes

$$\begin{cases} \frac{d^{\alpha}x}{dt^{\alpha}} = (25a+10)(y-x) \\ \frac{d^{\beta}y}{dt^{\beta}} = (28-35a)x - xz + (29a-1)y \\ \frac{d^{\gamma}z}{dt^{\gamma}} = xy - \frac{a+8}{3}z \end{cases}, \tag{8.69}$$

where α, β, and γ are fractional orders and system parameter $a \in [0,1]$. The fractional order for different variables x, y, and z may be different. The principle is the same for different fractional orders or the same fractional order. Here, we set $0 < \alpha = \beta = \gamma = q \leq 1$ to analyze fractional-order unified system.

8.5.2 Dynamic Simulation of Fractional-Order Unified Chaotic System

In this section, we focus on the dynamics of the fractional-order unified chaotic system by Simulink simulation

8.5.2.1 Simulation Model

When $a = 1$, eq. (8.69) becomes

$$\begin{cases} \frac{d^{q}x}{dt^{q}} = 35(y-x) \\ \frac{d^{q}y}{dt^{q}} = -7x - xz + 28y \\ \frac{d^{q}z}{dt^{q}} = xy - 3z \end{cases}. \tag{8.70}$$

According to eq. (8.70), when we design the model of fractional-order system, some modules are required, including state-space module, gain module, sum module, and produce module. Based on Simulink, we copy the required modules from module library to edit window, and then connecting these modules according to system (8.70) as shown in Figure 8.8.

8.5.2.2 Setting Simulation Parameter

After the model is set up simulation environment and parameters are required to be configured. By choosing menu "Simulation: Simulation Parameters" in the model window, the simulation configuration dialog will pop up. Then we modify simulation configuration. Almost all of the modules have dialog box for property setting. By double-clicking the module or clicking "Edit: BLOCK Propertied" after choosing the module, a basic properties dialog box will be opened. In this dialog, users can set the basic properties of the module.

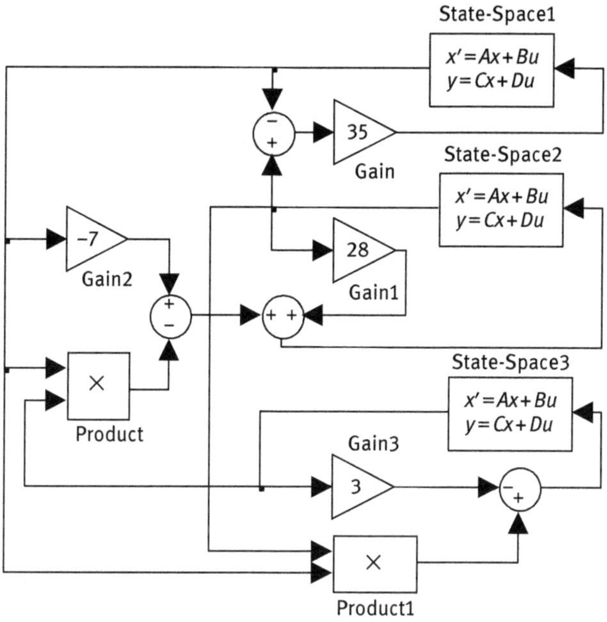

Figure 8.8: Simulation model of fractional-order unified system with $a = 1$.

In state-space module, the values of parameters A, B, C, and D are determined by the value of fractional order as explained in Section 8.4.1. The initial values of system are set by the initial condition. The other parameters are set as default values. Here, the initial values are set as (0.1, 0.1, 0.1) and simulation time is set as 30 s. The ODE45 differential equation solver is used to solve the equation.

8.5.3 Dynamics of Fractional-Order Unified System with Fixed Parameter

According to eq. (8.70), we analyze dynamics of the fractional-order unified system with $q = 0.1$–1, changing parameter or order.

8.5.3.1 Direct Observation Analysis

According to direct observation method, period motion in the phase space corresponds to a closed curve, while chaotic behavior corresponds to a never closed trajectory which is random separated in a certain area (strange attractor). After we finish the simulation of fractional-order chaotic system, the time sequence diagram and system phase diagram are plotted. Then we can preliminarily determine whether the fractional-order system is chaotic.

According to the fractional-order unified system model, the Scope module and XYgraph module are used to observe simulation results.

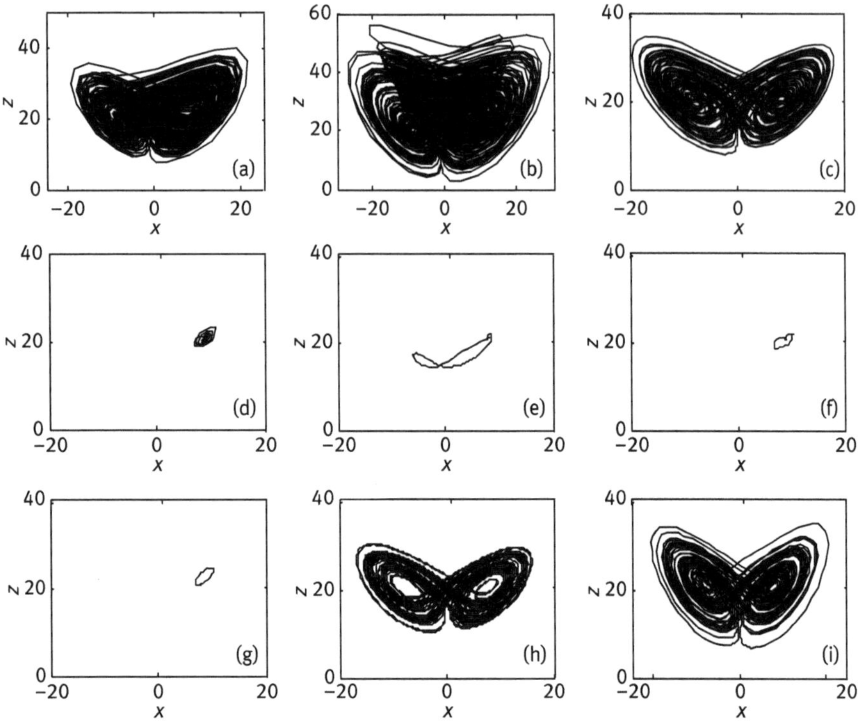

Figure 8.9: Attractors in the x–z plane of fractional-order unified system with different orders: (a) $q = 0.1$, (b) $q = 0.2$, (c) $q = 0.3$, (d) $q = 0.4$, (e) $q = 0.5$, (f) $q = 0.6$, (g) $q = 0.7$, (h) $q = 0.8$, and (i) $q = 0.9$.

Let $q = 0.1$–1 and set step size as 0.1. Simulation time is set to 40 s. For different fractional orders, we get the trajectory in x–z phase plane as shown in Figure 8.9. It can be seen that there exist the random separate states in the phase space when $q = 0.1$–0.3 and $q = 0.8$–1. It means chaotic attractor exists and the system is chaotic in this case. When $q = 0.4$–0.7, the phase curve finally converges to one point, which means the system is in periodic state.

By observing phase diagrams of the fractional-order unified system directly, we get the following conclusions. When $q = 0.1$–0.3, the system is chaotic. When $q = 0.4$–0.7, the system is periodic. When $q = 0.8$–1, the system shows chaotic behavior again. Therefore, fractional-order unified system is piecewise chaotic.

8.5.3.2 Power Spectrum Analysis

In this section, we will use power spectrum analysis method mentioned in Section 2.2 to analyze the fractional-order unified system. Power spectrum analysis can provide frequency information of signals. Discrete power spectrum corresponds to periodic behavior, while continuous power spectrum corresponds to chaotic behavior. So we

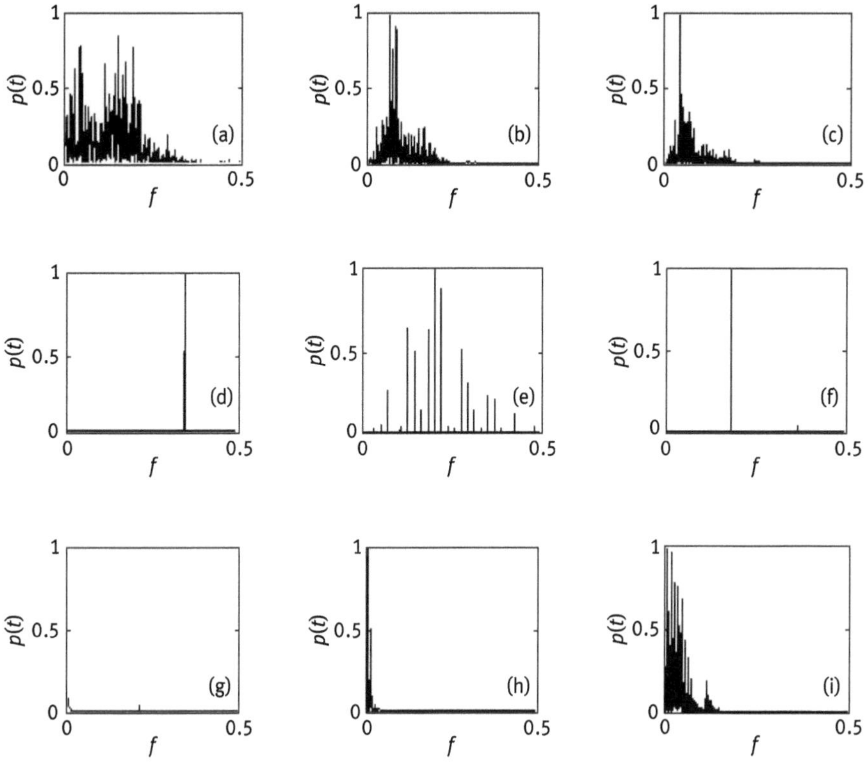

Figure 8.10: Normalized power spectrum density of fractional-order unified system with different orders: (a) $q = 0.1$, (b) $q = 0.2$, (c)$q = 0.3$, (d) $q = 0.4$, (e) $q = 0.5$, (f) $q = 0.6$, (g) $q = 0.7$, (h) $q = 0.8$, and (i) $q = 0.9$.

can tell whether the system is chaotic or not by observing power spectrum. Based on Matlab, we calculated normalized power spectrum when fractional order $q = 0.1–1$, which is shown in Figure 8.10.

From Figure 8.10, we get the following conclusions. When $q = 0.8–1$ and $q = 0.1–0.3$, the power spectrum is continuous, which means the system is chaotic. When $q = 0.4–0.7$, the power spectrum is discrete, which means the system is periodic. Therefore, by observing power spectrum, the fractional-order unified system is piecewise chaotic, which is consistent with the conclusion obtained by direct observation.

8.5.3.3 Analysis of the Largest Lyapunov Exponent

Lyapunov exponent represents the average ratio of convergence or divergence of the system variables in the phase space. It is an important parameter to determine whether the system is chaotic or not. If the largest Lyapunov exponent is greater than zero, the system is chaotic definitely.

Table 8.7: Largest Lyapunov exponent of unified system with different order q.

q	System order	The largest LE
0.1	0.3	0.2180
0.2	0.6	0.1616
0.3	0.9	0.1249
0.4	1.2	−0.0050
0.5	1.5	−1.4618
0.6	1.8	−0.0032
0.7	2.1	−0.0078
0.8	2.4	0.0199
0.9	2.7	0.0756
1	3	0.0263

Here, we use Wolf method [190] to calculate the largest Lyapunov exponent of fractional-order unified system. When the fractional order $q = 0.1–1$ with step size of 0.1, the results are shown in Table 8.7. When $q = 0.8–1$ and $q = 0.1–0.3$, the largest Lyapunov exponent is positive, which means the system is chaotic. When $q = 0.4–0.7$, the largest Lyapunov exponent is zero, which means the system is periodic. Thus, from the largest Lyapunov exponent, we can get the conclusion that the system is piecewise chaotic [191].

8.5.4 Dynamics of Fractional-Order Unified System with Different Parameter

8.5.4.1 Dynamics with Different Parameter and Order

We analyzed chaotic characteristic of fractional-order unified system with fixed parameter (a= 1). Next, we use the above methods to analyze dynamics of fractional-order unified system when parameter $a = 0–1$ and step size is 0.1. The variation of fractional order is still between 0.1 and 1 and its step size is 0.1. Dynamics of fractional-order unified system with different parameters is shown in Table 8.8. However, when $q = 1$, the system is integer-order unified system. When parameter varies from 0 to 1, the system is always chaotic. From Table 8.8, we can get the following conclusion. When $a = 0$, system changes from chaotic to nonchaotic with the increase of order q. More specifically, system is chaotic at $q = 1$ and nonchaotic at the range of $0 < q \leq 0.9$. When $a = 0.1–0.3$, the system is chaotic at $q = 1$ and $q = 0.1$. It is nonchaotic at the range of $0.2 \leq q \leq 0.9$. When $a = 0.4–0.7$, the system is chaotic at $q = 1$ and at the range of $0.1 \leq q \leq 0.3$. It is nonchaotic at the range of $0.4 \leq q \leq 0.9$. When $a = 0.8–1$, the system is chaotic at the range of $0.8 \leq q < 1$ and $0.1 \leq q \leq 0.3$, and it is nonchaotic at the range of $0.4 \leq q \leq 0.7$.

Obviously, dynamics of fractional-order unified chaotic system is related to both the system parameter and the fractional order. When the system parameter is fixed,

Table 8.8: Status of fractional-order unified system with different parameter.

Parameter a	Fractional order	Largest LE	Status
0	$0 < q \leq 0.9$	[−341.0837, −0.0459]	Nonchaotic
	$q = 1$	0.0521	Chaotic
	$q = 0.1$	[0.1418, 0.0605]	Chaotic
$0.1 \leq a \leq 0.3$	$0.2 \leq q \leq 0.9$	[−339.9469, −0.0005]	Nonchaotic
	$q = 1$	[0.0730, 0.0886]	Chaotic
	$0.1 \leq q \leq 0.3$	[0.2250, 0.0694]	Chaotic
$0.4 \leq a \leq 0.7$	$0.4 \leq q \leq 0.9$	[−331.3635, −0.0085]	Nonchaotic
	$q = 1$	[0.0509, 0.0742]	Chaotic
	$0.1 \leq q \leq 0.3$	[0.1206, 0.0686]	Chaotic
$0.8 \leq a \leq 1$	$0.4 \leq q \leq 0.7$	[−330.7911, −0.0028]	Nonchaotic
	$0.8 \leq q \leq 1$	[0.0104, 0.0756]	Chaotic

the system is piecewise chaotic. When the fractional order is fixed, it is possible for the system to change from nonchaotic to chaotic. Besides, it can be seen that with the increase of parameter and order, the chaotic range of the system expands, while the nonchaotic range of order narrows.

8.5.4.2 Routes to Chaos with Fixed Fractional Order and Parameter Varying

We take $q = 0.3$ as an example to discuss the routes to chaos of the fractional-order system. When the initial values are (−4.7508, −7.3886, 12.3605) and parameter $a = 0$–1, the phase diagram on the x–z plane of fractional-order unified system is shown in Figure 8.11. When the system parameter a is relatively small, the system

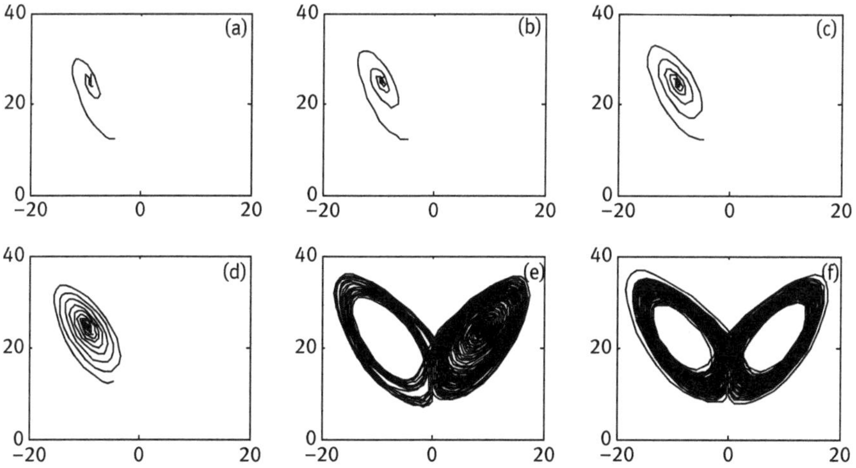

Figure 8.11: Attractors in the x–z phase of unified system with different parameter and $q = 0.3$: (a) $a = 0$, (b) $a = 0.1$, (c) $a = 0.2$, (d) $a = 0.3$, (e) $a = 0.45$, and (f) $a = 0.49$.

converges to the equilibrium point (as shown in Figure 8.11(a), 8.11(b), and 8.11(c)). However, with the increase of a, the evolution time of state variable tending to the equilibrium point becomes longer (as shown in Figure 8.11(d) and 8.11(e)). As a further increase to a certain value, such as $a = 0.49$, system reaches a border of attraction basin. At this time, system enters into chaos through boundary crisis bifurcation (as shown in Figure 8.11(f)). However, if we pursue accurate analysis for routes to chaos of fractional-order system, complex mathematical analysis and calculation are required.

By observing system phase diagrams and calculating power spectrum and the largest Lyapunov exponent, we analyzed dynamical characteristic of fractional-order unified system in detail. We also find that system shows piecewise chaotic behavior with the variation of system parameter and differential order.

8.6 Dynamics of Fractional-Order Diffusionless Lorenz System Based on Time Domain Method

8.6.1 Fractional-Order Diffusionless Lorenz System Model

The mathematic model of diffusionless Lorenz system is

$$\begin{cases} \dot{x} = -y - x \\ \dot{y} = -xz \\ \dot{z} = xy + R \end{cases}, \tag{8.71}$$

where R is the system parameter. When $R \in (0, 5)$, there exist equilibrium points $(x^*, y^*, z^*) = (\pm\sqrt{R}, \mp\sqrt{R}, 0)$. The eigenvalues of this system satisfy the equation $\lambda^3 + \lambda^2 + R\lambda + 2R = 0$. When $R = 3.4693$, the chaotic attractor of diffusionless Lorenz system is shown in Figure 8.12(a). Its maximum Kaplan–Yorke fractal is 2.23542 [192].

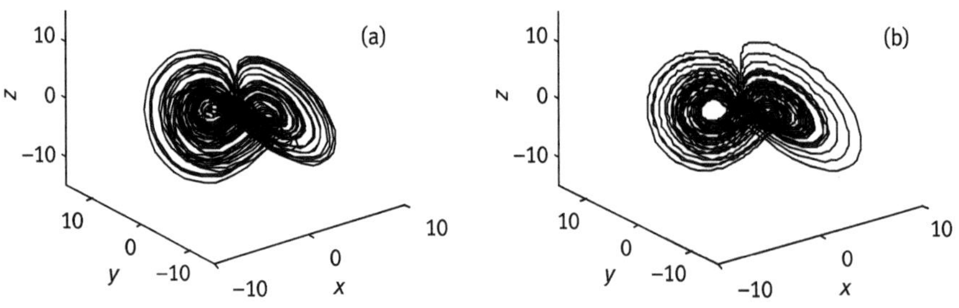

Figure 8.12: Attractors of diffusionless Lorenz system with $R = 3.4693$: (a) $\alpha = \beta = \gamma = 1$ and (b) $\alpha = \beta = \gamma = 0.95$.

Correspondingly, the fractional-order diffusionless Lorenz system is

$$\begin{cases} \frac{d^{\alpha}x}{dt^{\alpha}} = -y - x \\ \frac{d^{\beta}y}{dt^{\beta}} = -xz \\ \frac{d^{\gamma}z}{dt^{\gamma}} = xy + R \end{cases}, \tag{8.72}$$

where α, β, and γ are the fractional order and $0 < \alpha, \beta, \gamma \le 1$. When $\alpha = \beta = \gamma = 0.95$ and $R = 3.4693$, the chaotic attractor of diffusionless Lorenz system is shown in Figure 8.12(b).

According to the previous discussion about time domain solution algorithm of the fractional-order system, we can get the iterative calculation equation of fractional-order diffusionless system as shown in eq. (8.64). Next, we analyze its dynamic characteristic in detail.

8.6.2 Chaotic Characteristic of Fractional-Order Diffusionless Lorenz System

Here, we adopt Wolf algorithm [190] calculating the largest Lyapunov exponent. Let $\alpha = \beta = \gamma = q$ and control the variation of parameter, we can get the effective chaotic range of frictional-order diffusionless system as shown in Figure 8.13. Obviously, there exist three different dynamic statuses: limit circle, chaos, and convergence. If the integer-order system is chaotic, the corresponding chaotic range of fractional-order system will narrow slowly with the increase of system control parameter. If the integer-order system shows limit circle, the corresponding fractional-order system is chaotic. But the chaotic range will narrow quickly with the increase of system parameter. When $R = 8$, the largest Lyapunov exponent of fractional-order diffusionless Lorenz system

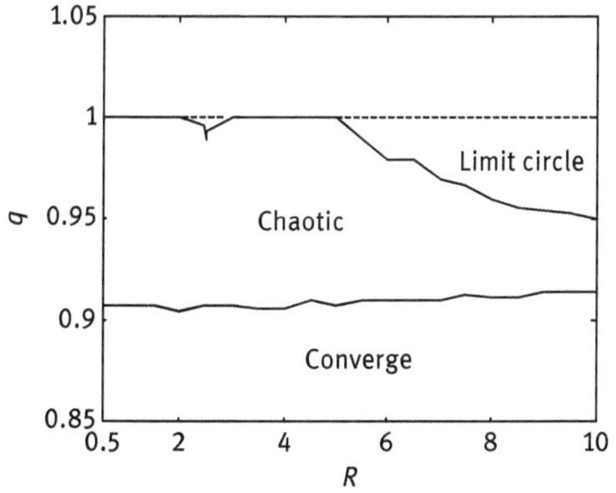

Figure 8.13: Chaos range of fractional-order diffusionless Lorenz system.

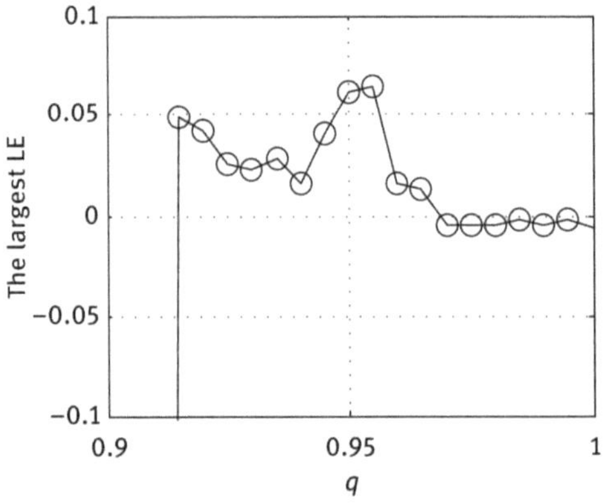

Figure 8.14: The largest Lyapunov exponent of fractional-order diffusionless Lorenz system with $R = 8$.

is shown in Figure 8.14. Obviously, its chaotic characteristic is consistent with the chaotic range as shown in Figure 8.13.

8.6.3 Bifurcation Analysis with Varying System Parameter *R*

Here, we set the fractional order $\alpha = \beta = y = 0.95$ and let the system control parameter R vary from 0.5 to 11 with the step size of 0.01. The initial values are $x(0) = -0.0249$, $y(0) = -0.1563$, and $z(0) = 0.9441$, and the simulation time is 140 s. The bifurcation diagram of fractional-order system is shown in Figure 8.15. Obviously, when the order of fractional-order diffusionless system is 2.85 and R varies, the system is chaotic with a periodic window. When the control parameter R decreased from 11, the

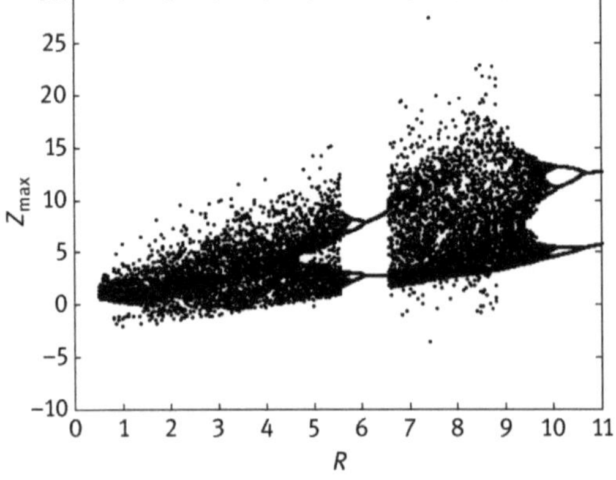

Figure 8.15: Bifurcation diagram of fractional-order diffusionless system with R varying at $q = 0.95$.

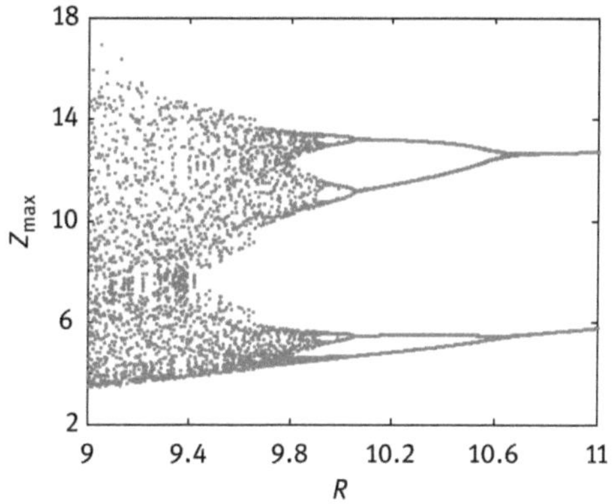

Figure 8.16: Periodic window of fractional-order diffusionless system with R varying at $q = 0.95$.

fractional-order diffusionless system enters into chaotic by period-doubling bifurcation as shown in Figure 8.16 (the step size is 0.005). In periodic window, several bifurcations are observed, such as an interior crisis bifurcation at $R \cong 5.57$, a flip bifurcation at $R \cong 6.02$, and a tangent bifurcation at $R \cong 6.55$. In the interior crisis, the chaotic attractor collides with an unstable periodic orbit or limit cycle within its basin of attraction. When the collision occurs, the attractor suddenly expands in size but remains bounded. For the tangent bifurcation, a saddle point and a stable node coalesce and annihilate one another, producing an orbit that has periods of chaos interspersed with periods of regular oscillation [193]. This same behavior occurs in the periodic windows of the logistic map, including the small windows within the larger windows [194]. In this periodic window, we also observe the route to chaos by a period-doubling bifurcation.

8.6.4 Bifurcation Analysis with Varying Fractional Order *q*

Now, we set $\alpha = \beta = \gamma = q$, $R = 8$ and q varying from 0.9 to 1. The initial values and running time kept the same as above, but the step size of R is 0.0005. The system bifurcation diagram is shown in Figure 8.17. The system enters into chaos by period-doubling bifurcation with decreasing fractional order. It is interesting to note that a chaotic transient is observed when q is less than 0.912. The state-space trajectory is shown in Figure 8.18(a) for $q = 0.911$, which suddenly switches to a pattern of oscillation that decays to the right equilibrium point. The time domain waveform of variation $z(t)$ shown in Figure 8.18(b) eventually converges to the fixed point. On the average, the transform from chaotic behavior into damped state requires about 70 oscillations. When q is relatively greater, its chaotic behavior will remain longer. Similar phenomenon has been reported in the standard Lorenz system [195]. The fractional-order system enters to chaos by a period-doubling bifurcation as shown

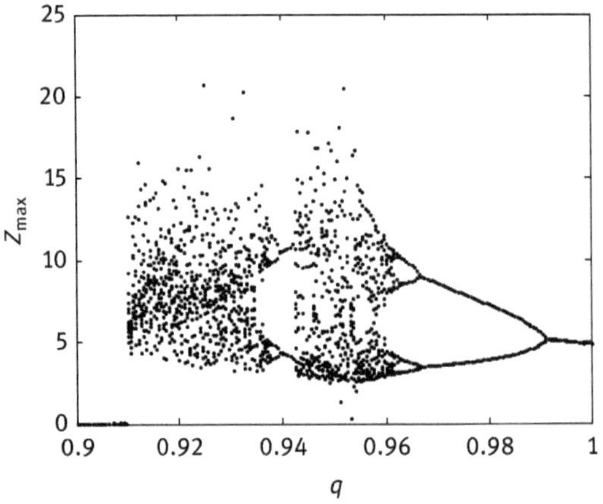

Figure 8.17: Bifurcation diagram of fractional-order diffusionless system with q varying at $R = 8$.

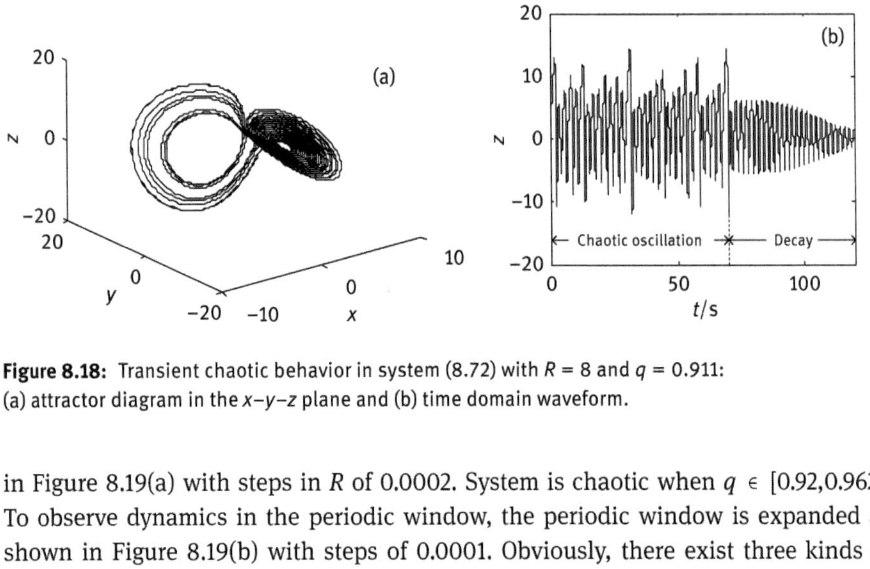

Figure 8.18: Transient chaotic behavior in system (8.72) with $R = 8$ and $q = 0.911$: (a) attractor diagram in the x–y–z plane and (b) time domain waveform.

in Figure 8.19(a) with steps in R of 0.0002. System is chaotic when $q \in [0.92, 0.962]$. To observe dynamics in the periodic window, the periodic window is expanded as shown in Figure 8.19(b) with steps of 0.0001. Obviously, there exist three kinds of bifurcations, such as a tangent bifurcation, a flip bifurcation, and an interior crisis. The fractional order parameter can be taken as a bifurcation parameter, just like the control parameter.

8.6.5 Bifurcations with Different Fractional Order

1. Fix $\beta = y = 1$, $R = 8$ and let α vary. The system is calculated numerically for $\alpha \in [0.4, 1]$ with an increment of 0.002. The bifurcation diagram is shown in Figure 8.20(a). Obviously, the fractional-order system is chaotic when $\alpha \in [0.43, 0.94)$

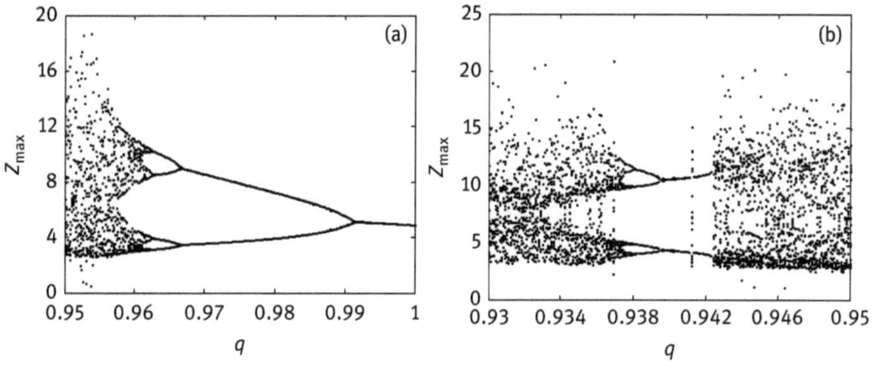

Figure 8.19: Bifurcation diagram of fractional-order diffusionless Lorenz system at $R = 8$: (a) $q \in [0.95, 1]$ and (b) $q \in [0.93, 0.95]$.

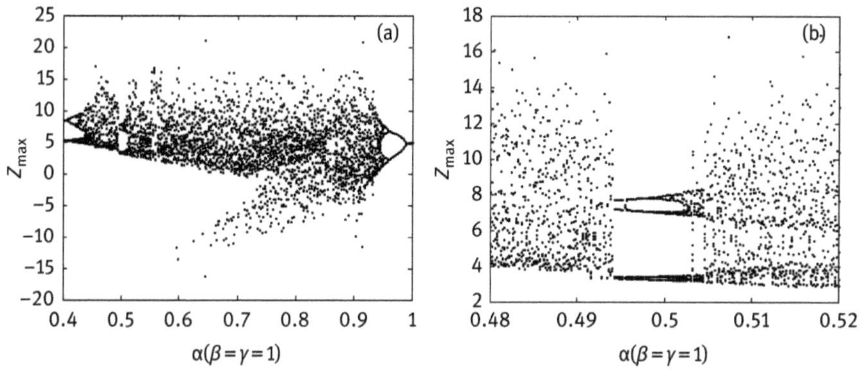

Figure 8.20: Bifurcation diagrams of fractional-order diffusionless Lorenz system with α at $\beta = \gamma = 1$: (a) $\alpha \in [0.4, 1]$ and (b) $\alpha \in [0.48, 0.52]$.

and it is periodic when $\alpha \in (0.484, 0.503)$. The periodic window is expanded as shown in Figure 8.20(b) with steps of 0.0002. When α increases from 0.4, or decreases from 1, a period-doubling bifurcation route to chaos is observed.

2. Fix $\alpha = \gamma = 1$, $R = 8$ and let β vary. The step of β is 0.001. The system is calculated numerically for $\beta \in [0.75, 1]$ with an increment of β equal to 0.001. The bifurcation diagram is shown in Figure 8.21. Obviously, when β decreased from 1, the fractional-order system enters into chaos by period-doubling bifurcation. When β is less than 0.8, the system converges to a fixed point. Thus, the minimum order of fractional-order diffusionless Lorenz system is 2.8 in this case.
3. Fix $\alpha = \beta = 1$, $R = 8$ and let γ vary. The step of γ is 0.001. The system is calculated numerically for $\gamma \in [0.8, 1]$. The bifurcation diagram is shown in Figure 8.22(a). It has similar dynamics with the previous two cases, which enters into chaotic by period-doubling bifurcation. Finally, the system converges to a fixed point at

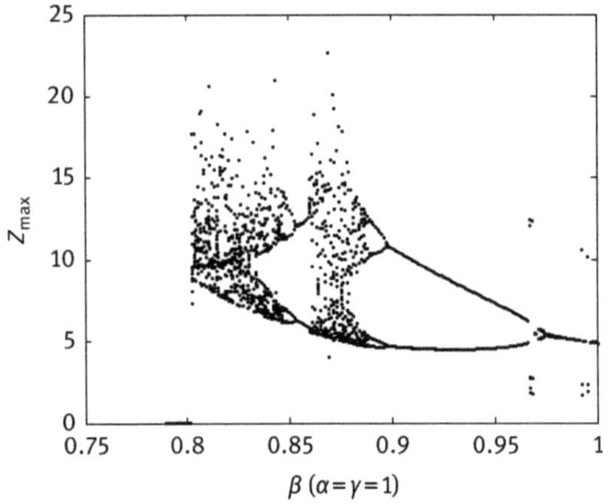

Figure 8.21: Bifurcation diagram of fractional-order diffusionless system with β at $\alpha = \gamma = 1$.

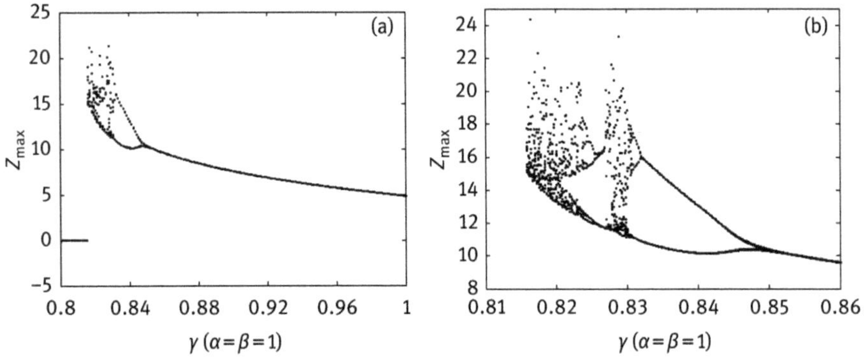

Figure 8.22: Bifurcation diagrams of fractional-order diffusionless Lorenz system with γ at $\alpha = \beta = 1$: (a) $\gamma \in [0.8, 1]$ and (b) $\gamma \in [0.81, 0.86]$.

$\gamma = 0.816$. However, its range of chaotic order is smaller. The dynamic behaviors in the periodic window are similar with the case of (2) as shown in Figure 8.22(b) with steps of 0.0002.

The analysis results mentioned above show the fractional-order diffusionless Lorenz system has rich dynamical behaviors. It provides a theoretical and simulation basis for practical application of fractional-order chaotic system.

Questions

1. What is the difference and relationship between fractional-order chaotic system and integer-order system?

2. For fractional-order chaotic system, which solving method is the best?
3. What are the methods of dynamic analysis for fractional-order chaotic system?
4. Compare characteristics among different simulation methods for fractional-order chaotic system.
5. Choose an integer-order chaotic system to analyze its corresponding fractional-order dynamic characteristic.

9 Simulation and Hardware Implementation of Chaotic System

Chaos system can be applied in many fields, such as secure communications, information encryption, system control, system optimization, and so on. To develop the applications of chaos theory, the hardware design and implementation of chaotic system must be solved, including the design and implementation of chaotic analog circuit and chaotic digital circuit. In this chapter, we focus on the simulation and hardware implementation of chaotic system.

9.1 Dynamic Simulation of Chaotic System

9.1.1 Dynamic Simulation Steps on Simulink

Dynamic simulation steps of Simulink are shown in Figure 9.1. The steps include establishing simulation model, setting system parameters and observation window, analyzing simulation results, and modifying model and parameters [188]. A detailed description is presented as follows:

1. Establishing simulation mathematical model. It is the key for the simulation of chaos system. According to the chaotic system equation, the whole system is simplified to the source system. Functions of the system are determined, and each part of the function is modeled. We should find out the relationship between the various parts and draw the system flow diagram model. With the increase of the modules and complexity of the system, the model will become more and more complicated. To reduce the number of modules in the model window, we can group the same function modules and simplify the complex problem by packaging subsystem. For example, to observe the time evolution of the synchronization error signal ($|e| = \sqrt{(x_2 - x_1)^2 + (y_2 - y_1)^2 + (z_2 - z_1)^2}$), the module is constructed as shown in Figure 9.2(a), which can be packaged into a subsystem as shown in Figure 9.2(b), which simplifies the structure of the simulation system.
2. Building simulation system model. According to the established mathematical model, the function module is dragged into the editing window from the Simulink model library. After the modules are connected, the simulation model of chaotic system is constructed.
3. Setting simulation parameters. It is an important part of the simulation process, which may directly affect the simulation times and results. Parameters setting include setting operating system parameters (such as the system running time, the integration step size, the tolerable error, the integral algorithm) and the operation

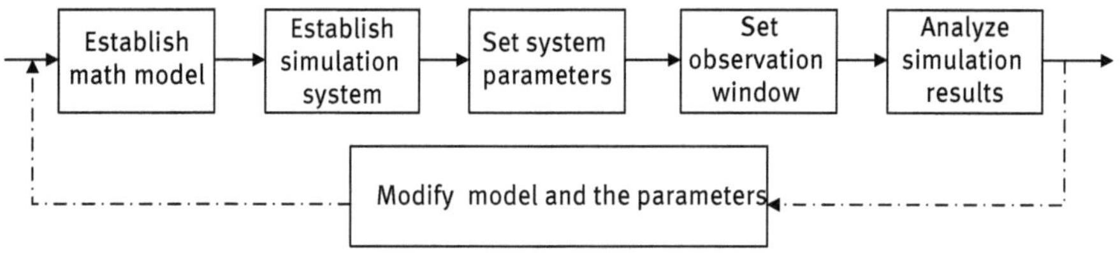

Figure 9.1: Matlab/Simulink simulation steps.

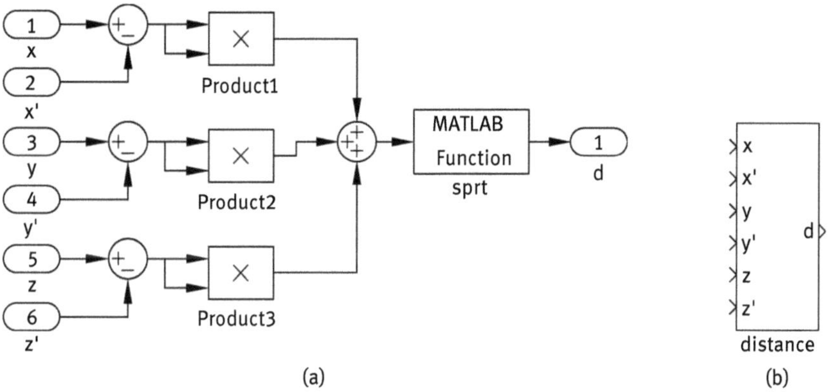

Figure 9.2: Error calculation submodule for synchronization system: (a) error signal function module and (b) packaged subsystem.

parameters of the function modules (such as the initial value, signal amplitude, frequency). The former is finished in the parameter submenu of the Simulink menu, and the latter is finished in the pop-up window by double-clicking the module.

4. Setting observation window. The observation module is used to observe the operation of the simulation system, so as to adjust the parameters and analyze the results. The observation modules include the oscilloscope (Scope), XYgraph, Display, to workspace, and so on.
5. Analyzing simulation results and correcting model or parameter. We can judge whether the simulation results are consistent with that of the theoretical analysis by analyzing the output waveform and output data of the simulation system. Generally, it is difficult to succeed at a time. Many factors affect the simulation results. First, you need to check whether the simulation system is consistent with the theoretical model or whether the parameter settings are correct. If the correct simulation results cannot be obtained by checking and modifying the simulation parameters, you need to consider whether the simulation model is correct. We need to modify the model and then simulate it until the results are satisfactory.

9.1.2 Simulation Result Analysis and Performance Improvement

Outputting simulation results (graphics) to the word document is an important way to observe the results. First, the output signal is written into the return variable. That is to say, the Out module in the sink library is connected as the output. After choosing the variable time (I/O) and output (Output) box and setting the corresponding time variable tout() and output variable yout(), the simulation is operated. Finally, we can run the Matlab command plot(tout, yout) in the command window. In addition, by using "Figure Copy" in the Edit Windows of the submenu, one can get the copied graph, and then paste it into the drawing software window. To improve the resolution, the graph is generally processed by graphics processing software and then be pasted into the document.

Note that the simulation time is different from the time of simulation running. The simulation time is defined as the difference between the Start time and the Stop time. The time of simulation running is the time to run a simulation. In general, the two are not equal. The time required to run a simulation depends on many factors, including the complexity of the model, the maximum and minimum step size, computer performance, and so on.

Several reasons will lead to simulation time too long. For example, the simulation step size and the tolerance error are too small. When the model contains Fcn Matlab module, the simulation of each step needs to explain the program, which will greatly reduce the simulation speed, so it is best to use the S-function M file. S-function of M file also allows Simulink to call the Matlab interpreter at each simulation time, so it can be replaced by the C-mex function or equivalent subsystem.

9.2 Circuit Simulation of Chaotic System

Circuit simulation experiment method is used to simulate the chaotic system by the electronic circuit diagram and to study the characteristics of the chaotic system by analyzing the experimental results. For example, the Electronics Workbench (EWB) software platform is used to design the chaotic system.

EWB is a virtualization working platform that simulates the analog circuit. It has a more perfect model library and several common analytical instruments. It can be used to analyze the electronic circuit and directly output the designed circuit files to common electronic circuit layout software, such as Protel. The experimental analysis steps of the chaotic system are shown in Figure 9.3 [188].

1. According to the chaotic system, design the EWB simulation circuit of a chaotic system.
2. Choose the components in the window of EWB and connect them.
3. Determine the component parameters, including the label of the components and its parameter values.

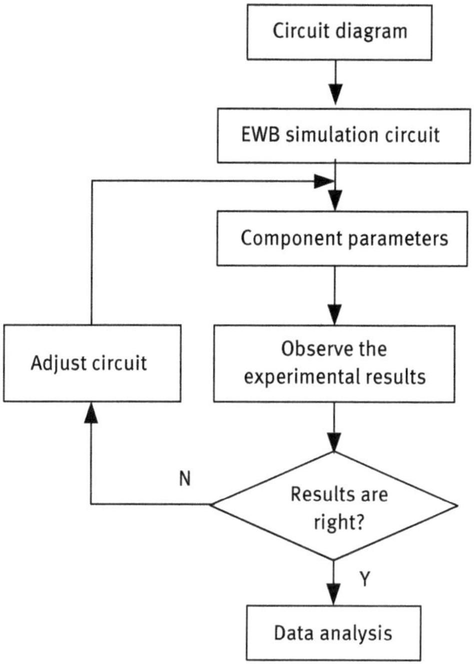

Figure 9.3: EWB simulation procedure of chaotic system.

4. Use EWB virtual instrument to observe the experimental results, analyze data, and compare whether the results are consistent with that of the theoretical analysis. Otherwise we should adjust the circuit.

9.3 Design of Chaotic Analog Circuit

9.3.1 Procedure of Chaotic Circuit Design

There are three kinds of chaotic circuit design methods, such as individual design, modular design, and improved modular design. The advantage of the individual design is that the number of circuit components is the least, but it needs a rich prior knowledge and circuit design skills. The modular design is based on the nondimensional state equation, which is universal but needs more components. The improved modular design is applied to determine the system parameters by comparing the state differential equation with the actual circuit state differential equation so that the number of components is the minimum. Chaotic circuit modular design flowchart is shown in Figure 9.4. First, based on the dimensionless chaotic state equation, the phase

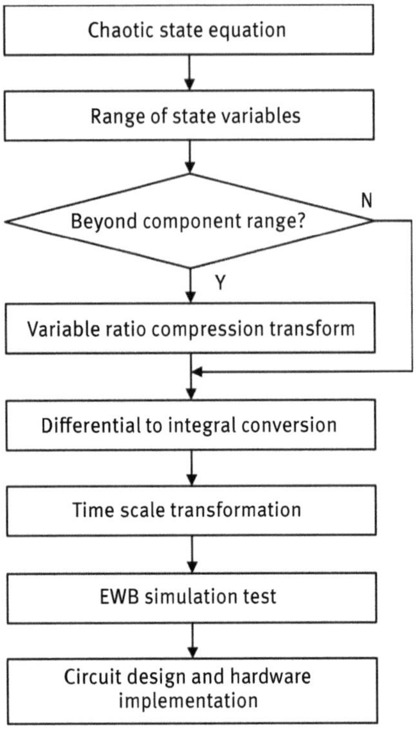

Figure 9.4: Chaotic circuit modular design flowchart.

diagrams are calculated to determine the range of the state variables. If the variable is too large to beyond the scope of the dynamic variables of components, the proportion of compression transform is needed, and all variables will be reduced N times. If the variables are within the range, it is not necessary to carry out the compression transformation. Then the EWB circuit simulation is carried out, and the circuit is designed finally.

Circuit design process mainly includes following five steps.

1. Variable ratio compression transformation. Because the common power supply voltage is 15 V, the linear dynamic range of an operational amplifier (OA) is 13.5 V. According to the simulation graph, one needs to analyze whether the variable range is beyond. If it is beyond, it needs to do compression transformation.
2. Do timescale transformation.
3. Do differential integral transformation.
4. The equation was standardized to adapt to the inverse phase adder in the module circuit.
5. According to the standard integral equation, the corresponding modular circuit is designed.

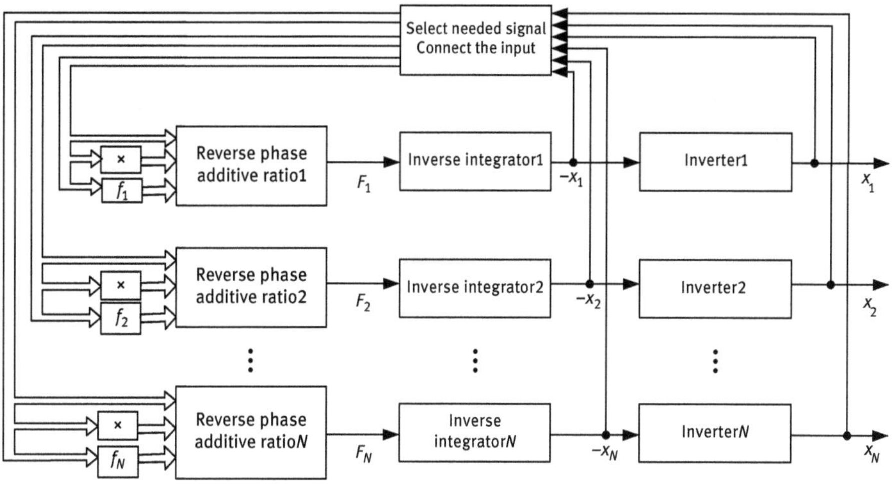

Figure 9.5: Block diagram of chaotic circuit design.

9.3.2 Modular Design of Chaotic Circuit

The modular design of chaotic circuit is designed mainly based on dimensionless equation. The general principle block diagram is shown in Figure 9.5 [196]. The circuit consists of four modules, including nonlinear function, reverse phase additive proportional arithmetic unit, reverse integrator, and inverter. A nonlinear function includes multiplication function, the absolute value function, and so on. Circuits of the latter three modules mainly are built based on amplifiers.

9.3.3 Basic Components of Chaotic Circuit

In general, a chaotic circuit contains a nonlinear unit and a linear unit. A nonlinear unit circuit includes analog multiplier, Chua's diode circuit, the absolute value of the arithmetic circuit and index operation circuit, and so on. Based on the OA device, the common linear basic unit circuits are shown in Figure 9.6, and the input and output expressions of each unit circuit are presented as follows.

1. Reverse phase proportion amplifier: $v_o = -\frac{R_2}{R_1}v_i$ as shown in Figure 9.6(a).
2. Reverse phase adder: $v_o = -\left(\frac{R_3}{R_1}v_{i1} + \frac{R_3}{R_2}v_{i2}\right)$ as shown in Figure 9.6(b).
3. Subtractor: $v_o = \frac{R_2}{R_1}(v_{i2} - v_{i1})$ as shown in Figure 9.6(c).
4. Proportional amplifier: $v_o = \frac{R_1+R_2}{R_1}v_i$ as shown in Figure 9.6(d).
5. Reverse phase differentiator: $v_o = -RC\frac{dv_i}{dt}$ as shown in Figure 9.6(e).
6. Reverse phase integrator: $v_o = -\frac{1}{RC}\int v_i dt$ as shown in Figure 9.6(f).

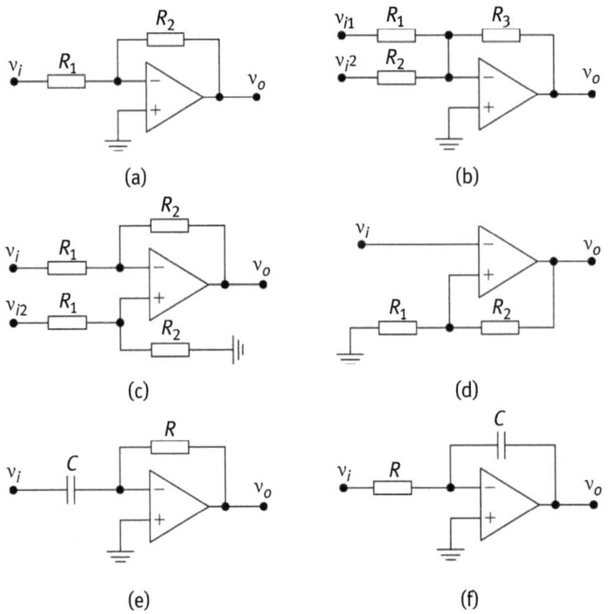

Figure 9.6: Linear unit circuits: (a) reverse phase proportion amplifier; (b) reverse phase adder; (c) subtractor; (d) proportional amplifier; (e) inverse phase differentiator; and (f) reverse phase integrator.

Now taking the simplified Lorenz system as an example, the modular design process is described in detail. The mathematical model of simplified Lorenz system is [56]

$$\begin{cases} \dot{x} = 10(y - x) \\ \dot{y} = -xz + (24 - 4c)x + cy \, , \\ \dot{z} = xy - 8z/3 \end{cases} \tag{9.1}$$

where c is the system parameter. When $c \in [-1.59, 7.75]$, it is chaotic.

1. When $c = 5$, by numerical simulation, all the state variables of the system are beyond the dynamic range of the device, so it is necessary to change the system variables. Let $x' = 0.25x$, $y' = 0.25y$, and $z' = 0.25z$, then system (9.1) becomes to

$$\begin{cases} \dot{x}' = 10(y' - x') \\ \dot{y}' = -4x'z' + 4x' + 5y' \, . \\ \dot{z}' = 4x'y' - 8z'/3 \end{cases} \tag{9.2}$$

2. Do timescale transformation to system (9.2) and transform t to $\tau_0 t'$, where τ_0 is the time-scaling factor. Let $\tau_0 = 1/(R_0 C_0)$, then we have

$$\begin{cases} \dot{x}' = \frac{1}{R_0C_0}10(y' - x') \\ \dot{y}' = \frac{1}{R_0C_0}(-4x'z' + 4x' + 5y') \\ \dot{z}' = \frac{1}{R_0C_0}(4x'y' - 8z'/3) \end{cases} . \tag{9.3}$$

3. After the differential is transformed to integral, the system becomes

$$\begin{cases} x' = -\int \frac{1}{R_0C_0}10((-y') + x')dt' \\ y' = -\int \frac{1}{R_0C_0}(4x'z' + 4(-x') + 5(-y'))dt' \\ z' = -\int \frac{1}{R_0C_0}(-4x'y' + 8z'/3)dt' \end{cases} . \tag{9.4}$$

4. According to system (9.4), the basic circuit units include inverter, inverting adder, integrator, and multiplier. The circuit is shown in Figure 9.7.

In Figure 9.7, x, y, and z represent the output voltage values, respectively. According to the circuit theory, the state equation of the circuit is obtained:

$$\begin{cases} \dot{x} = \frac{1}{R_4C_1}\left(\frac{R_3}{R_2}y - \frac{R_3}{R_1}x\right) \\ \dot{y} = \frac{1}{R_{11}C_2}\left(-\frac{R_{10}}{R_9}0.1xz + \frac{R_{10}}{R_7}x + \frac{R_{10}}{R_8}y\right) \\ \dot{z} = \frac{1}{R_{17}C_3}\left(\frac{R_{16}}{R_{14}}0.1xy - \frac{R_{16}}{R_{15}}z\right) \end{cases} . \tag{9.5}$$

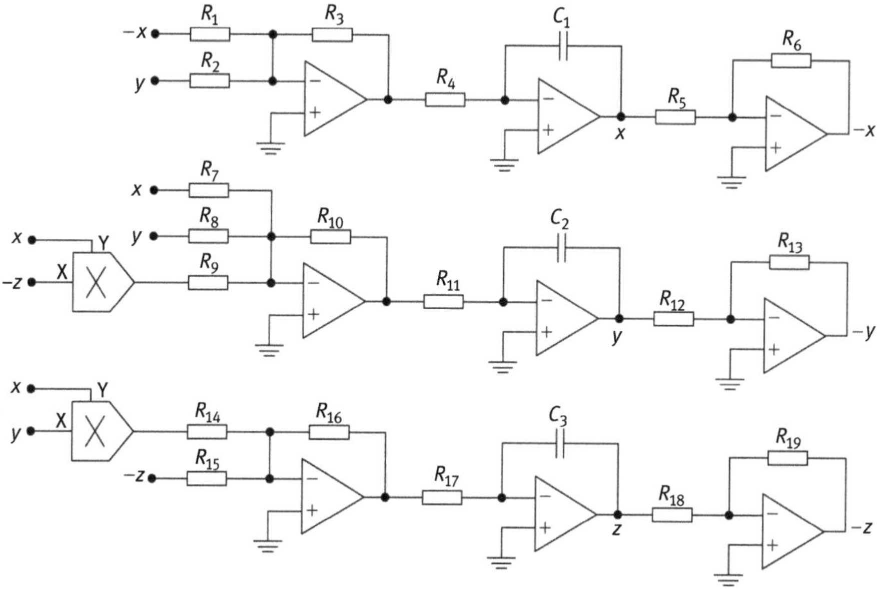

Figure 9.7: Circuit diagram of simplified Lorenz system.

Comparing eq. (9.4) with eq. (9.5), the value of the components is obtained as $R_i = 10\ \text{k}\Omega$ ($i = 1, 2, 4, 5, 6, 11, 12, 13, 17, 18, 19$), $R_i = 100\ \text{k}\Omega$ ($i = 3, 10, 16$), $R_7 = 20\ \text{k}\Omega$, $R_8 = 25\ \text{k}\Omega$, $R_9 = R_{14} = 2.5\ \text{k}\Omega$, $R_{15} = 37.5\ \text{k}\Omega$, $C_i = 100\ \text{nF}$ ($i = 1, 2, 3$).

9.4 Improved Modular Design of Chaotic Circuit

The modular design of chaotic circuit is intuitive, simple, and easy to debug, but it has a large number of components. The circuit based on the improved modular design has fewer components, and its precision is higher than that of the standard-type chaotic circuit. The basic unit circuit module is shown in Figure 9.8. Here taking the multiscroll Jerk system as an example, we illustrate the improved modular design in detail.

The equation of multiscroll Jerk system is [197]

$$\begin{cases} \dot{x} = y \\ \dot{y} = z \\ \dot{z} = -x - y - \alpha z + f(x) \end{cases}, \tag{9.6}$$

where α is the system parameter. When $\alpha = 0.6$ and $f(x) = \text{sgn}(x)+\text{sgn}(x+2)+\text{sgn}(x-2)$, it can produce a four-scroll chaotic attractor. According to the results of the numerical simulation, the system variables are not beyond the dynamic range of the device, so it is not necessary to do proportion transformation for variables.

Do timescale transformation for system (9.6), let $t = \tau_0 t'$, $\tau_0 = 10,000$, we obtain

$$\begin{cases} \dfrac{dx}{dt'} = 10,000y \\ \dfrac{dy}{dt'} = 10,000z \\ \dfrac{dz}{dt'} = -10,000x - 10,000y - 6,000z + 10,000f(x) \end{cases}. \tag{9.7}$$

The circuit units for system (9.7) include basic unit circuit module, inverter module, and nonlinear circuit module. Next, we design the system circuit based on OA and current conveyor (CC).

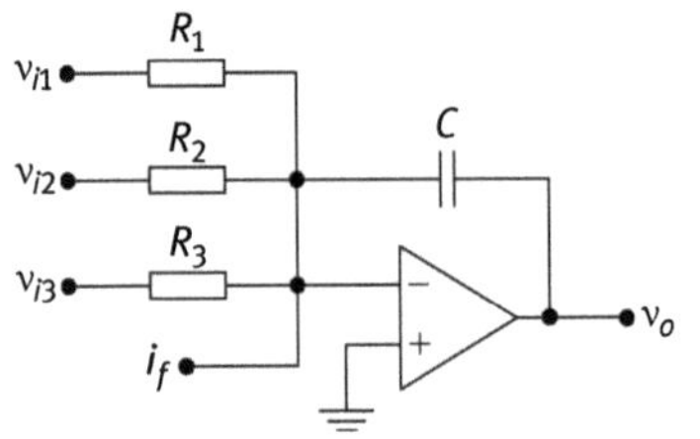

Figure 9.8: General basic unit circuit module based on improved modular design.

9.4.1 Design of Multiscroll Jerk Circuit Based on OA

Four-scroll Jerk circuit based on OA devices is shown in Figure 9.9.

In Figure 9.9, x, y, and z represent the output voltage, respectively, the output current value of the subcircuit $f(x)$ is: $i(x) = -\text{sgn}(x-2)/R_{x1} - \text{sgn}(x)/R_{x2} - \text{sgn}(x+2)/R_{x3}$. According to the circuit theory, we can obtain the state equation of the circuit

$$\begin{cases} \dot{x} = \frac{1}{R_1 C_1} y \\ \dot{y} = \frac{1}{R_2 C_2} z \\ \dot{z} = -\frac{1}{C_3}\left(\frac{x}{R_5} + \frac{y}{R_6} + \frac{z}{R_7} + i(x)\right) \end{cases}. \tag{9.8}$$

Comparing eq. (9.7) with eq. (9.8), the values of the components are obtained as follows: $R_i = 10\ \text{k}\Omega$ ($i = 1, 2, 3, 4, 5, 6$), $R_{xi} = 135\ \text{k}\Omega$ ($i = 1, 2, 3$), $R_7 = 16.7\ \text{k}\Omega$, $C_i = 10$ nF ($i = 1, 2, 3$). According to Figure 9.9, we do simulations on EWB, and the results are shown in Figure 9.10.

9.4.2 Design of Multiscroll Jerk Circuit Based on CC

The CC is the active device commonly used in current mode circuit, and it has strong versatility, high precision, less external devices, and good high-frequency characteristics. It has been widely used in active power filter, analog signal processing, and so on. Among the CCs, the second-generation current conveyor (CCII) is widely applied. Its circuit symbol is shown in Figure 9.11. Characteristics of the ports are $i_y = 0$, $V_x = V_y$, and $i_z = \pm i_x$. When $i_z = i_x$, it is the second-generation current conveyor (CCII+). When $i_z = -i_x$, it is the reverse second-generation current conveyor (CCII−).

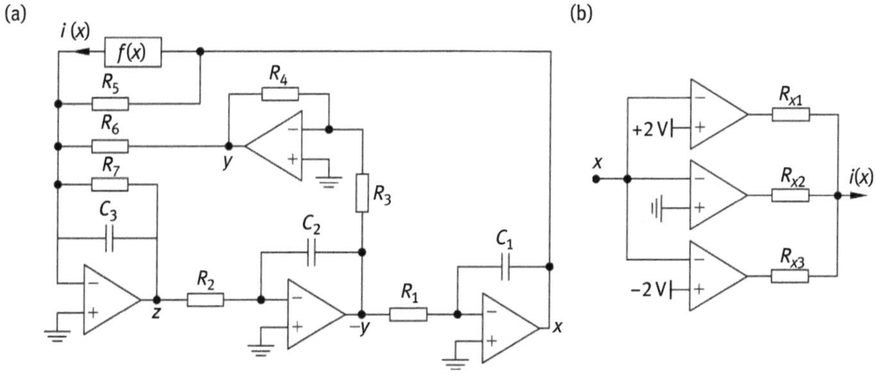

Figure 9.9: Multiscroll Jerk circuit based on OA devices: (a) main circuit and (b) subcircuit of function $f(x)$.

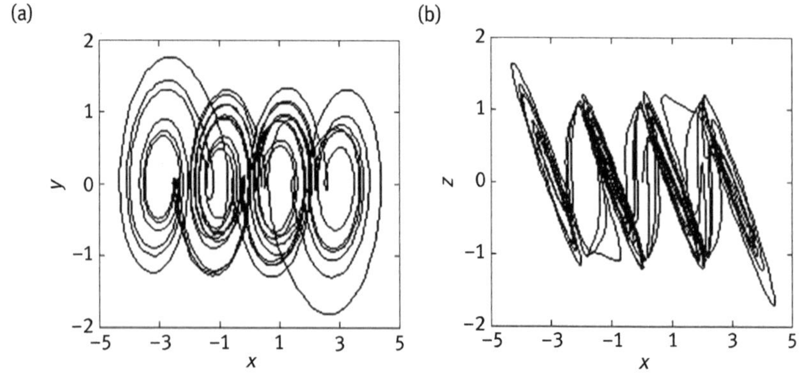

Figure 9.10: Jerk circuit simulation with four scrolls based on the OA: (a) phase diagram on x–y plane and (b) phase diagram on x–z plane.

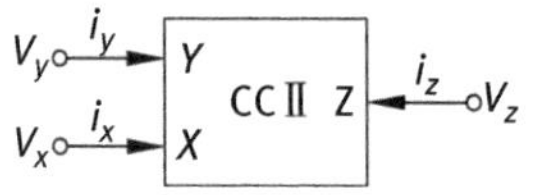

Figure 9.11: CCII circuit symbol.

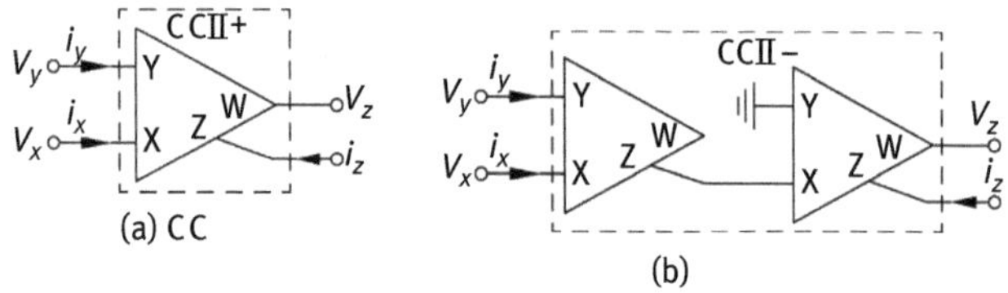

Figure 9.12: CCII implementation circuit : (a) CCII+ and (b) CCII–.

CCII+ can be implemented by employing an AD844 device, while CCII– can be realized by cascading two AD844 devices as shown in Figure 9.12.

Based on CCII+, the common linear basic unit circuits are shown in Figure 9.13; the input and output expressions of each unit circuit are shown as follows [198]:

1. Proportional amplifier: $v_o = \frac{R_2}{R_1} v_i$, as shown in Figure 9.13(a).
2. Reverse proportional amplifier: $v_o = -\frac{R_2}{R_1} v_i$, as shown in Figure 9.13(b).
3. Inverting differential device: $v_o = -RC\frac{dv_i}{dt}$, as shown in Figure 9.13(c).
4. Inverting adder: $v_o = -\left(\frac{R_3}{R_1} v_{i1} + \frac{R_3}{R_2} v_{i2}\right)$, as shown in Figure 9.13(d).
5. Reverse integrator: $v_o = -\frac{1}{RC}\int v_i dt$, as shown in Figure 9.13(e).
6. Integrator: $v_o = \frac{1}{RC}\int v_i dt$, as shown in Figure 9.13(f).

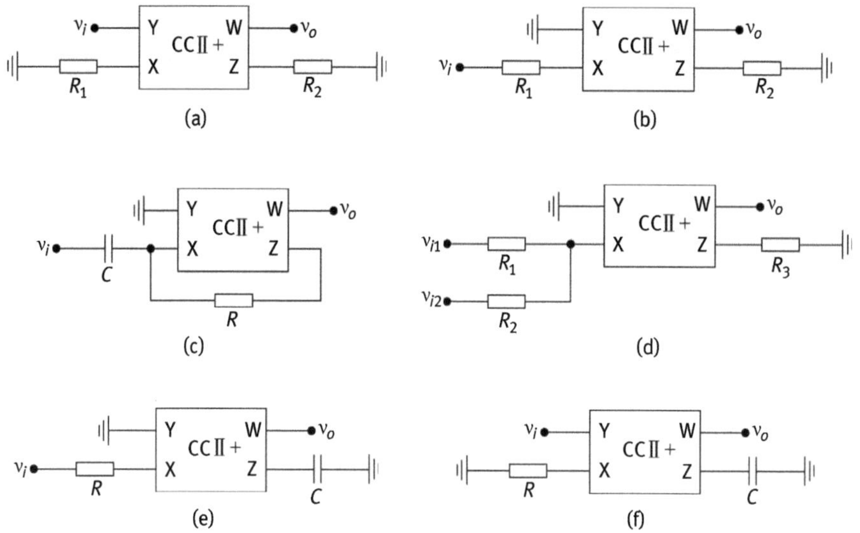

Figure 9.13: Basic unit circuits: (a) proportional amplifier; (b) reverse proportional amplifier; (c) reverse differentiator; (d) inverting adder; (e) reverse integrator; and (f) integrator.

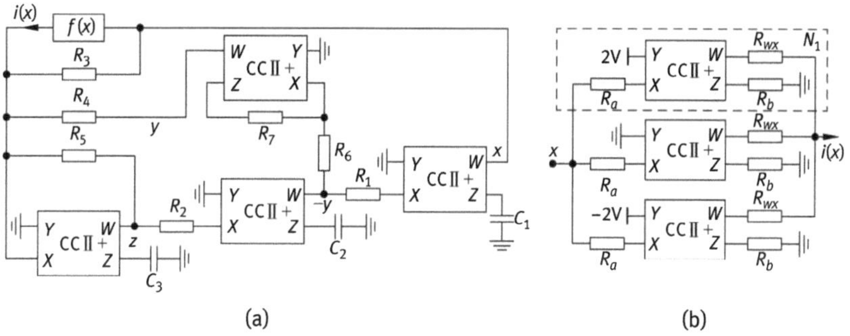

Figure 9.14: Four-scroll Jerk circuit based on CCII device: (a) main circuit and (b) subcircuit $f(x)$.

Based on the mathematical models and system (9.7), the four-scroll Jerk circuit is designed based on CCII as shown in Figure 9.14.

Nonlinear function $f(x)$ is a step function, and a saturation function is achieved based on the N_1 network as shown in Figure 9.14(b). The saturation function sequence is realized by parallel network. If R_b/R_a is large enough, the sequence of saturation function can be approximated by infinite step function. The output current is $i(x) = -[\mathrm{sgn}(x-2)+\mathrm{sgn}(x)+\mathrm{sgn}(x+2)]/R_{wx}$. According to the circuit theory, the state equation of the circuit is

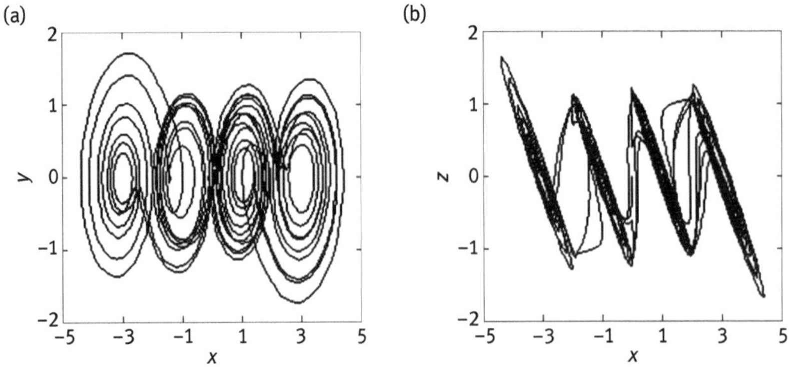

Figure 9.15: Four-scroll Jerk chaotic attractor based on CCII: (a) phase diagram on x–y plane and (b) phase diagram on x–z plane.

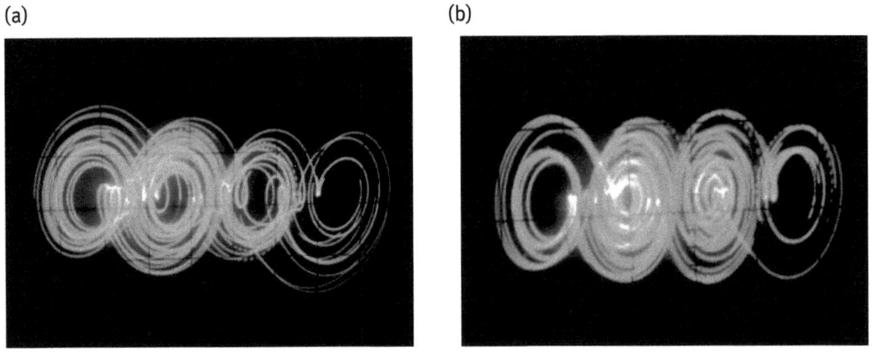

Figure 9.16: Phase diagram on x–y plane: (a) circuit implement based on OA and (b) circuit implement based on CC.

$$\begin{cases} \dot{x} = \frac{1}{R_1C_1}y \\ \dot{y} = \frac{1}{R_2C_2}z \\ \dot{z} = -\frac{1}{C_3}\left(\frac{x}{R_3} + \frac{y}{R_4} + \frac{z}{R_5} + i(x)\right) \end{cases}. \tag{9.9}$$

When comparing eq. (9.7) with eq. (9.9), the value of each component was obtained as follows: R_i = 10 kΩ (i = 1, 2, 3, 4, 6), R_5 = 16.7 kΩ, R_7 = 20 kΩ, R_{wx} = 102 kΩ, R_a = 470 Ω, R_b = 1 MΩ, C_i = 10 nF (i = 1, 2, 3). The circuit in Figure 9.14 is simulated on EWB. According to Figure 9.14, we do simulations on EWB, and the results are shown in Figure 9.15.

9.4.3 Multiscroll Jerk Circuit Implementation

According to Figures 9.9 and 9.14, the four-scroll Jerk circuit is designed based on CC and OA devices, and the phase diagram is observed by oscilloscope as shown in

Figure 9.16. When compared with Figures 9.10 and 9.15, respectively, we concluded that the results of actual circuit experiment and circuit simulation experiment are consistent, and it verified the correctness of the circuit design and the physical realization of the four-scroll Jerk system.

9.5 Digital Signal Processor Design and Implementation of Chaotic Systems

Hardware implementation of chaotic system is a key problem for the applications of chaotic system in engineering practice. With the development of microelectronics and digital signal processing technology, the digital circuit realization of chaotic system has become a research hotspot [199–201]. Compared with the analog devices, digital signal processor (DSP) device has the characteristics of stability, repeatability, programmable, and easy to implement. It brings about great developments of digital signal processing in the fields of aviation, aerospace, radar, sonar, communication, domestic appliance, and so on and becomes the core of electronic system. With increase in the calculation speed of DSP, the bandwidth of the real-time processing is greatly increased, and the DSP is widely used in the application of chaotic system.

Here, taking simplified Lorenz system and fractional-order simplified Lorenz system as examples, we will study the DSP implementation of the chaotic system based on DSP TMS320F2812.

9.5.1 DSP Experimental Platform of Chaotic System

The hardware platform to implement chaotic system based on DSP is shown in Figure 9.17, in which the D/A converter is the 16-bit dual-channel DAC8552 chip, and the interface chip is MAX3232. Their pin distribution is shown in Figure 9.18.

According to the system requirements, we can choose DSP chip with fixed point or floating point. The main frequency of DSP chip will affect the operation speed, and consequently affect the generating speed of the chaotic sequence. In addition, the width of the data bus will affect the numerical accuracy of the chaotic sequence. We need to consider the DSP chip peripherals to connect with other parts of the

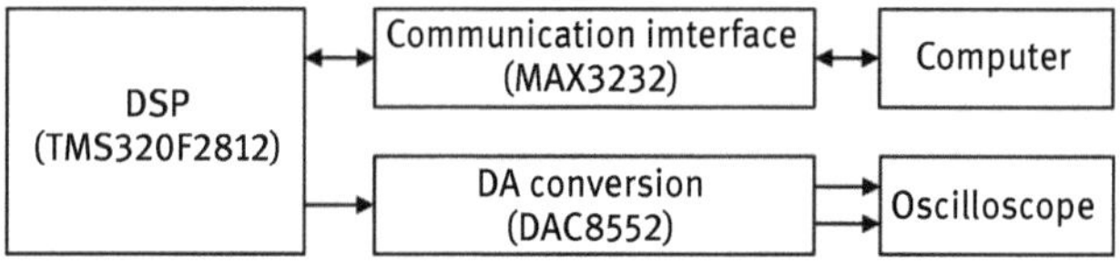

Figure 9.17: Hardware block diagram of chaotic system based on DSP.

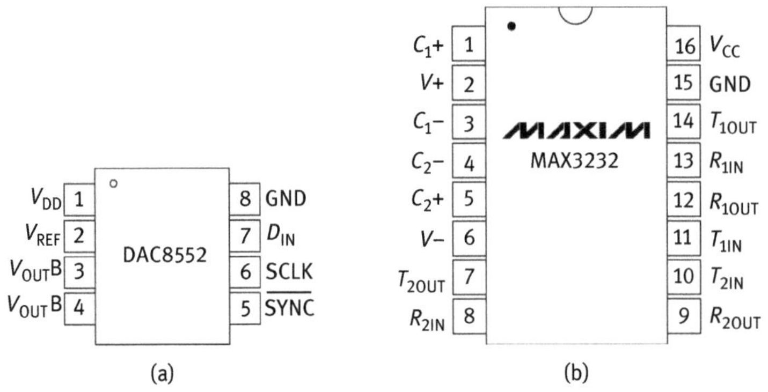

Figure 9.18: Pin distribution of DAC8552 and MAX3232.

system conveniently. Here, we choose TMS320F2812 with 32-bit fixed point DSP chip, amount of peripheral interfaces, and 150 MHz main frequency. Universal Asynchronous Receiver/Transmitter (UART) serial interface is generally used in communication interface, which transmits chaos sequences to the computer in real time. The D/A converter converts the chaotic sequence generated by DSP into analog signals, which can be observed on the oscilloscope. So it demands that the D/A conversion has the dual-channel conversion function for synchronous output, and its conversion speed and precision are high enough. The operation command words of DSP for the DAC8552 are shown in Figure 9.19, where DB15–DB0 is the converted digital signal.

By employing CCS4.2 as the DSP integrated development environment, the experimental platform of chaotic system based on DSP is shown in Figure 9.20.

The software flowchart for the realization of the chaos system on DSP is shown in Figure 9.21, which includes DSP initialization, setting of the clock frequency, initialization of the related interface, and so on. Different chaotic system has different system parameters, and the iteration step size, iteration number, and iteration initial value also need to be determined. According to different solution algorithm, we can derive the iterative calculation formula. Each iteration calculation results needs to be stored as the initial value of the next iteration. If data output to D/A or PC, the corresponding data processing is needed, as shown in Figures 9.22 and 9.23.

24 bits in the *A* data buffer written by the first step

Res	Res	LDB	LDA	DC	Buffer select	PD1	PD0	DB15	—	DB1	DB0
0	0	0	0	×	0	0	0	D15	—	D1	D0

24 bits in the *B* data buffer written by the second step

Res	Res	LDB	LDA	DC	Buffer select	PD1	PD0	DB15	—	DB1	DB0
0	0	1	1	×	1	0	0	D15	—	D1	D0

Figure 9.19: Control word outputted by DAC8552 dual channel.

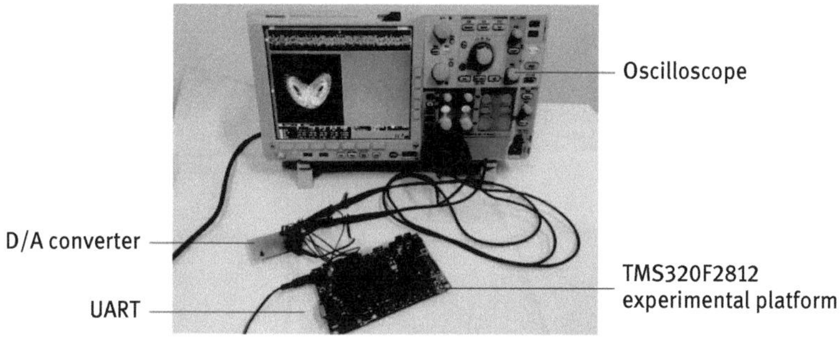

Figure 9.20: Experimental platform of chaotic system based on DSP.

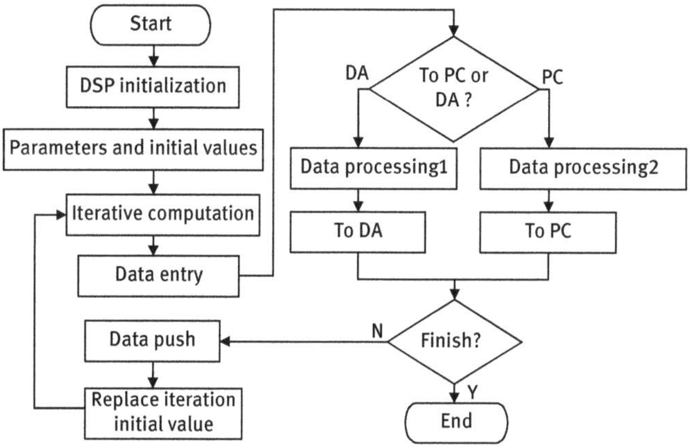

Figure 9.21: Software flow of chaos system based on DSP.

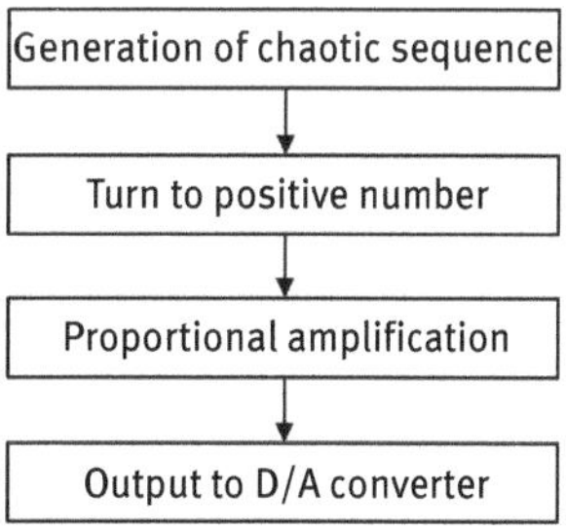

Figure 9.22: Data processing 1.

1. Data processing 1: To realize the D/A conversion correctly, a large enough positive number is added to all of the output value of chaotic sequence, which is equivalent to coordinate translation on the output of the system. DAC8552 is a 16-bit D/A converter, and the converted digital quantity must be positive integers between 0

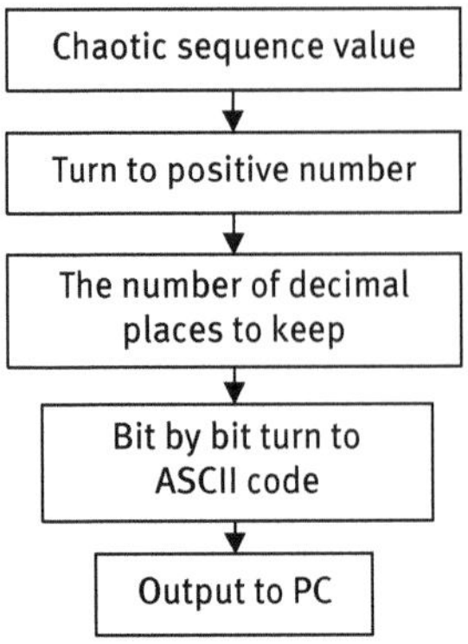

Figure 9.23: Data processing 2.

and 65,535, while the majority value of the chaos sequence generated by DSP is too small. So we need to multiply the sequence value with a positive integer so that the digital numbers are distributed in the whole range of conversion as far as possible.

2. Data processing 2: Similarly, the sequence value of variable is turned to positive. To facilitate the computer display and process, each bit of the chaotic sequence value is converted to ASCII code. Namely, each decimal adds 48 and transmits to the computer bit by bit. Each complete sequence is separated with commas (ASCII $0 \times 2C$). The length of the data can be chosen. When we analyzing it on computer, the chaotic sequence value can be restored to the values calculated by DSP.

The DSP implementation steps are presented by taking integer-order simplified Lorenz system and its fractional-order counterpart as examples.

9.5.2 DSP Implementation of Integer-Order Chaotic System

Taking the simplified Lorenz system as an example, the DSP implementation of the chaotic system based on the fourth-order Runge–Kutta algorithm is described. The simplified Lorenz system equation is [56]

$$\begin{cases} \dot{x} = 10(y - x) \\ \dot{y} = (24 - 4c)x - xz + cy \\ \dot{z} = xy - 8z/3 \end{cases}, \tag{9.10}$$

where c is the system parameter. When $c \in [-1.59, 7.75]$, it is chaotic. By solving the differential equations based on the fourth-order Runge–Kutta method, we obtain

$$\begin{cases} x_{n+1} = x_n + (K_{11} + 2K_{12} + 2K_{13} + K_{14})/6 \\ y_{n+1} = y_n + (K_{21} + 2K_{22} + 2K_{23} + K_{24})/6 \\ z_{n+1} = z_n + (K_{31} + 2K_{32} + 2K_{33} + K_{34})/6 \end{cases} . \tag{9.11}$$

The recursive parameters are as follows:

$$\begin{cases} K_{11} = 10[(y_n - x_n)]T \\ K_{21} = [(24 - 4c)x_n - x_n z_n + c y_n]T \\ K_{31} = [x_n y_n - 8z_n/3]T \end{cases} , \tag{9.12}$$

$$\begin{cases} K_{12} = 10[(y_n + 0.5K_{21}) - (x_n + 0.5K_{11})]T \\ K_{22} = [(24 - 4c)(x_n + 0.5K_{11}) - (x_n + 0.5K_{11})(z_n + 0.5K_{31}) + c(y_n + 0.5K_{21})]T \\ K_{32} = [(x_n + 0.5K_{11})(y_n + 0.5K_{21}) - 8(z_n + 0.5K_{31})/3]T \end{cases} , \tag{9.13}$$

$$\begin{cases} K_{13} = 10[(y_n + 0.5K_{22}) - (x_n + 0.5K_{12})]T \\ K_{23} = [(24 - 4c)(x_n + 0.5K_{12}) - (x_n + 0.5K_{12})(z_n + 0.5K_{32}) + c(y_n + 0.5K_{22})]T \\ K_{33} = [(x_n + 0.5K_{12})(y_n + 0.5K_{22}) - 8(z_n + 0.5K_{32})/3]T \end{cases} , \tag{9.14}$$

$$\begin{cases} K_{14} = 10[(y_n + K_{23}) - (x_n + K_{13})]T \\ K_{24} = [(24 - 4c)(x_n + K_{13}) - (x_n + K_{13})(z_n + K_{33}) + c(y_n + K_{23})]T \\ K_{34} = [(x_n + K_{13})(y_n + K_{23}) - 8(z_n + K_{33})/3]T \end{cases} , \tag{9.15}$$

where T is the iteration step size. According to eqs (9.11)–(9.15), setting initial values $x = 0.1$, $y = 0.2$, $z = 0.3$, $c = 5$, and $T = 0.01$, the attractor phase diagram is generated by the hardware implementation as shown in Figure 9.24.

Taking 4 bits after decimal point of the chaotic sequence generated by DSP, and transmitting a total of 10,000 sequence values between 1,001 and 20,000 to the computer, the phase diagram of the attractor is plotted as shown in Figure 9.25.

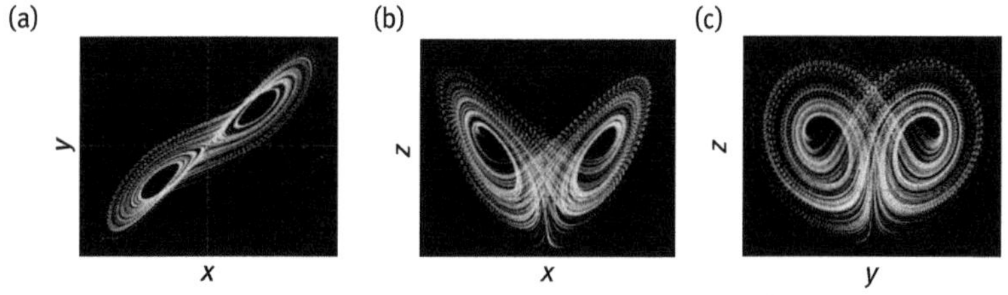

Figure 9.24: Attractors in an oscilloscope for the Lorenz system based on the DSP: (a) x–y plane; (b) x–z plane; and (c) y–z plane.

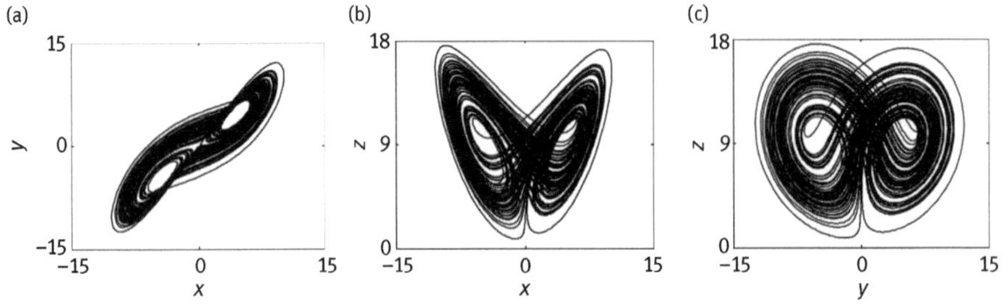

Figure 9.25: Attractors of the simplified Lorenz system by experiment data from DSP: (a) x–y plane; (b) x–z plane; and (c) y–z plane.

9.5.3 DSP Implementation of Fractional-Order Chaotic System

Except the difference of the iterative algorithm, the realization of fractional-order chaotic system based on DSP is similar to that of the integer-order chaotic system based on DSP. Fractional-order simplified Lorenz system is described as [186]

$$\begin{cases} {}^{*}D_{t0}^{q}x = 10(y-x) \\ {}^{*}D_{t0}^{q}y = (24-4c)x - xz + cy \\ {}^{*}D_{t0}^{q}z = xy - 8z/3 \end{cases}, \tag{9.16}$$

where q is the order of fractional differential and c is the system parameter. Choosing suitable q and c, it is chaotic. According to the Adomian decomposition algorithm, the fractional differential equations are solved, and the first seven items are obtained as follows [202]:

$$\begin{cases} C_{10} = x_m \\ C_{20} = y_m \\ C_{30} = z_m \end{cases}, \tag{9.17}$$

$$\begin{cases} C_{11} = 10(C_{20} - C_{10}) \\ C_{21} = (24-4c)C_{10} + cC_{20} - C_{10}C_{30} \\ C_{31} = -\frac{8}{3}C_{30} + C_{10}C_{20} \end{cases}, \tag{9.18}$$

$$\begin{cases} C_{12} = 10(C_{21} - C_{11}) \\ C_{22} = (24-4c)C_{11} + cC_{21} - C_{11}C_{30} - C_{10}C_{31} \\ C_{32} = -\frac{8}{3}C_{31} + C_{11}C_{20} + C_{10}C_{21} \end{cases}, \tag{9.19}$$

$$\begin{cases} C_{13} = 10(C_{22} - C_{12}) \\ C_{23} = (24 - 4c)C_{12} + cC_{22} - C_{12}C_{30} - C_{11}C_{31}\frac{\Gamma(2q+1)}{\Gamma^2(q+1)} - C_{10}C_{32} \\ C_{33} = -\frac{8}{3}C_{32} + C_{12}C_{20} + C_{11}C_{21}\frac{\Gamma(2q+1)}{\Gamma^2(q+1)} + C_{10}C_{22} \end{cases}, \tag{9.20}$$

$$\begin{cases} C_{14} = 10(C_{23} - C_{13}) \\ C_{24} = (24 - 4c)C_{13} + cC_{23} - C_{13}C_{30} - (C_{12}C_{31} + C_{11}C_{32})\frac{\Gamma(3q+1)}{\Gamma(q+1)\Gamma(2q+1)} - C_{10}C_{33} \\ C_{34} = -\frac{8}{3}C_{33} + C_{13}C_{20} + (C_{12}C_{21} + C_{11}C_{22})\frac{\Gamma(3q+1)}{\Gamma(q+1)\Gamma(2q+1)} + C_{10}C_{23} \end{cases}, \tag{9.21}$$

$$\begin{cases} C_{15} = 10(C_{24} - C_{14}) \\ C_{25} = (24 - 4c)C_{14} + cC_{24} - C_{14}C_{30} - (C_{13}C_{31} + C_{11}C_{33})\frac{\Gamma(4q+1)}{\Gamma(q+1)\Gamma(3q+1)} \\ -C_{12}C_{32}\frac{\Gamma(4q+1)}{\Gamma^2(2q+1)} - C_{10}C_{34} \\ C_{35} = -\frac{8}{3}C_{34} + C_{14}C_{20} + (C_{13}C_{21} + C_{11}C_{23})\frac{\Gamma(4q+1)}{\Gamma(q+1)\Gamma(3q+1)} \\ +C_{12}C_{22}\frac{\Gamma(4q+1)}{\Gamma^2(2q+1)} + C_{10}C_{24} \end{cases}, \tag{9.22}$$

$$\begin{cases} C_{16} = 10(C_{25} - C_{15}) \\ C_{26} = (24 - 4c)C_{15} + cC_{25} - C_{15}C_{30} - (C_{14}C_{31} + C_{11}C_{34})\frac{\Gamma(5q+1)}{\Gamma(q+1)\Gamma(4q+1)} \\ -(C_{13}C_{32} + C_{12}C_{33})\frac{\Gamma(5q+1)}{\Gamma(2q+1)\Gamma(3q+1)} - C_{10}C_{35} \\ C_{36} = -\frac{8}{3}C_{35} + C_{15}C_{20} + (C_{14}C_{21} + C_{11}C_{24})\frac{\Gamma(5q+1)}{\Gamma(q+1)\Gamma(4q+1)} \\ +(C_{13}C_{22} + C_{12}C_{23})\frac{\Gamma(5q+1)}{\Gamma(2q+1)\Gamma(3q+1)} + C_{10}C_{25} \end{cases}, \tag{9.23}$$

$$\begin{bmatrix} x_{m+1} \\ y_{m+1} \\ z_{m+1} \end{bmatrix} = \begin{bmatrix} C_{10} & C_{11} & C_{12} & C_{13} & C_{14} & C_{15} & C_{16} \\ C_{20} & C_{21} & C_{22} & C_{23} & C_{24} & C_{25} & C_{26} \\ C_{30} & C_{31} & C_{32} & C_{33} & C_{34} & C_{35} & C_{36} \end{bmatrix} \left[1 \ \frac{h^q}{\Gamma(q+1)} \ \frac{h^{2q}}{\Gamma(2q+1)} \ \frac{h^{3q}}{\Gamma(3q+1)} \ \frac{h^{4q}}{\Gamma(4q+1)} \ \frac{h^{5q}}{\Gamma(5q+1)} \ \frac{h^{6q}}{\Gamma(6q+1)} \right]^{\mathrm{T}}. \tag{9.24}$$

For all the above-mentioned equations, h is the iterative step size and $\Gamma(\cdot)$ is the gamma function. According to the calculating process, $\Gamma(\cdot)$ and h^{nq} are unchanged in the iterative process, so to speed up the calculation, we need to calculate the relevant $\Gamma(\cdot)$ and h^{nq} before iteration, taking them as constants in iterative process. The initial values are $x_0 = 0.1$, $y_0 = 0.2$, $z_0 = 0.3$, $c = 5$, $q = 0.9$, and $h = 0.01$, and hardware implementation of the attractor phase diagram is shown in Figure 9.26.

Taking 4 bits after decimal point of the chaotic sequence generated by DSP and transmitting a total of 10,000 sequence values between 10,001 and 20,000 to the computer, the phase diagram of the attractor is plotted as shown in Figure 9.27.

In this chapter, we study the simulation and circuit design techniques of the chaotic system, including the dynamic simulation, circuit simulation, and realization of the chaotic system. The realization of analog circuit and digital circuit of chaotic system has laid an experimental foundation for the applications of chaotic system.

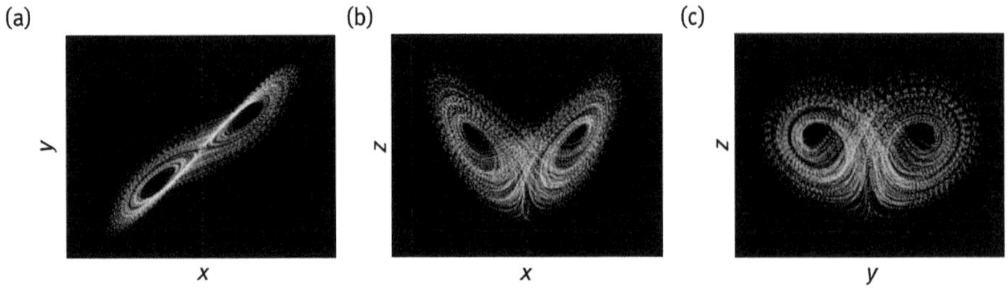

Figure 9.26: Attractors in an oscilloscope for the fractional-order simplified Lorenz system based on the DSP: (a) x–y plane; (b) x–z plane; and (c) y–z plane.

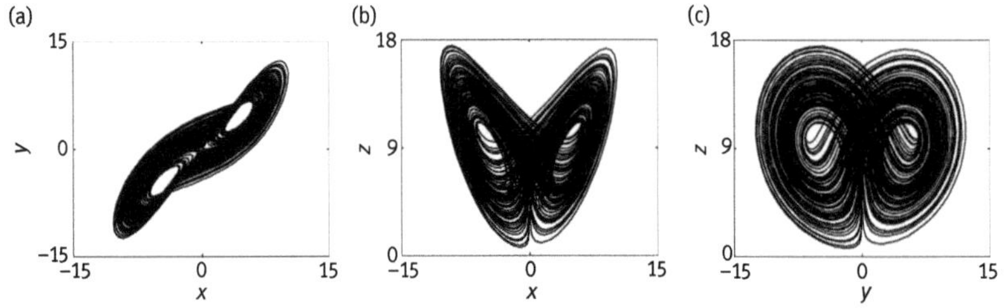

Figure 9.27: Attractors of the fractional-order simplified Lorenz system by experiment data from DSP: (a) x–y plane; (b) x–z plane; and (c) y–z plane.

Questions

1. Try to study a chaotic system by using different simulation methods and compare the results.
2. What are the similarities and differences of the chaotic systems based on the OA device and the CC device? What is the difference between the chaotic signals generated by them?
3. To achieve the digital circuit implementation of a chaotic system, what are the demands for solving algorithm of differential equations?
4. Is there finite precision effect in DSP implementation of the chaotic system, and how to overcome it?
5. Design and implement a chaotic system circuit.

Appendix A: Important Academic Journals in the Field of Chaos

No.	Journals	ESI subject	Influence factor
1	*Physical Review Letter*	Physics	7.943
2	*Communications in Nonlinear Science and Numerical Simulation*	Physics	2.773
3	*Physical Review E*	Physics	2.313
4	*Chaos*	Physics	2.188
5	*Physics Letters A*	Physics	1.766
6	*Physica D – Nonlinear Phenomena*	Physics	1.669
7	*Chaos, Solitons and Fractals*	Physics	1.246
8	*Chinese Physics B*	Physics	1.148
9	*Chinese Physics Letters*	Physics	0.811
10	*Journal of Nonlinear Mathematical Physics*	Physics	0.569
11	*Acta Physica Sinica*	Physics	1.016
12	*Nonlinear Analysis – Theory, Methods & Applications*	Mathematics	1.64
13	*Nonlinearity*	Mathematics	1.602
14	*International Journal of Bifurcation and Chaos*	Mathematics	0.921
15	*Nonlinear Analysis: Modelling and Control*	Mathematics	0.861
16	*Advanced Nonlinear Studies*	Mathematics	0.538
17	*Nonlinear Dynamics*	Engineering	3.009
18	*IEEE Transactions on Circuits and Systems I/II*	Engineering	2.24/1.327
19	*Electronics Letters*	Engineering	1.038
20	*International Journal of Nonlinear Sciences and Numerical Simulation*	Engineering	0.622

Appendix B: Matlab Source Programs of Chaos Characteristic Analysis

Taking the simplified Lorenz system as an example, Matlab simulation programs are presented. First, the analysis method of integer-order system is introduced, including Runge–Kutta solving algorithm, Euler solving algorithm, Lyapunov exponents (LE) calculation, plotting phase diagram, bifurcation diagram, Poincaré section, and 0–1 test method. Second, the programs about solving and analyzing algorithm of the fractional-order chaotic system are designed, including frequency domain approach (Simulink simulation), Adams–Bashforth–Moulton approach, Adomian decomposition method, C_0 complexity algorithm, and chaos diagram method. Finally, Matlab programs about chaotic synchronization algorithm are introduced for readers' reference. Of course, readers can modify the relevant codes according to their own needs to meet the system or research.

If a document begins with "function," then it is a function. The reader needs to be entered and saved in a separate .m file, so that you can call it when you need it. Design program by employing function belongs to the module design approach, which is beneficial to improve the programming efficiency.

B.1 Mode of the Simplified Lorenz System

The simplified Lorenz system is defined by

$$\begin{cases} \dot{x} = 10\,(y - x) \\ \dot{y} = (24 - 4c)\,x - xz + cy \\ \dot{z} = xy - 8z/3 \end{cases}, \tag{B.1}$$

where c is the system parameter. When $c \in [-1.59\ 7.75]$, the system is chaotic.

The fractional-order simplified Lorenz system is defined as

$$\begin{cases} {}^{*}D_{t0}^{q}x = 10(y - x) \\ {}^{*}D_{t0}^{q}y = (24 - 4c)x - xz + cy \\ {}^{*}D_{t0}^{q}z = xy - 8z/3 \end{cases}. \tag{B.2}$$

In this system, there are two parameters, order q and control parameter c.

B.2 Programs about Characteristic Analysis of Integer-Order Simplified Lorenz System

B.2.1 Simulation Codes for Matlab Toolbox with Runge–Kutta Algorithm

First, program the system function according to system (B.1):

```
function f=chao_SimpleLorenz(t,y)
global c
f=zeros(3,1);
f(1)=10*(y(2)-y(1));
f(2)=(24-4*c)*y(1)+c*y(2)-y(1)*y(3);
f(3)=y(1)*y(2)-8/3*y(3);
```

Running the following codes in Matlab window, the attractor phase diagram is obtained as shown in Figure B.1.

```
clc
clear
global c
c=2;
[T,Y]=ode45(@chao_SimpleLorenz,0:0.01:500,[1 2 3 ]);
figure
plot(Y(30000:end,1),Y(30000:end,3))
xlim([-15 15])
xlabel('\itx')
ylabel('\itz')
```

B.2.2 Solution Based on Euler and Runge–Kutta Algorithm

In practical applications, for example, digital signal processing or Field Programmable Gate Array (FPGA) digital circuit implementation, we need to prepare the

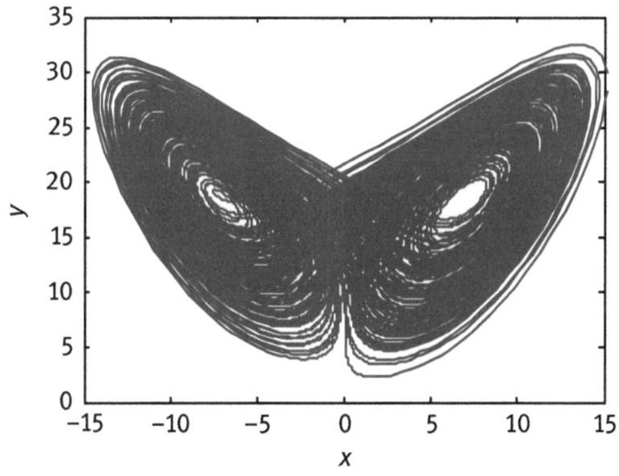

Figure B.1: Attractor of the simplified Lorenz system (ode45).

relevant code to solve the numerical solution algorithm, and cannot use the Matlab toolbox program. So solution algorithm based on the Euler algorithm and Runge–Kutta method has a good application value.

1. Numerical solution program based on Euler algorithm:

```
function [x y z]=EulerSolve(c,x0,h,N)
% c is the system parameter; x0 is the ini-
tial value; h is the
iteration step size; N is the iteration number.
x=zeros(1,N);
y=zeros(1,N);
z=zeros(1,N);
x(1)=x0(1)+h*(10*(x0(2)-x0(1)));
y(1)=x0(2)+h*((24-4*c)*x0(1)-x0(1)*x0(3)+c*x0(2));
z(1)=x0(3)+h*(x0(1)*x0(2)-(8/3)*x0(3));
for i=2:N
        x(i)=x(i-1)+h*(10*(y(i-1)-x(i-1)));
        y(i)=y(i-1)+h*((24-4*c)*x(i-1)-x(i-1)*z(i-
1)+c*y(i-1));
        z(i)=z(i-1)+h*(x(i-1)*y(i-1)-(8/3)*z(i-1));
end
```

Running the following codes in Matlab window, the attractor phase diagram is obtained as shown in Figure B.1.

```
[x y z]= EulerSolve (2,[1 2 3],0.01,20000);
plot(x(10000:end),z(10000:end))
xlabel('\itx')
ylabel('\itz')
```

2. Numerical solution program based on fourth-order Runge–Kutta algorithm:

```
function [x y z]=RungeSolve(c,xin,h,N)
% c is the system parameter; x0 is the ini-
tial value; h is the
iteration step size; N is the iteration number.
x=zeros(1,N);
y=zeros(1,N);
z=zeros(1,N);
x0=xin(1);y0=xin(2);z0=xin(3);

k11=10*(-x0+y0);
k21=(24-4*c)*x0-x0*z0+c*y0;
k31=x0*y0-(8/3)*z0;
```

```
k12=10*(-(x0+0.5*h*k11)+(y0+0.5*h*k21));
k22=(24-4*c)*(x0+0.5*h*k11)-(x0+0.5*h*k11)*(z0+0.5*h*k31)
                    +c*(y0+0.5*h*k21);
k32=(x0+0.5*h*k11)*(y0+0.5*h*k21)-(8/3)*(z0+0.5*h*k31);

k13=10*(-(x0+0.5*h*k12)+(y0+0.5*h*k22));
k23=(24-4*c)*(x0+0.5*h*k12)-(x0+0.5*h*k12)*(z0+0.5*h*k32)
     +c*(y0+0.5*h*k22);
k33=(x0+0.5*h*k12)*(y0+0.5*h*k22)-(8/3)*(z0+0.5*h*k32);
k14=10*(-(x0+h*k13)+(y0+h*k23));
k24=(24-4*c)*(x0+h*k13)-(x0+h*k13)*(z0+h*k33)+c*(y0+h*k23);
k34=(x0+h*k13)*(y0+h*k23)-(8/3)*(z0+h*k33);

x0=x0+h*(k11+2*k12+2*k13+k14)/6;
y0=y0+h*(k21+2*k22+2*k23+k24)/6;
z0=z0+h*(k31+2*k32+2*k33+k34)/6;
x(1)=x0;y(1)=y0;z(1)=z0;
for i=2:N
        k11=10*(-x0+y0);
        k21=(24-4*c)*x0-x0*z0+c*y0;
        k31=x0*y0-(8/3)*z0;

        k12=10*(-(x0+0.5*h*k11)+(y0+0.5*h*k21));
        k22=(24-4*c)*(x0+0.5*h*k11)-(x0+0.5*h*k11)
             *(z0+0.5*h*k31)+c*(y0+0.5*h*k21);
        k32=(x0+0.5*h*k11)*(y0+0.5*h*k21)-
             (8/3)*(z0+0.5*h*k31);

        k13=10*(-(x0+0.5*h*k12)+(y0+0.5*h*k22));
        k23=(24-4*c)*(x0+0.5*h*k12)-(x0+0.5*h*k12)
             *(z0+0.5*h*k32)+c*(y0+0.5*h*k22);
        k33=(x0+0.5*h*k12)*(y0+0.5*h*k22)-
             (8/3)*(z0+0.5*h*k32);

        k14=10*(-(x0+h*k13)+(y0+h*k23));
        k24=(24-4*c)*(x0+h*k13)-(x0+h*k13)*(z0+h*k33)
             +c*(y0+h*k23);
        k34=(x0+h*k13)*(y0+h*k23)-(8/3)*(z0+h*k33);

        x0=x0+h*(k11+2*k12+2*k13+k14)/6;
        y0=y0+h*(k21+2*k22+2*k23+k24)/6;
        z0=z0+h*(k31+2*k32+2*k33+k34)/6;
```

```
        x(i)=x0;y(i)=y0;z(i)=z0;
end
```

Running the following codes in Matlab window, the attractor phase diagram is obtained as shown in Figure B.1.

```
[x y z]=RungeSolve(2,[1 2 3],0.01,20000);
plot(x(10000:end),z(10000:end))
xlabel('\itx')
ylabel('\itz')
```

B.2.3 Program of Calculating LE

1. Program based on tool box:
   The web site of the tool box to download is http://www.math.rsu.ru/mexmat/kvm/matds/. Download the toolbox, as long as preparation of the following functions according to the simplified Lorenz system.

```
function f=SimLorenz_ly(t,X)
global c
x=X(1); y=X(2); z=X(3);
Y= [X(4), X(7), X(10);
    X(5), X(8), X(11);
    X(6), X(9), X(12)];
f=zeros(12,1);
f(1)=10*(y-x);
f(2)=-x*z+(24-4*c)*x+c*y;
f(3)=x*y-8/3*z;
 Jac=[-10,10,0;
        24-4*c-z,   c,    -x;
        y,    x,       -8/3];
 f(4:12)=Jac*Y;
```

You can calculate the system's LE spectrum by building a new .m file, and write codes as follows:

```
clc
clear
global c
c=2;
[T,Res]=lyapunov(3,@SimLorenz_ly,@ode45,0,0.5,500,
        [ 0.1 0 0.1],5);=Ly=Res(end,:)
```

The running result is

```
Ly =
        0.8486    0.0014    -11.5161
```

2. Program of calculating LEs based on Euler and QR decomposition:

```
function ly=oulaJianhua(c)
h=0.01;
x0=1;y0=2;z0=3;
Q=[1 0 0;0 1 0; 0 0 1];
ly1=0;ly2=0;ly3=0;m=20000;
I=eye(3);
for i=1:m
      x1=x0+h*(10*(y0-x0));
      y1=y0+h*((24-4*c)*x0-x0*z0+c*y0);
      z1=z0+h*(x0*y0-(8/3)*z0);
      x0=x1;y0=y1;z0=z1;
      J=I+h*[-10 10 0; (24-4*c)-z1 c -x1; y1 x1 -8/3];
      B=J*Q;
      [Q,R]=qr(B);
      ly1=ly1+log(abs(R(1,1)));
      ly2=ly2+log(abs(R(2,2)));
      ly3=ly3+log(abs(R(3,3)));
end
ly=[ly1/(m*h) ly2/(m*h) ly3/(m*h)];
```

Running on Matlab

```
ly=oulaJianhua(c)
```

The result is

```
ly =
      0.9996   -0.0013   -11.7466
```

3. Program of calculating LEs based on Runge–Kutta and QR decomposition:

```
function ly=RungeJianhua(c)
h=0.01;
x0=1;y0=1;z0=2;
Q=eye(3);
ly1=0;ly2=0;ly3=0;m=50000;
I=eye(3);
for i=1:m
      k11=10*(-x0+y0);
      k21=(24-4*c)*x0-x0*z0+c*y0;
      k31=x0*y0-(8/3)*z0;

      k12=10*(-(x0+0.5*h*k11)+(y0+0.5*h*k21));
      k22=(24-4*c)*(x0+0.5*h*k11)-(x0+0.5*h*k11)
           *(z0+0.5*h*k31)+c*(y0+0.5*h*k21);
    k32=(x0+0.5*h*k11)*(y0+0.5*h*k21)-(8/3)
        *(z0+0.5*h*k31);
```

```
        k13=10*(-(x0+0.5*h*k12)+(y0+0.5*h*k22));
        k23=(24-4*c)*(x0+0.5*h*k12)-(x0+0.5*h*k12)
             *(z0+0.5*h*k32)+c*(y0+0.5*h*k22);
        k33=(x0+0.5*h*k12)*(y0+0.5*h*k22)-
            (8/3)*(z0+0.5*h*k32);

        k14=10*(-(x0+h*k13)+(y0+h*k23));
        k24=(24-4*c)*(x0+h*k13)-(x0+h*k13)*(z0+h*k33)
            +c*(y0+h*k23);
        k34=(x0+h*k13)*(y0+h*k23)-(8/3)*(z0+h*k33);

        x0=x0+h*(k11+2*k12+2*k13+k14)/6;
        y0=y0+h*(k21+2*k22+2*k23+k24)/6;
        z0=z0+h*(k31+2*k32+2*k33+k34)/6;
        J1=[-10 10 0;(24-4*c)-z0 c -x0;y0 x0 -8/3];
        J2=[-10 10 0;(24-4*c)-(z0+0.5*h*k31) c -
            (x0+0.5*h*k11);y0+0.5*h*k21x0+0.5*h*k11 -8/3];%
        J3=[-10 10 0;  (24-4*c)-(z0+0.5*h*k32) c -
            (x0+0.5*h*k12); y0+0.5*h*k22x0+0.5*h*k12 -8/3];%
        J4=[-10 10 0;(24-4*c)-(z0+h*k33) c -
            (x0+h*k13);y0+h*k23 x0+h*k13-8/3];%

J=I+h*(J1+2*J2*(I+0.5*h*J1)+2*J3*(I+0.5*h*J2*(I+0.5*h*J1))
  +J4*(I+h*J3*(I+0.5*h*J2*(I+0.5*h*J1))))/6;
      B=J*Q;
      [Q,R]=qr(B);
      ly1=ly1+log(abs(R(1,1)));
      ly2=ly2+log(abs(R(2,2)));
      ly3=ly3+log(abs(R(3,3)));
end
ly=[ly1/(m*h) ly2/(m*h) ly3/(m*h)];
```

Running in the Matlab window

```
ly=RungeJianhua(2)
```

The result is

```
ly=
    0.8620    -0.0041 -11.5245
```

4. Program of calculating LEs with parameter varying:
   Here, program is presented for the calculation of LEs of the simplified Lorenz system with parameter *c*. LE diagram as shown in Figure B.2 is obtained by employing Lyapunov.m:

```
clc
clear
```

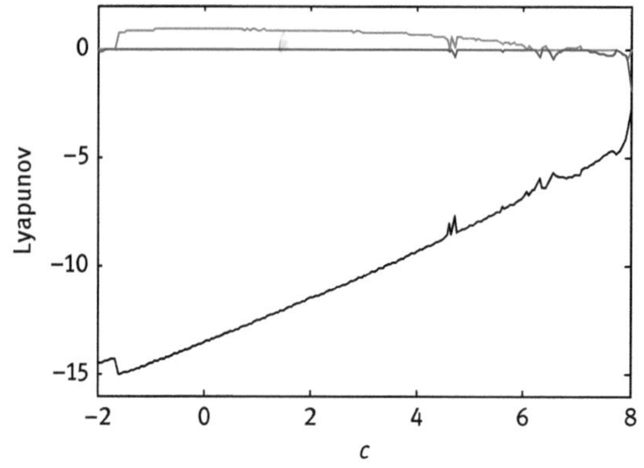

Figure B.2: Lyapunov exponents of the simplified Lorenz system.

```
global c
C=-2:0.05:8;
L=length(C);
ly=zeros(3,L);
for i=1:L
        c=C(i);
        [T,Res]=lyapunov(3,@SimLorenz_ly,@ode45,0,0.5,500,
        [ 0.1 0 0.1],5);ly(:,i)=Res(end,:);
end
plot(C,ly)
xlabel('\itc')
ylabel('Lyapunov')
```

B.2.4 Program of Calculating the Maximum LE

The maximum LE of chaotic system has several calculation methods, and this program is one of them. It must be used with f=chao_SimpleLorenz(t,y), and the specific code is as follows:

```
function Ly=MaxLypunov(cc)
% simplified Lorenz system
global c;
c=cc;
T,Y]=ode45(@chao_SimpleLorenz,200:0.01:202,[0.1 0.1 0.1]);
lyap=0;
tao=2;       % time interval
x0=Y(end,:);
y0=[x0(1)-3e-2,x0(2)-4e-2,x0(3)-5e-2];
d0=sqrt((y0(1)-x0(1))^2+(y0(2)-x0(2))^2+(y0(3)-x0(3))^2);
```

```
N=1000;
d=zeros(1,N);
for i=1:N
       [T,Y]=ode45(@chao_SimpleLorenz,0:0.01:tao,x0);
       x0=Y(end,:);
       [T,Y]=ode45(@chao_SimpleLorenz,0:0.01:tao,y0);
       y1=Y(end,:);
       d(i)=sqrt((y1(1)-x0(1))^2+(y1(2)-x0(2))^2+(y1(3)-
            x0(3))^2);
       z=d0/d(i);
       y0=x0+(y1-x0)*z;
       if mod(i,40)==0
             lyap1=lyap;
             lyap=0;
             for j=1:i
                 lyap=lyap+log(d(i)/d0);
             end
             lyap=lyap/(j*tao);
             if abs(lyap-lyap1)<0.001
                break;
           end
      end
end
Ly=0;
for j=1:i
    Ly=Ly+log(d(i)/d0);
end
Ly=Ly/(j*tao);
```

Input in the Matlab window

```
Ly=MaxLypunov(2)
```

The result is

```
Ly =
0.6559
```

B.2.5 Program for Calculating Bifurcation Diagram

```
clc
clear
global c
C=-2:0.025:8;
L=length(C);
for i=1:L
```

```
        c=C(i);
        [T,Y]=ode45(@chao_SimpleLorenz,0:0.01:200,[1 2 3]);
        data=Y(10000:end,3);
        for j=2:(length(data)-1)
            if data(j)>data(j-1)&&data(j)>data(j+1)
                plot(c,data(j),'.','markersize',1);
                hold on;
                if j==20
                    break;
                      end
                end
        end
end
xlabel('\itc')
ylabel('\itzmax')
```

Running the above codes, the bifurcation diagram of the simplified Lorenz system can be obtained as shown in Figure B.3. In practice, the step size of parameter *c* is appropriate to take a smaller one.

B.2.6 Program for Plotting Poincaré Section

```
clc
clear
global c
c=2;
[t,Y]=ode45(@chao_SimpleLorenz,0:0.01:500,[1 2 3]);
A=3;  % z
B=2;  % y
D=1;  % x
m=22;  %value of the section
```

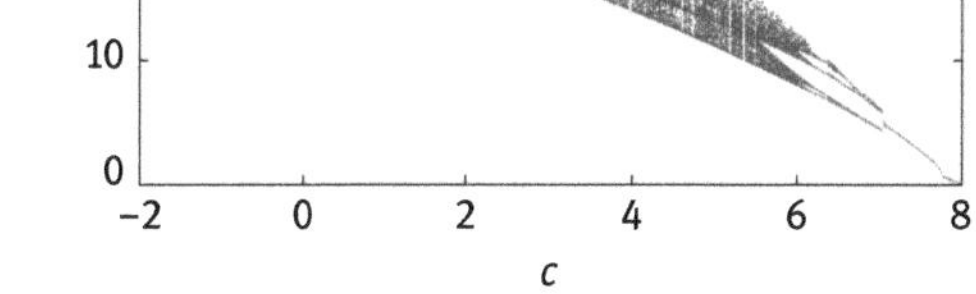

Figure B.3: Bifurcation of the simplified Lorenz system.

```
N=length(Y);
for k=1:N-1
    if (Y(k,A)>m&&Y(k+1,A)<m)||(Y(k,A)<m&&Y(k+1,A)>m)
       %choose a plane on z axis
        a=(Y(k+1,A)-Y(k,A))/(t(k+1)-t(k));
          % the chosen plane
        b=(Y(k,A)*t(k+1)-Y(k+1,A)*t(k))/(t(k+1)-t(k));
          % t=0, Intercept of z line (coordinates)
        c=(Y(k+1,B)-Y(k,B))/(t(k+1)-t(k));
          % slope of y line
        d=(Y(k,B)*t(k+1)-Y(k+1,B)*t(k))/(t(k+1)-t(k));
          % t=0 Intercept of y line (coordinates)
        e=(Y(k+1,D)-Y(k,D))/(t(k+1)-t(k));
          % slope of x line
        f=(Y(k,D)*t(k+1)-Y(k+1,D)*t(k))/(t(k+1)-t(k));
          % t=0 Intercept of x line (coordinates)x
        g=(m-b)/a; % z=10, the time needed
        q=e*g+f; % z=10, the value of x
        p=c*g+d; % z=10, the value of y plot(q,p);hold on;
     end
end
xlabel('y')
ylabel('x')
```

Running the above codes, the Poincaré section of the simplified Lorenz system can be obtained as shown in Figure B.4.

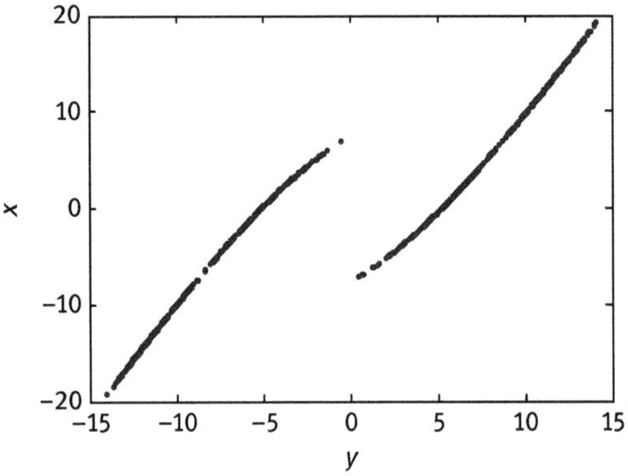

Figure B.4: Poincaré section of the simplified Lorenz system.

B.2.7 Program for 0–1 Test

Here is the program for plotting p–s diagram of 0–1 test:

```
clc
clear
global c
c=2;
[T,Y]=ode45(@chao_SimpleLorenz,0:0.01:1500,[1 2 3]);
xt=Y(10000:end,3);
N=length(xt)-5;
n=floor(N/15);
th=zeros(1,N+1);
p=th;
s=p;
o=p;
M=p;
for i=2:N+1
    th(i)=c+th(i-1)+xt(i);%
end
for i=2:N+1
    p(i)=p(i-1)+xt(i)*cos(th(i));
end
for i=2:N+1
    s(i)=s(i-1)+xt(i)*sin(th(i));
end
plot(p,s)
xlabel('p')
ylabel('s')
```

Running the above codes, the p–s diagram of the simplified Lorenz system with 0–1 test can be obtained as shown in Figure B.5.

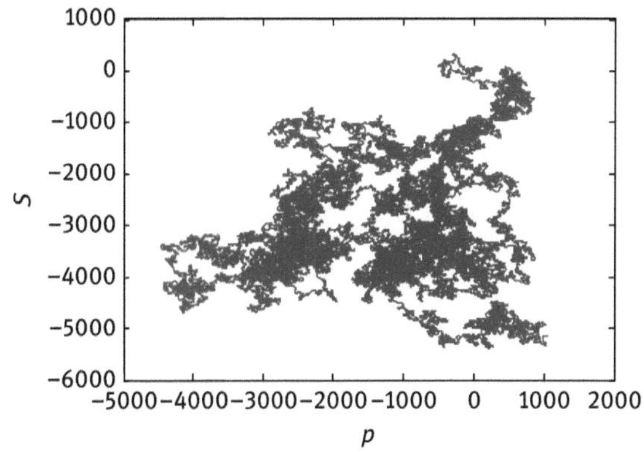

Figure B.5: p–s diagram of the simplified Lorenz system with 0–1 test.

B.3 Solution Program for the Fraction-Order Simplified Lorenz System

B.3.1 Solution Program Based on Predictor–Corrector Approach

```
function [x y z]=jianhua(c,q1,q2,q3)
if nargin==0
        c=5;
        q1=0.98;q2=0.98;q3=9.98;
end
h=0.01;    % iteration step size
N=5000;    % iteration number
x0=-6.7781;y0=-8.3059;z0=9.2133; %   initial val-
ues of the system values;
M1=0;M2=0;M3=0;
x(N+1)=[0];y(N+1)=[0];z(N+1)=[0];
x1(N+1)=[0];y1(N+1)=[0];z1(N+1)=[0];
x1(1)=x0+h^q1*10*(y0-x0)/(gamma(q1)*q1);
y1(1)=y0+h^q2*((24-4*c)*x0-x0*z0+c*y0)/(gamma(q2)*q2);
z1(1)=z0+h^q3*(x0*y0-8*z0/3)/(gamma(q3)*q3);
x(1)=x0+10*h^q1*(y1(1)-x1(1)+q1*(y0-x0))/gamma(q1+2);
y(1)=y0+h^q2*((24-4*c)*x1(1)-x1(1)*z1(1)+c*y1(1)+q2*((24-
4*c)*x0-x0*z0+c*y0))/gamma(q2+2);
z(1)=z0+h^q3*(x1(1)*y1(1)-8*z1(1)/3+q3*(x0*y0-
8*z0/3))/gamma(q3+2);
for n=1:N
            M1=(n^(q1+1)-(n-q1)*(n+1)^q1)*10*(y0-x0);
            M2=(n^(q2+1)-(n-q2)*(n+1)^q2)*((24-4*c)
               *x0-x0*z0+c*y0);
            M3=(n^(q3+1)-(n-q3)*(n+1)^q3)*(x0*y0-8*z0/3);
            N1=((n+1)^q1-n^q1)*10*(y0-x0);
            N2=((n+1)^q2-n^q2)*((24-4*c)*x0-x0*z0+c*y0);
            N3=((n+1)^q3-n^q3)*(x0*y0-8*z0/3);
      for j=1:n
M1=M1+((n-j+2)^(q1+1)+(n-j)^(q1+1)-2*(n-j+1)^(q1+1))
   *10*(y(j)-x(j));
M2=M2+((n-j+2)^(q2+1)+(n-j)^(q2+1)-2*(n-j+1)^(q2+1))
   *((24-4*c)*x(j)-x(j)*z(j)+c*y(j));
M3=M3+((n-j+2)^(q3+1)+(n-j)^(q3+1)-2*(n-j+1)^(q3+1))
   *(x(j)*y(j)-8*z(j)/3);
             N1=N1+((n-j+1) q1-(n-j)^q1)*10*(y(j)-x(j));
             N2=N2+((n-j+1)^q2-(n-j)^q2)*((24-4*c)*x(j)
                -x(j)*z(j)+c*y(j));
```

```
            N3=N3+((n-j+1) q3-(n-j)^q3)*(x(j)*y(j)-8*z(j)/3);
        end
        x1(n+1)=x0+h^q1*N1/(gamma(q1)*q1);
        y1(n+1)=y0+h^q2*N2/(gamma(q2)*q2);
        z1(n+1)=z0+h^q3*N3/(gamma(q3)*q3);
        x(n+1)=x0+h^q1*(10*(y1(n+1)-x1(n+1))+M1)/gamma(q1+2);
        y(n+1)=y0+h^q2*((24-4*c)*x1(n+1)-x1(n+1)*z1(n+1)
              +c*y1(n+1)+M2)/gamma(q2+2);
        z(n+1)=z0+h^q3*(x1(n+1)*y1(n+1)-8*z1(n+1)/3+M3)
        /gamma(q3+2);
end
```

Running following codes in Matlab window:

```
[x y z]=jianhua(2,0.98,0.98,0.98);
plot(x,z)
xlabel('\itx')
ylabel('\itz')
```

the phase diagram of the fractional-order simplified Lorenz system is obtained as shown in Figure B.6.

B.3.2 Solution Program Based on Adomian Decomposition Approach

```
function dy=SimpleFratral(x0,h,q,c)
dy=zeros(1,3);
b=8/3;
a=10;
if q<=1 && q>0
    c10=x0(1);
    c20=x0(2);
```

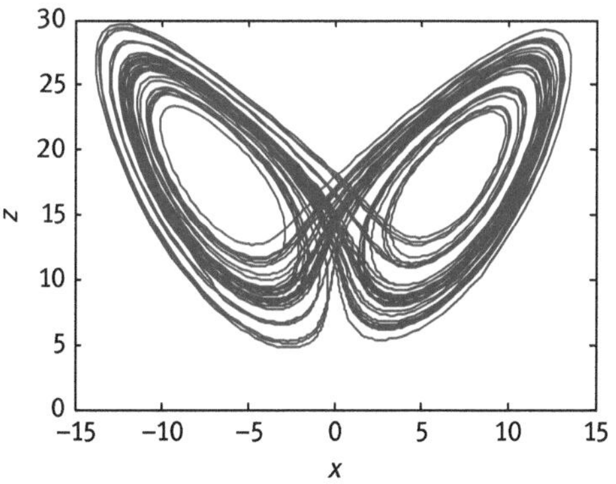

Figure B.6: Phase diagram of the fractional-order simplified Lorenz system with $q = 0.98$, $c = 2$.

```
    c30=x0(3);
end
if q>1&& q<2
   c10=x0(1)+x0(1)*h;
   c20=x0(2)+x0(2)*h;
   c30=x0(3)+x0(3)*h;
end
c11=a*(c20-c10);
c21=(24-4*c)*c10+c*c20-c10*c30;
c31=-b*c30+c10*c20;

c12=a*(c21-c11);
c22=(24-4*c)*c11+c*c21-c11*c30-c10*c31;
c32=-b*c31+c11*c20+c10*c21;

c13=a*(c22-c12);
c23=(24-4*c)*c12+c*c22-c12*c30-c11*c31*(gamma(2*q+1)
    /gamma(q+1)^2)-c10*c32;
c33=-b*c32+c12*c20+c11*c21*(gamma(2*q+1)/gamma(q+1)^2)+c10*c22;

c14=a*(c23-c13);
c24=(24-4*c)*c13+c*c23-c13*c30-(c12*c31+c11*c32)*(gamma(3*q+1)
    /(gamma(q+1)*gamma(2*q+1)))-c10*c33;
c34=-b*c33+c13*c20+(c12*c21+c11*c22)*(gamma(3*q+1)/(gamma(q+1)
    *gamma(2*q+1)))+c10*c23;

c15=a*(c24-c14);
c25=(24-4*c)*c14+c*c24-c14*c30-(c13*c31+c11*c33)*(gamma(4*q+1)
    /(gamma(q+1)*gamma(3*q+1)))-c12*c32*(gamma(4*q+1)
    /gamma(2*q+1)^2)-c10*c34;
c35=-b*c34+c14*c20+(c13*c21+c11*c23)*(gamma(4*q+1)
     /(gamma(q+1)*gamma(3*q+1)))+c12*c22*(gamma(4*q+1)
     /gamma(2*q+1)^2)+c10*c24;

c16=a*(c25-c15);
c26=(24-4*c)*c15+c*c25-c15*c30-(c14*c31+c11*c34)
    *(gamma(5*q+1)/(gamma(q+1)*gamma(4*q+1)))
    -(c13*c32+c12*c33)
    *(gamma(5*q+1)/gamma(2*q+1)*gamma(3*q+1))-c10*c35;
c36=-b*c35+c15*c20+(c14*c21+c11*c24)*(gamma(5*q+1)
     /(gamma(q+1)*gamma(4*q+1)))+(c13*c22+c12*c23)
     *(gamma(5*q+1)/gamma(2*q+1)*gamma(3*q+1))+c10*c25;
```

```
dy(1)=c10+c11*(h^q/gamma(q+1))+c12*h^(2*q)/gamma(2*q+1)+c13
      *h^(3*q)/gamma(3*q+1)+c14*h^(4*q)/gamma(4*q+1)+c15*h^(5*q)
      /gamma(5*q+1)+c16*h^(6*q)/gamma(6*q+1);
dy(2)=c20+c21*(h^q/gamma(q+1))+c22*h^(2*q)/gamma(2*q+1)+c23
      *h^(3*q)/gamma(3*q+1)+c24*h^(4*q)/gamma(4*q+1)+c25*h^(5*q)
      /gamma(5*q+1)+c26*h^(6*q)/gamma(6*q+1);
dy(3)=c30+c31*(h^q/gamma(q+1))+c32*h^(2*q)/gamma(2*q+1)
      +c33*h^(3*q)/gamma(3*q+1)+c34*h^(4*q)/gamma(4*q+1)+c35
      *h^(5*q)/gamma(5*q+1)+c36*h^(6*q)/gamma(6*q+1);
```

```
function [t,y]=FratalSim(h,N,x0,q,c)
y=zeros(3,N);
t=zeros(1,N);
tt=0;
for i=1:N
      t(i)=tt;
      dy=SimpleFratral(x0,h,q,c);
      y(:,i)=dy;
      x0=dy;
      tt=tt+h;
end
```

Running in Matlab window,

```
clc
clear
h=0.01;
c=5;
N=10000;
q=0.95;
x0=[0.1 0.2 0.3];
[t,y]=FratalSim(h,N,x0,q,c);
plot(y(1,:),y(3,:))
xlim([-12 12])
xlabel('\itx')
ylabel('\itz')
```

The attractor (similar to Figure B.6) and the chaotic sequences of the fractional-order simplified Lorenz system are obtained.

B.3.3 Analysis Program of Complexity and Chaos Diagram

1. Program for calculating C_0 complexity:

```
function CO=COFuZadu(x,r)
% function name: COFuZadu
% function: calculating C0 complexity
% input parameter x and r. r is the toler-
ance limit ; x is the chaotic sequence
% output parameters: C0.
Y=fft(x);
Gn=mean(abs(Y).^2);
YY=zeros(1,length(x));
for i=1:length(x)
    if abs(Y(i))^2>r*Gn
        YY(i)=Y(i);
      end
end
xx=ifft(YY);
Ssum=0;
for i=1:length(x)
    Ssum=Ssum+(xx(i)-x(i))^2;
end
CO=Ssum/sum(x.^2);
```

2. Program for calculating spectral entropy (SE) complexity:

```
function SE=SEShannon(x)
%function name: SEShannon
%function: calculating the SE complex-
ity of chaotic sequence
%input parameter x. x is the chaotic sequence
%output parameter: SE.
N=length(x);
flag=0;
x=x-mean(x);
for i=1:N
    if x(i)~=0
        flag=1;
      end
end
if flag==0
    SE=0;
    return;
```

```
end
Y=fft(x);
Xk=(abs(Y).^ 2)./N;
Xk=Xk(1:(floor(N/2)));
ptot=sum(Xk);
PK=Xk./ptot;
P=0;
for i=1:N/2
        if PK(i)~=0
           P=P-PK(i)*log(PK(i));
       end
end
se2=sum(P);
SE=se2/log(N/2);
```

3. Program for complexity with order *q* varying

```
clc
clear
L=201;
c=5;
Q=linspace(0.5,1,L);
C0=zeros(1,L);
SE=zeros(1,L);
for i=1:L
      q=Q(i);
      [t,y]=FratalSim(0.01,11000,[1 2 3],q,c);
      temp=y(1,1001:end);
      C0(i)=C0FuZadu(temp,15);
      SE(i)=SEShannon(temp);
      disp(i)
end
figure
plot(Q,SE)
xlabel('q')
ylabel('SE')
figure
plot(Q,C0)
xlabel('q')
ylabel('C0')
```

Running the above codes, the complexity of the fractional-order simplified Lorenz system with varying *q* can be obtained as shown in Figure B.7.

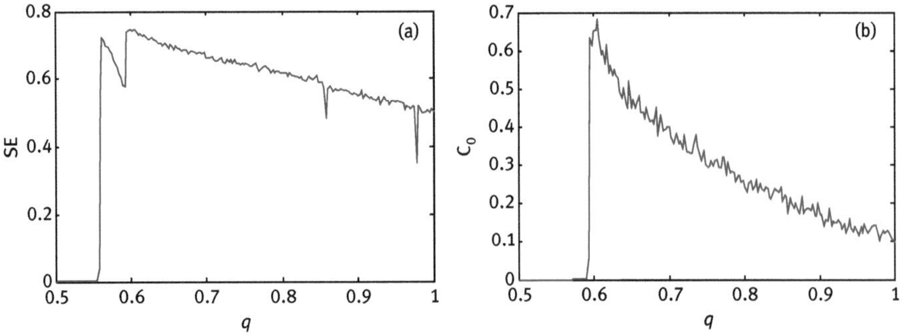

Figure B.7: Complexity of the fractional-order simplified Lorenz system with varying $q(c = 2)$: (a) SE complexity and (b) C_0 complexity.

4. Program for complexity with order c varying:

```
clc
clear
L=201;
q=0.98;
C=linspace(-2,8,L);
C0=zeros(1,L);
SE=zeros(1,L);
for i=1:L
        c=C(i);
        [t,y]=FratalSim(0.01,11000,[1 2 3],q,c);
        temp=y(1,1001:end);
        C0(i)=C0FuZadu(temp,15);
        SE(i)=SEShannon(temp);
        disp(i)
end
figure
plot(C,SE)
xlabel('c')
ylabel('SE')
figure
plot(C,C0)
xlabel('c')
ylabel('C0')
```

Running the above codes, the complexity of the fractional-order simplified Lorenz system with c varying can be obtained as shown in Figure B.8.

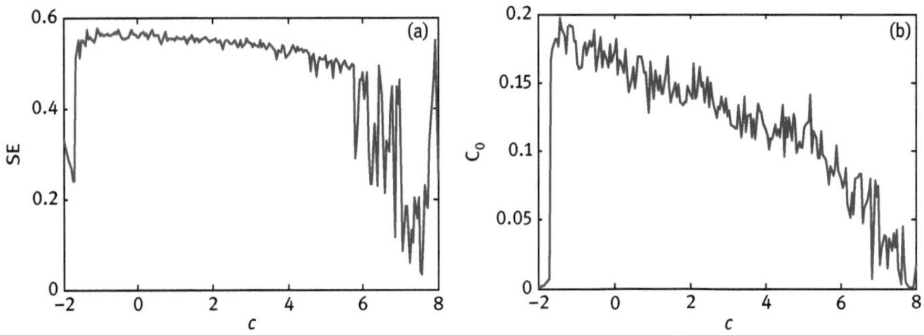

Figure B.8: Complexity of the fractional-order simplified Lorenz system with c varying (q= 0.98): (a) SE complexity and (b) C_0complexity.

5. Program for complexity-based chaos diagram:

```
clc
clear
L=11;
Q=linspace(0.65,1,L);
C=linspace(-2,8,L);
C0=zeros(L,L);
SE=zeros(L,L);
for i=1:L
    q=Q(i);
    for j=1:L
        c=C(j);
          [t,y]=FratalSim(0.01,12000,[1 2 3],q,c);
          temp=y(1,2001:end);
          C0(i,j)=COFuZadu(temp,15);
     SE(i,j)=SEShannon (temp);
end
disp(i)
end
[X,Y]=meshgrid(C,Q);
figure
colormap(flipud(hot));
contourf(X,Y,SE);
xlabel('c')
ylabel('q')
figure
colormap(flipud(hot));
contourf(X,Y,C0);
```

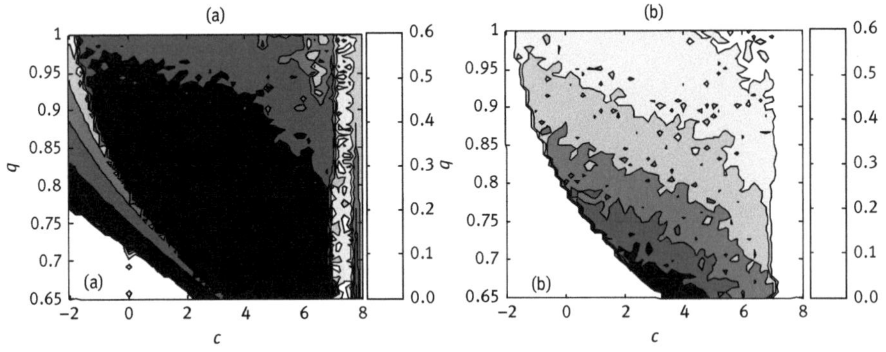

Figure B.9: Chaos diagram of the fractional-order simplified Lorenz system on q–c plane: (a) based on SE complexity and (b) based on C_0 complexity.

```
xlabel('c')
ylabel('q')
```

Running the above codes, the complexity-based chaos diagram on q–c plane of the fractional-order simplified Lorenz system can be obtained as shown in Figure B.9.

It is worth pointing out that the greater the value of L, the more the mesh of the parameters plane, and the more details can be reflected. But the code running time is relatively long, and it needs to wait patiently. We suggest to choose $L = 101$ at least. At the same time, the length of the sequence is also important when we calculate the complexity. Because both the SE algorithm and C_0 algorithm are based on fast Fourier transformation, the length of the sequence is suggested to choose $N = 104$. For test, we choose $L = 51$ and $N = 104$.

B.4 Simulation of Chaotic Synchronization

B.4.1 Drive–Response Synchronization

$$\text{Drive system:} \begin{cases} \dot{x} = 10\,(y - x) \\ \dot{y} = (24 - 4c)\,x - xz + cy \\ \dot{z} = xy - 8z/3 \end{cases}, \tag{B.3}$$

$$\text{Response system:} \begin{cases} \dot{x}_1 = 10\,(y_1 - x_1) \\ \dot{y}_1 = (24 - 4c)\,x - xz_1 + cy_1 \\ \dot{z}_1 = xy_1 - 8z_1/3 \end{cases}, \tag{B.4}$$

$$\text{Error system:}\begin{cases}\dot{e}_1 = \dot{x}_1 - \dot{x} \\ \dot{e}_2 = \dot{y}_1 - \dot{y} \\ \dot{e}_3 = \dot{z}_1 - \dot{z}\end{cases}. \tag{B.5}$$

It will be very convenient to solve this problem by using the Matlab Ode45 function, if the drive system and the response system are considered as a big system. Of course, readers can also solve the drive system and input the drive system sequence into the response system, but the program may be relatively complex:

```
function dy = SimplePC(t,y)
%Function name: SimplePC
%function: the drive-response synchronization of the simpli-
fied Lorenz system
global c;
dy = zeros(6,1);
dy(1)=10*(y(2)-y(1));
dy(2)=(24-4*c)*y(4)-y(4)*y(3)+c*y(2);
dy(3)=y(4)*y(2)-8*y(3)/3;
dy(4)=10*(y(5)-y(4));
dy(5)=(24-4*c)*y(4)-y(4)*y(6)+c*y(5);
dy(6)=y(4)*y(5)-8*y(6)/3;
```

Setting $c = -1$, initial values of the drive system are [5, 8, −10] and the initial values of the response system are [5, −5, 20]. Running the below Matlab codes, the synchronization error curves of the synchronization system are obtained as shown in Figure B.10:

```
clc
clear
global c
c=-1;
[T,Y]=ode45(@SimplePC,0:0.01:15,[5, -5, 20, 15, 8, -10 ]);
figure
plot(T,Y(:,1)-Y(:,4))
xlabel('t')
ylabel('e1')
figure
plot(T,Y(:,2)-Y(:,5))
xlabel('t')
ylabel('e2')
figure
plot(T,Y(:,3)-Y(:,6))
xlabel('t')
ylabel('e3')
```

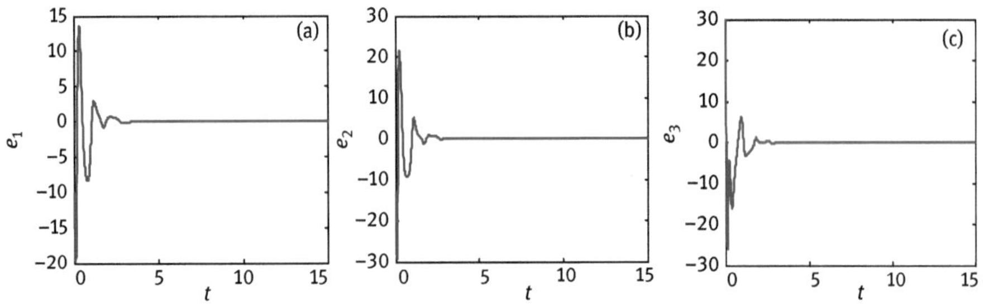

Figure B.10: Synchronization error curves of the simplified Lorenz system: (a) error curve of sequence x; (b) error curve of sequence y; and (c) error curve of sequence z.

B.4.2 Coupling Synchronization

$$\text{Drive system:} \begin{cases} \dot{x} = 10\,(y - x) \\ \dot{y} = (24 - 4c)\,x - xz + cy \\ \dot{z} = xy - 8z/3 \end{cases}, \tag{B.6}$$

$$\text{Response system:} \begin{cases} \dot{x}_1 = 10\,(y_1 - x_1) + k_1(x - x_1) \\ \dot{y}_1 = (24 - 4c)\,x_1 - x_1 z_1 + cy_1 + k_2(y - y_1) \\ \dot{z}_1 = x_1 y_1 - 8z_1/3 + k_3(z - z_1) \end{cases}, \tag{B.7}$$

$$\text{Error system:} \begin{cases} \dot{e}_1 = \dot{x}_1 - \dot{x} \\ \dot{e}_2 = \dot{y}_1 - \dot{y} \\ \dot{e}_3 = \dot{z}_1 - \dot{z} \end{cases}. \tag{B.8}$$

```
function dy = SimpleOH(t,y)
%function name: SimpleOH %function: the coupling synchroniza-
tion of the simplified Lorenz system
global c;
global k1;
global k2;
global k3;
dy = zeros(6,1);
dy(1)=10*(y(2)-y(1))+k1*(y(4)-y(1));
dy(2)=(24-4*c)*y(1)-y(1)*y(3)+c*y(2)+k2*(y(5)-y(2));
dy(3)=y(1)*y(2)-8*y(3)/3+k3*(y(6)-y(3));
dy(4)=10*(y(5)-y(4));
dy(5)=(24-4*c)*y(4)-y(4)*y(6)+c*y(5);
dy(6)=y(4)*y(5)-8*y(6)/3;
```

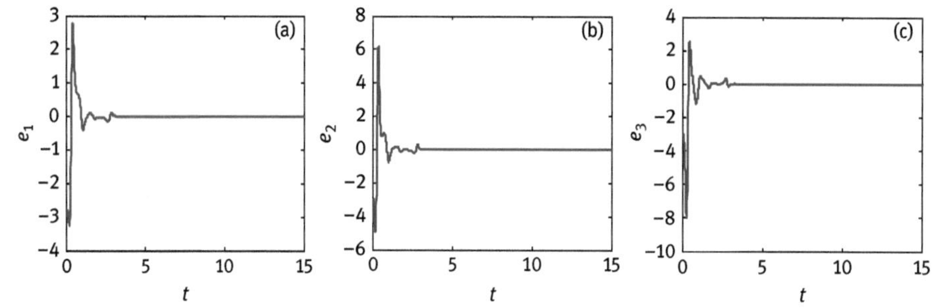

Figure B.11: Coupling synchronization error curves of the simplified Lorenz system: (a) error curve of sequence *x*; (b) error curve of sequence *y*; and (c) error curve of sequence *z*.

Setting $c = 2$, $k_1 = 4$, $k_2 = 2$, and $k_3 = 3$, the initial values of the drive system are [4, 5, 6] and the initial values of the coupling system are [1, 2, 3]. Running the Matlab codes below, the synchronization error curves of the synchronization system are obtained as shown in Figure B.11:

```
clc
clear
global c
global k1
global k2
global k3
c=2;
k1=4;k2=2;k3=3;
[T,Y]=ode45(@SimpleOH,0:0.01:15,[1 2 3 4 5 6 ]);
figure
plot(T,Y(:,1)-Y(:,4))
xlabel('t')
ylabel('e1')
figure
plot(T,Y(:,2)-Y(:,5))
xlabel('t')
ylabel('e2')
figure
plot(T,Y(:,3)-Y(:,6))
xlabel('t')
ylabel('e3')
```

B.4.3 Tracking Synchronization

$$\begin{cases} \dot{x} = 10\,(y - x) + u_1 \\ \dot{y} = (24 - 4c)\,x - xz + cy + u_2 \\ \dot{z} = xy - 8z/3 + u_3 \end{cases} . \tag{B.9}$$

Design the controller:

$$\begin{cases} u_1 = \dot{r}_1 + r_1 - 10y + 9x \\ u_2 = \dot{r}_2 + r_2 - (24 - 4c)\,x + xz - (c + 1)y \\ u_3 = \dot{r}_3 + r_3 - xy + 5z/3 \end{cases} , \tag{B.10}$$

where r_1, r_2, and r_3 are signals to be tracked.

Next, we let the simplified Lorenz system to track the sin curve, the constant curve, and the x sequence of the hyperchaotic Lorenz system, respectively,

$$\begin{cases} r_1 = \sin t \\ r_2 = 4 \\ r_3 = x_1 \end{cases} \quad \text{and} \quad \begin{cases} \dot{r}_1 = \cos t \\ \dot{r}_2 = 0 \\ \dot{r}_3 = 10(y_1 - x_1) \end{cases} , \tag{B.11}$$

$$\begin{cases} \dot{x}_1 = 10\,(y_1 - x_1) \\ \dot{y}_1 = 28x_1 - x_1 z_1 + y_1 - w_1 \\ \dot{z}_1 = x_1 y_1 - 8z_1/3 \\ \dot{w}_1 = k y_1 z_1 \end{cases} , \tag{B.12}$$

```
clc; clear
L=10000;
c=2;h=0.001;k=0.21;
y0=[1 2 1];
yy0=[5 6 7 8];
data=zeros(3,L);
T=zeros(1,L);
R1=zeros(1,L);
R2=zeros(1,L);
R3=zeros(1,L);
t=0;
for i=1:L
    t=t+h;
    yy0(1)=yy0(1)+h*(10*(yy0(2)-yy0(1)));
    yy0(2)=yy0(2)+h*(-yy0(1)*yy0(3)+28*yy0(1)+yy0(2)-yy0(4));
    yy0(3)=yy0(3)+h*(yy0(1)*yy0(2)-8/3*yy0(3));
```

```
    yy0(4)=yy0(4)+h*(k*yy0(2)*yy0(3));
    r1=sin(t);dr1=cos(t);
    r2=4;dr2=0;
    r3=yy0(1);dr3=10*(yy0(2)-yy0(1));

    u1=dr1+r1-10*y0(2)+9*y0(1);
    u2=dr2+r2-(24-4*c)*y0(1)+y0(1)*y0(3)-(c+1)*y0(2);
    u3=dr3+r3-y0(1)*y0(2)+5*y0(3)/3;

    y0(1)=y0(1)+h*(10*(y0(2)-y0(1))+u1);
    y0(2)=y0(2)+h*((24-4*c)*y0(1)-y0(1)*y0(3)+c*y0(2)+u2);
    y0(3)=y0(3)+h*(y0(1)*y0(2)-8*y0(3)/3+u3);

    data(1,i)=y0(1);data(2,i)=y0(2);data(3,i)=y0(3);
    T(i)=t;
    R1(i)=r1;
    R2(i)=r2;
    R3(i)=r3;
end
figure
        plot(T,R1,'-.')
        hold on
        plot(T,data(1,:),'--')
        hold off
        legend('r1','x')
        xlabel('t')
        ylabel('r1,x')
figure
        plot(T,R2,'-.')
        hold on
        plot(T,data(2,:),'--')
        hold off
        legend('r2','x')
        xlabel('t')
        ylabel('r2,x')
figure
        plot(T,R3,'-.')
        hold on
        plot(T,data(3,:),'--')
        hold off
        legend('r3','x')
        xlabel('t')
```

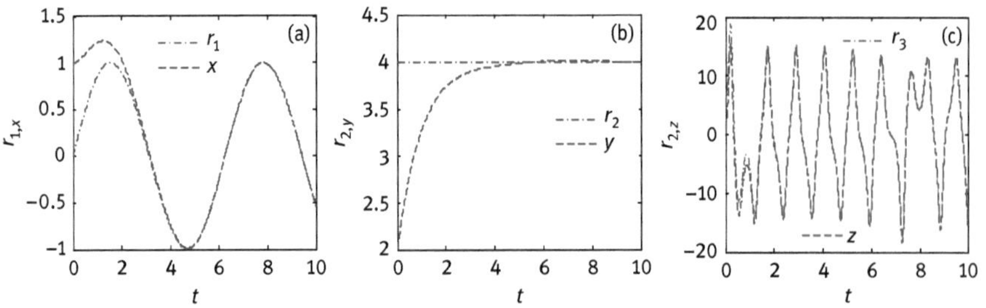

Figure B.12: Tracking curves of the simplified Lorenz system: (a) sequence x; (b) sequence y; and (c) sequence z.

```
ylabel('r3,x')
```

Running the above Matlab codes, the tracking curves of the simplified Lorenz system are obtained as shown in Figure B.12. Of course, readers can also plot the tracking error curves.

References

[1] Ford J. Chaos: Solving the unsolvable, predicting the unpredictable, in chaotic dynamic and fractal. New York, Academic Press, 1986.

[2] Lorenz EN. Deterministic nonperiodic flows. J Atmos Sci 1963, 20, 130–41.

[3] Li TY, Yorke JA. Period three implies chaos. Am Math Mon 1975, 82(10), 985–92.

[4] May R. Simple mathematical models with very complicated dynamics. Nature 1976, 261, 459–67.

[5] Feigenbaum MJ. Quantitative universality for a class of non-linear transformations. J Stat Phys 1978, 19(1), 25–52.

[6] Feigenbaum MJ. The universal metric properties of non-linear transformations. J Stat Phys 1979, 21(6), 669–706.

[7] Takens F. "Detecting strange attractors in turbulence" in Dynamical systems and turhulence, Warwick, 1980, Lecture Notes in Mathematics, vol. 898. D. A. Rand and L. S. Young, Eds. Berlin, Springer, 1981, 366–81.

[8] Glass L, Guevara MR, Shrier A, Perez R. Bifurcation and chaos in a periodically stimulated cardiac oscillator. Physica D 1983, 7, 89–101.

[9] Grassberger P. Toward a quantitative theory of self-generated complexity. Int J Theor Phys 1986, 25(9), 907–38.

[10] Pecora LM, Carroll TL. Synchronization in chaotic systems. Phys Rev Lett 1990, 64(8), 821–4.

[11] Pecora LM, Carroll TL. Driving systems with chaotic signals. Phys Rev A 1991, 44(4), 2374–83.

[12] Ott E, Grebogi C, Yorke JA. Controlling chaos. Phys Rev Lett 1990, 64(11), 1196–9.

[13] Ding J. Puzzles of the wise men-a brief history of chaos and fractals. Beijing, Higher Education Press, 2013.

[14] Liang ML, Wang ZK. Talks about chaos and balance. Dalian, Dalian University of Technology Press, 2008.

[15] Matsumoto T, Chua LO, Tanaka S. Simplest chaotic nonautonomous circuit. Phys Rev A 1984, 30, 1155–7.

[16] Chen GR, Ueta T. Yet another chaotic attractor. Int J Bifurcation Chaos 1999, 9(7), 1465–6.

[17] Wu CW, Chua LO. A unified framework for synchronization and control of dynamical systems. Int J Bifurcation Chaos 1994, 4(4), 979–98.

[18] Fang JQ. Controlling chaos and developing high and new technology. Beijing, Atomic Energy Press, 2002.

[19] Cuomo KM, Oppenheim AV, Strogatz SH. Synchronization of Lorenz-based chaotic circuits with applications to communications. IEEE Trans Circuits Syst II 1993, 40(10), 626–33.

[20] Carroll TL, Pecora LM. Cascading synchronized chaotic systems. Physica D 1993, 67(1), 126–40.

[21] Kocarev L, Parlitz U. General synchronization with application to communication. Phys Rev Lett 1995, 74(25), 5028–31.

[22] Kapitaniak T, Chua LO. Hyperchaotic attractors of unidirectionally-coupled Chua's circuits. Int J Bifurcation Chaos 1994, 4(2), 477–82.

[23] Pyragas K, Tamasevicius A. Experimental control of chaos by delayed self-controlling feedback. Phys Lett A 1993, 180(1–2), 99–102.

[24] Huberman BA, Lumer E. Dynamics of adaptive systems. IEEE Trans CAS-I 1990, 37(4), 547–50.

[25] John JK, Amritkar RE. Synchronization of unstable orbits using adaptive control. Phys Rev E 1994, 49(6), 4843–8.

[26] Yang T, Chua LO. Impulsive stabilization for control and synchronization of chaotic systems: Theory and application to secure communication. IEEE Trans CAS-I 1997, 44(10), 976–88.

[27] Yang T, Chua LO. Impulsive control and synchronization of nonlinear dynamical systems and application to secure communication. Int J Bifurcation Chaos 1997, 7(3), 645–64.

[28] Mainieri R, Rehacek J. Projective synchronization in three-dimensional chaotic systems. Phys Rev Lett 1999, 82(15), 3042–5.
[29] Du HY, Zeng QS, Wang CH. Function projective synchronization of different chaotic systems with uncertain parameters. Phys Lett A 2008, 372, 5402–10.
[30] Park JH. Adaptive control for modified projective synchronization of a four-dimensional chaotic system with uncertain parameters. J Comput Appl Math 2008, 213, 288–93.
[31] Angeli AD, Genesio R, Tesi A. Dead-beat chaos synchronization in discrete time system. IEEE Trans CAS-I 1995, 42(1), 54–6.
[32] Vasegh N, Khellat F. Takagi-Sugeno fuzzy modeling and chaos control of partial differential systems. Chaos 2013, 23(4), 042101.
[33] Shieh CS, Hung RT. Hybrid control for synchronizing a chaotic system. Appl Math Model 2011, 35(8), 3751–8.
[34] Sun KH, Mou J, Zhou JL. Synchronization control and its application for the unified chaotic system based on chaos observer. IET Control Theory Appl 2008, 25(4), 794–8.
[35] Sun KH, Zhang TS. Synchronization control method of hyperchaotic systems by multi variables driving and error feedback. J Cent South Univ (Science and Technology) 2005, 36(4), 653–7.
[36] Kennedy M, Kolumbon G. Special issue on noncoherent chaotic communication. IEEE Trans CAS-I 2000, 47(12), 1661–732.
[37] Kocarev L, Maggio G, Ogorzalerk M. Special issue on applications of chaos in modern communication systems. IEEE Trans CAS-I 2001, 48(12), 1385–527.
[38] Hasler M, Mazzini G. Special issue on applications of nonlinear dynamics to electronic and information engineering. Proc IEEE 2002, 90(5), 631–40.
[39] Wang XY. Synchronization of chaotic system and its application in secure communication. Beijing, Science Press, 2012.
[40] Yin Y. Experimental demonstration of chaotic synchronization in the modified Chua's Oscillators. Int J Bifurcation Chaos 1997, 7(6), 1401–10.
[41] Dedieu HD, Kennedy MP, Hasler M. Chaos shift-keying: Modulation and demodulation of a chaotic carrier using self-synchronizing Chua's circuits. IEEE Trans CAS-II 1993, 40(10), 634–42.
[42] Yang T, Chua LO. Secure communication via chaotic parameter modulation. IEEE Trans CAS-I 1996, 43(9), 817–9.
[43] Batini GH, McGillem CD. A chaotic direct sequence spread spectrum communication system. IEEE Trans Commun 1994, 42(2, 3, 4), 1524–7.
[44] Itoh M. Spread spectrum communication via chaos. Int J Bifurcation Chaos 1999, 9(1), 155–213.
[45] Cuomo KM, Oppenheim AV. Circuit implementation of synchronized chaos with application to communications. Phys Rev Lett 1993, 71, 65–8.
[46] Kennedy MP, Kolumban G. Digital communications using chaos. CRC PRESS'99, 1–24.
[47] Kolumban G, Kis G, Jako Z, Kennedy MP. FM-DCSK: A robust modulation scheme for chaotic communications. IEICE Trans Fundam 1998, E81–A, 1798–1802.
[48] Xu WK, Wang L, Kolumban G. A novel differential chaos shift keying modulation scheme. Int J Bifurcation Chaos 2011, 21(3), 799–814.
[49] Mathews R. On the derivation of a chaotic encryption algorithm. Cryptrologia 1989, 29–42.
[50] Baptista MS. Cryptography with chaos. Phys Lett A 1998, 240, 50–4.
[51] Wong WK. A fast chaotic cryptographic with dynamic look-up table. Phys Lett A 2002, 298(4), 238–42.
[52] Wong WK, Ho SW, Yung CK. A chaotic cryptography scheme for generating short ciphertext. Phys Lett A 2003, 310, 67–73.
[53] Akgul A, Calgan H, Koyuncu I, Pehlivan I, Istanbullu A. Chaos-based engineering applications with a 3D chaotic system without equilibrium points. Nonlinear Dynamics 2016, 84(2): 481–495.

[54] Utami D, Suwastio H, Sumadjudin BB. FPGA implementation of digital chaotic cryptography. Lect Notes Comput Sci 2002, 25(10), 239–47.
[55] Sun KH, Sprott JC. Periodically forced chaotic system with signum nonlinearity. Int J Bifurcation Chaos 2010, 20(5), 1499–507.
[56] Sun KH, Sprott JC. Dynamics of a simplified Lorenz system. Int J Bifurcation Chaos 2009, 19(4), 1357–66.
[57] Gottwald GA, Melbourne I. A new test for chaos in deterministic systems. Proc R Soc Med 2004, 460, 603–11.
[58] Gottwald GA, Melbourne I. Testing for chaos in deterministic systems with noise. Physica D 2005, 212, 100–10.
[59] Sun KH, Liu X, Zhu CX. The 0–1 test algorithm for chaos and its applications. Chin Phys B 2010, 19(11), 110510.
[60] Horgan J. From complexity to perplexity. Sci Am 1995, 272(6), 104–9.
[61] Lempel A, Ziv J. On the complexity of finite sequence. IEEE Trans 1976, IT-22, 75–9.
[62] Pincus SM. Approximate entropy (ApEn) as a complexity measure. Chaos 1995, 5(1), 110–17.
[63] Sun KH, Tan GQ, Sheng LY. Complexity analysis of TD-ERCS discrete chaotic pseudo-random sequences. Acta Phys Sin 2008, 57(6), 3359–68.
[64] Chen WT, Zhuang J, Yu WX, Wang ZZ. Measuring complexity using FuzzyEn, ApEn, and SampEn. Med Eng Phys 2009, 31(1), 61–8.
[65] Sun KH, He SB, Yin LZ, A Di-Li·Duo Li-Kun. Application of FuzzyEn algorithm to the analysis of complexity of chaotic sequence. Acta Phys Sin 2012, 61(13), 130507.
[66] Larrondo HA, González CM, Martin MT, Plastino A, Rosso OA. Intensive statistical complexity measure of pseudorandom number generators. Phys A Stat Mech Appl 2005, 356, 133–8.
[67] Sun KH, He SB, Sheng LY. Complexity analysis of chaotic sequence based on the intensive statistical complexity algorithm. Acta Phys Sin 2011, 60(2), 020505
[68] Rajeev KA, Subba JR, Ramakrishna Ramaswamy. Information-entropic analysis of chaotic time series: Determination of time-delays and dynamical coupling. Chaos Solitons Fractals 2002, 14(4), 633–41.
[69] Malihe S, Serajeddin K, Reza B. Entropy and complexity measures for EEG signal classification of schizophrenic and control participants. Artif Intell Med 2009, 47(3), 263–74.
[70] Abdulnasir Y, Mehmet A, Mustafa P. Application of adaptive neuro-fuzzy inference system for vigilance level estimation by using wavelet-entropy feature extraction. Expert Syst Appl 2009, 36(4), 7390–9.
[71] Chen F, Xu JH, Gu FJ. Dynamic process of information transmission complexity in human brains. Biol Cybern 2000, 83 (4), 355–66.
[72] Shen EH, Cai ZJ, Gu FJ. Mathematical foundation of a new complexity measure. Appl Math Mech 2005, 26(9), 1188–96.
[73] Cai ZJ, Sun J. Modified C_0 complexity and applications. Journal of Fudan University (Natural Science), 2008, 47(6), 791–96
[74] Sun KH, He SB, He Y, Yin LZ. Complexity analysis of chaotic pseudo-random sequences based on spectral entropy algorithm. Acta Phys Sin 2013, 62(1), 010501.
[75] Sun KH, He SB, Zhu CX, He Y. Analysis of chaotic complexity characteristics based on C_0 algorithm. Acta Electronica Sin 2013, 41(9), 1765–71
[76] Hénon M. A two-dimensional mapping with a strange attractor. Commun Math Phys 1976, 50, 291–312.
[77] Sheng LY, Sun KH, Li CB. Study of a discrete chaotic system based on tangent-delay for elliptic reflecting cavity and its properties. Acta Phys Sin 2004, 53(9), 2871–6.
[78] Duffing G. Erzwungene Schwingungen bei Veränderlicher Eigenfrequenz. Braunschweig, Vieweg, 1918.

[79] van der Pol B. On relaxation-oscillations. The London, Edinburgh, and Dublin Philosophical Magazine and Journal of Sciences Series 1926, 7(2), 978–92.
[80] Rössler OE. An equation for continuous chaos. Phys Lett A 1976, 57(5), 397–8.
[81] Zhong GQ, Tang KS. Circuitry implementation and synchronization of Chen's attractor. Int J Bifurcation Chaos 2002, 12(6), 1423–7.
[82] Lü JH, Chen GR. A new chaotic attractor coined. Int J Bifurcation Chaos 2002, 12(3), 659–61.
[83] Vaněček A, Čelikovský S. Control systems: From linear analysis to synthesis of chaos. London, Pretice-Hall, 1996.
[84] Lü JH, Chen GR, Cheng DZ, Čelikovský S. Bridge the gap between the Lorenz system and the Chen system. Int J Bifurcation Chaos 2002, 12(12), 2917–26.
[85] Lü JH, Zhou TS, Chen GR, Zhang SC. The compound structure of Chen's attractor. Int J Bifurcation Chaos 2002, 12(4), 855–8.
[86] Zhang ZQ, Park J H; Shao HY. Adaptive synchronization of uncertain unified chaotic systems via novel feedback controls. Nonlinear Dyn 2015, 81(1–2), 695–706.
[87] Cheng CJ. Robust synchronization of uncertain unified chaotic systems subject to noise and its application to secure communication. Appl Math Comput 2012, 219(5), 2698–712.
[88] Sun KH, Yang JL, Ding JF, Sheng LY. Circuit design and implementation of Lorenz chaotic system with one parameter. Acta Phys Sin 2010, 59(12), 8385–92.
[89] Sprott JC. A proposed standard for the publication of new chaotic systems. Int J Bifurcation Chaos 2011, 21(9), 2391–4.
[90] Saltzman B. Finite amplitude free connection as an initial value problem-I. J Atmos Sci 1962 19, 329–41.
[91] Rössler OE. An equation for hyperchaos. Phys Lett A 1979, 71(2–3), 155–7.
[92] Li Y, Chen G, Tang KS. Generating hyperchaos via state feedback control. Int J Bifurcation Chaos 2005, 15(10), 3367–75.
[93] Zgliczynski P. Symbolic dynamics for the Rössler folded towel map. Banach Center Publ 1999, 47(1), 253–8.
[94] Carroll TL, Pecora LM. Synchronization chaotic circuits. IEEE Trans CAS-I 1991, 38(4), 453–6.
[95] Fang JQ. Control and synchronization of chaos in nonlinear systems and its application prospect (I). Prog Phys 1996, 16(1), 1–74.
[96] Fang JQ. Control and synchronization of chaos in nonlinear systems and its application prospect (II). Prog Phys 1996, 16(2), 137–201.
[97] Sun KH. Chaos synchronization control theory and its application in information encryption. Central South University, 2005.
[98] He R, Vaidya PG. Analysis and synthesis of synchronous periodic and chaotic systems. Phys Rev A 1992, 46(12), 7387–92.
[99] Zhou CS, Lai CH. Analysis of spurious synchronization with positive conditional Lyapunov exponents in computer simulations. Physica D 2000, 135(1–2), 1–23.
[100] Sun KH, Shang F, Zhong K, Sheng Z. Synchronous control between unified chaotic system and its modified system based on intermittent feedback. J Electron Inf Technol 2009, 31(1), 71–4.
[101] Ma J, Liao GH, Mo XH. Hyperchaos synchronization and control using intermittent feedback. Acta Phys Sin 2005, 54(12), 5585–90.
[102] Guan XP, Fang ZP, Chen CL, Hua CC. Chaos control and its application in secure communication. Beijing, National Defence Industry Press, 2002.
[103] Zhou P. Observers for a class of 3D continuous chaotic systems. Acta Phys Sin 2003, 52(5), 1108–11.
[104] Yang XS. On observability of 3D continuous-time autonomous chaotic systems based on scalar output measurement. Int J Bifurcation Chaos 2002, 12(5), 1159–62.
[105] Xu DL, Li ZG. Controlled projective synchronization in nonpartially-linear chaotic systems. Int J Bifurcation Chaos 2002, 12(6), 1395–402.

[106] Li GH. Modified projective synchronization of chaotic system. Chaos Solitons & Fractals 2007, 32 (5), 1786–90.
[107] Cafagna D, Grassi G. Observer-based projective synchronization of fractional systems via a scalar signal: Application to hyperchaotic Rössler systems. Nonlinear Dyn 2012, 68(1–2), 117–28.
[108] Farivar F, Nekoui MA, Shoorehdeli MA, Teshnehlab M. Modified projective synchronization of chaotic dissipative gyroscope systems via backstepping control. Indian J Phys 2012, 86(10), 901–6.
[109] Wang SB, Wang XY, Zhou YF. A memristor-based complex Lorenz system and its modified projective synchronization. Entropy 2015, 17(11), 7628–44.
[110] Kim DW, Chang PH, Kim SH. A new chaotic attractor and its robust function projective synchronization. Nonlinear Dyn 2013, 73(3), 1883–93.
[111] Liu J, Liu ST, Sprott JC. Adaptive complex modified hybrid function projective synchronization of different dimensional complex chaos with uncertain complex parameters. Nonlinear Dyn 2016, 83(1–2), 1109–21.
[112] Santo B. Chaos synchronization and cryptography for secure communications. New York, Hershey, 2010.
[113] Cuomo KM, Oppengeim AV, Strogatz SH. Synchronization of Lorenz-based chaotic circuits with applications to communications. IEEE Trans CAS-II 1993, 140(10), 626–33.
[114] Mianovic V, Zaghloul ME. Improved masking algorithm for chaotic communication systems. Electron Lett 1996, 32(1), 11–12.
[115] Carroll TC, Heagy JF, Pecora LM. Transforming signals with chaotic synchronization. Phys Rev E 1996, 54(5), 4676–80.
[116] Parlitz U, Chua LO, Kocarev L, Halle KS, Shang A. Transmission of digital signals by chaotic synchronization. Int J Bifurcation Chaos 1992, 2(2), 973–7.
[117] Kolumban G, Kennedy MP, Jako Z, Kis G. Chaotic communications with correlator receivers: Theory and performance limits. Proc IEEE 2002, 90(5), 711–32.
[118] Kolnmban G, Kis G, Kennedy MP, Já Z. FM-DCSK: A new and robust solution to chaos communications. In: Proceedings of International Symposium Nonlinear Theory and Its Applications, Hawaii, 1997, 117–20.
[119] Galias Z, Maggio GM. Quadrature chaos-shift keying: Theory and performance analysis. IEEE Trans CAS-I 2001, 48(12), 1510–19.
[120] Kennedy MP, Kolumbán G. Digital communications using chaos. Signal Process 2000, 80(7), 1307–20.
[121] Rulkov NF, Sushchik MM, Tsimring LS, et al. Digital communication using chaotic-pulse-position modulation. IEEE Trans CAS-I 2001, 48(12), 1436–44.
[122] Heidari-Bateni G, McGillem CD. A chaotic direct-sequence spread-spectrum communication system. IEEE Trans Commun 1994, 42(2/3/4), 1524–7.
[123] Sun KH, Zhou JL, Mou J. Design and performance analysis of multi-user chaotic sequence spread-spectrum communication system. J Electron Inf Technol 2007, 29(10), 2436–40.
[124] Liao XF, Xiao D, Chen Y, Xiang T. Principle of chaos cryptography and its application. Beijing, Science Press, 2009.
[125] Li CB, He B. Chaotic encryption algorithm and hash function construction research. Beijing, Publishing House of Electronic Industry, 2011.
[126] Su M, Wen WY, Zhang Y S Security evaluation of bilateral-diffusion based image encryption algorithm. Nonlinear Dyn 2014, 77(1–2), 243–6
[127] Dong C. Color image encryption using one-time keys and coupled chaotic systems. Signal Process Image Commun 2014, 29(5), 628–40.

[128] Yang YX, Lin XD. Coding cryptology. Beijing, People's Posts and Telecommunications Publishing House, 1992.
[129] Roland S. Use of chaotic dynamical systems in cryptography. J Franklin Inst 2001, 338(4), 429–41.
[130] Li P, Zhang W, Li Z. FPGA implementation of a coupled-map-lattice-based cryptosystem. Int J Circuit Theor Appl 2010, 38(1), 85–98.
[131] Wang FL. A new pseudo-random number generator and application to a digital secure communication scheme based on compound symbolic chaos. Acta Phys Sin 2011, 60(11), 110517.
[132] Hu YH, Luo J, Su H. A new method for improving the limited precision effect of chaotic sequences. Syst Simul 2012, 24 (11), 2349–52.
[133] Wang XS. A chaotic sequence encryption method. Chin J Comput 2002, (4), 351–6.
[134] Liu WH, Sun, KH, Zhu CX. A fast image encryption algorithm based on chaotic map. Optics and Lasers in Engineering 2016, 84: 26–36.
[135] Wang L. A novel chaos-based pseudo-random number generator. Acta Phys Sin 2006, 55(8), 3964–8.
[136] Wang XY, Bao XM. A novel block cryptosystem based on the coupled chaotic map lattice. Nonlinear Dyn 2013, 72(4), 707–15.
[137] Sheng LY, Wen J, Cao LL, Xiao YY. Differential cryptanalysis of TD-ERCS chaos. Acta Phys Sin 2007, 56(1), 78–83.
[138] Sheng LY, Xiao YY, Sheng Z. A universal algorithm for transforming chaotic sequences into uniform pseudo-random sequences. Acta Phys Sin 2008, 57(7), 4007–13
[139] McLauchlan L, Mehruebeoglu M. DWT and DCT embedded watermarking using Chaos theory. Conference on Mathematics of Data/Image Coding, Compression, and Encryption with Applications XII, San Diego, CA, 2010, 8.
[140] Millerioux G. Crypyanalysis of hybrid cryptosystems. Int J Bifurcation Chaos 2013, 23(10), 1350173.
[141] Gangadhar C, Rao KD. Hyperchaos based image encryption. Int J Bifurcation Chaos 2009, 19(11), 3833–9.
[142] Cao GH. Image scrambling algorithm based on chaotic weighted sampling theory and sorting transformation. J Beijing Univ Aeronaut Astronaut 2013, 39(1), 67–72
[143] Li HJ, Zhang JS. Embedding adaptive arithmetic coder in chaos-based cryptography. Chin Phys B 2010, 19(5), 050508.
[144] Rahman SMM, Hossain MA, Mouftah H. Chaos-cryptography based privacy preservation technique for video surveillance. Multimedia Syst 2012, 18(2), 145–55.
[145] Vidal G, Baptista MS, Mancini H. Fundamentals of a classical chaos-based Cryptosystem with some quantum cryptography features. Int J Bifurcation Chaos 2012, 22(10), 1250243.
[146] Fouda JSAE, Effa JY, Sabat SL, Ali M. A fast chaotic block cipher for image encryption. Commun Nonlinear Sci Numer Simul 2014, 19(3), 578–88.
[147] Wang YM, Liu JW. Security of communication network-theory and technology. Xi'an, Xi'an Electronic and Science University Press, 2000.
[148] Zhang FX, Zheng YJ. Communication secure technology. Beijing, National Defence Industry Press, 2003.
[149] Feng DG. Research status and development trend of domestic and foreign cryptography. J Commun 2002, 23(5), 18–26.
[150] American National Standards Institute, Triple Data Encryption Algorithm Modes of Operation, ANSI standard X9.52, 1998.
[151] Kocarev L. Chaos-based cryptography: A brief overview. IEEE Circuit Syst Mag 2001, 1(3), 6–21.
[152] Dachselt F, Kelber K, Schwartz W. Discrete-time chaotic encryption system-part III: Cryptographical analysis. IEEE Trans CAS-I 1998, 45(9), 1–19.

[153] Yang T, Yang LB, Yang CM. Cryptanalyzing chaos secure communication using return maps. Phys Lett A 1998, 245(6), 495–510.
[154] Zhou LH, Peng ZW, Feng ZJ, Zhong TX. Security property of chaotic encryption systems. J Shanghai Jiaotong Univ 2001, 35(1), 133–8.
[155] Syyid Umar. The adaptive communication environment: "ACE". Hughes network systems, 1998.
[156] Bhargava B, Shi CG, Wang SY. MPEG video encryption algorithms. Multimedia Tool Appl 2004, 24(1), 57–79.
[157] Zeng W, Lei S. Efficient frequency domain selective scrambling of digital video. IEEE Trans Multimedia 2003, 5(1), 118–29.
[158] Liu Z, Li X, Dong ZY. Enhancing security of frequency domain video encryption. Proceedings of ACM Multimedia, New York, USA, Poster Session, 2004, 304–7.
[159] Liu G, Ikenaga T, Goto S, Baba T. A selective video encryption scheme for MPEG compression standard. IEICE Trans Fundam Electron Commun Comput Sci 2006, E89–A(1), 194–202.
[160] Shang F, Sun K H, Cai YQ. A new efficient MPEG video encryption system based on chaotic maps. Proceedings-1st International Congress on Image and Signal Processing, CISP, 2008, v 3, 12–16, Haikou.
[161] Hilfer R. Applications of fractional calculus in physics. New Jersey, World Scientific, 2001.
[162] Mandelbort BB. The fractal geometry of nature. New York, Freeman, 1983.
[163] Hartley TT, Lorenzo CF, Qammer HK. Chaos in a fractional order Chua's system. IEEE Trans Circuits Syst- 1995, 42(8), 485–90.
[164] Arena P, Caponetto R, Fortuna L, Porto D. Chaos in fractional order Duffing system. Proceedings of the European Conference on Circuit Theory and Design. (ECCTD, 97), Budapest. Hungary, September 1997, 1258–72.
[165] Ge ZM, Ou CY. Chaos in a fractional order modified Duffing system. Chaos Solitons Fractals 2007, 34(2), 262–91.
[166] Ahmad WM, Sprott JC. Chaos in fractional-order autonomous nonlinear system. Chaos Solitons Fractals 2003, 16(2), 339–51.
[167] Li CG, Chen GR. Chaos and hyperchaos in the fractional-order Rössler equations. Phys A: Stat Mech Appl 2004, 341(1–4), 55–61.
[168] Podlubny I. Fractional differential equations. New York, Academic Press, 1999.
[169] Li CP, Deng WH. Remarks on fractional derivatives. Appl Math Comput 2007, 187(2), 777–84.
[170] Charef A, Sun HH. Fractal system as represented by singularity function. IEEE Trans Autom Control 1992, 37(9), 1465–70.
[171] Kai D. An algorithm for the numerical solution of differential equations of fractional order. Electron Trans Numer Anal 1998, 5(Mar.), 612–7.
[172] Kai D, Ford NJ, Freed AD. A predictor-corrector approach for the numerical solution of fractional differential equations. Nonlinear Dyn 2002, 29(1–4), 3–22.
[173] Kai D, Ford NJ. Analysis of fractional differential equations. J Math Anal Appl 2002, 265(2), 229–48.
[174] Ahmad W. Hyperchaos in fractional order nonlinear systems. Chaos Solitons Fractals 2005, 26(5), 1459–65.
[175] Arman KB, Kia F, Naser P, Henry L. A chaotic secure communication scheme using fractional chaotic systems based on an extended fractional Kalman filter. Commun Nonlinear Sci Numer Simul 2009, 14(3), 863–79.
[176] Bohannan GW, Hurst SK, Spangler L. Electrical component with fractional order impedance. 2006, No. PCT/US2006/008821.
[177] Wang FQ, Liu CX. Study on the critical chaotic system with fractional order and circuit experiment. Acta Phys Sin 2006, 55(8), 3922–7

[178] Liu CX. A hyperchaotic system and its fractional-order circuit simulation. Acta Phys Sin 2007, 56(12), 6865–73.
[179] Chen XR, Liu CX, Wang FQ. Study on the fractional-order Liu chaotic system with circuit experiment and its control. Acta Phys Sin 2008, 57(3), 1416–22.
[180] Zhu CX, Zou Y. Summary of research on fractional-order control. Control Decis 2009, 24(2), 161–9.
[181] Liu SD, Shi SY, Liu SS. Bridge between weather and climate: Fractional derivatives. Meteor Sci Technol 2007, 35(1), 15–19.
[182] Zhao CN, Li YS, Lu T.Analysis and design of fractional-order system. Beijing, National Defense Industry Press, 2011.
[183] Adomian G. A new approach to nonlinear partial differential equations. J Math Anal Appl 1984, 102(2), 420–34.
[184] Cafagna D, Grassi G. Bifurcation and chaos in the fractional-order Chen system via a time-domain approach. Int J Bifurcation Chaos 2008, 18(18), 1845–63.
[185] Cafagna D, Grassi G. Hyperchaos in the fractional-order Rössler system with lowest-order. Int J Bifurcation Chaos 2009, 19(1), 338–47.
[186] Sun KH, Wang X, Sprott JC. Bifurcations and chaos in fractional-order simplified Lorenz system. Int J Bifurcation Chaos 2010, 20(4), 1208–19.
[187] He SB, Sun KH, Wang HH. Solution of the fractional-order chaotic system based on Adomian decomposition algorithm and its complexity analysis. Acta Phys Sin 2014, 63(3), 030502.
[188] Sun KH, Yang JL, Qiu SS. Study of Simulation approaches for fractional-order chaotic systems. J Syst Simul 2011, 23(11), 2361–5, 2370.
[189] Chen GR, Lü JH. Dynamic analysis, control and synchronization of lorenz family. Beijing, Science Press, 2006.
[190] Wolf A, Swift JB, Swinney HL, Vastano JA. Determining Lyapunov exponents from a time series. Physica D 1985, 16(3), 285–317.
[191] Sun KH, Ren J, Qiu SS. Chaotic dynamic characteristics of fractional-order unified system. J South China Univ Technol (Natural Science Edition) 2008, 36(8), 6–10.
[192] Sun KH, Sprott JC. Bifurcations of fractional-order diffusionless Lorenz system. Electron J Theor Phys 2009, 6(22), 123–34.
[193] Pomeau Y, Manneville P. Intermittent transition to turbulence in dissipative dynamical systems. Commun Math Phys 1980, 74(2), 189–97.
[194] Sprott JC. Chaos and time-series analysis. New York, Oxford University Press, 2003.
[195] Yorke JA, Yorke ED. Metastable chaos: The transition to sustained chaotic behavior in the Lorenz model. J Stat Phys 1979, 21(3), 263–77.
[196] Yang JL. Research on circuit design and implementation of simplified Lorenz system. Central South University, 2011.
[197] Sprott JC. Some simple chaotic Jerk functions. Am J Phys 1997, 65(6), 537–43.
[198] Zuo T, Sun KH, Ai XX, Wang HH. High-order grid multi-scroll chaotic attractors generated by the second generation current conveyor circuit. IEEE Trans Circuits Syst-II 2014, 61(10), 818–22.
[199] Merah L, Ali-Pacha A, Said N H, Mamat M. Design and FPGA implementation of Lorenz chaotic system for information security issues. Appl Math Sci 2013, 7(5), 237–46.
[200] Xu X, Yu S, Xu L. Chaotic digital image encryption and its hardware implementation based on DSP technology. Electron Sci Technol 2011, 24(3), 1–5.
[201] Yu SM. Chaotic systems and chaotic circuits: Principle, design and its application in communications. Xi'an, Xi'an Electronic and Science University Press, 2011.
[202] Wang HH, Sun KH, He SB. Characteristic analysis and DSP realization of fractional-order simplified Lorenz system based on Adomain decomposition method. Int J Bifurcation Chaos 2015, 25(6), 155008.

Index

Bei Fragen zur Produktsicherheit wenden Sie sich bitte an:
If you have any questions regarding product safety,
please contact:

Walter de Gruyter GmbH
Genthiner Straße 13
10785 Berlin
productsafety@degruyterbrill.com